Sample Size Calculations
for Clustered and
Longitudinal Outcomes
in Clinical Research

Chapman & Hall/CRC Biostatistics Series

Published Titles

**Adaptive Design Methods in
Clinical Trials, Second Edition**
Shein-Chung Chow and Mark Chang

**Adaptive Design Theory and
Implementation Using SAS and R,
Second Edition**
Mark Chang

**Advanced Bayesian Methods for Medical
Test Accuracy**
Lyle D. Broemeling

Advances in Clinical Trial Biostatistics
Nancy L. Geller

Applied Meta-Analysis with R
Ding-Geng (Din) Chen and Karl E. Peace

**Basic Statistics and Pharmaceutical
Statistical Applications, Second Edition**
James E. De Muth

**Bayesian Adaptive Methods for
Clinical Trials**
Scott M. Berry, Bradley P. Carlin,
J. Jack Lee, and Peter Muller

**Bayesian Analysis Made Simple: An Excel
GUI for WinBUGS**
Phil Woodward

**Bayesian Methods for Measures of
Agreement**
Lyle D. Broemeling

Bayesian Methods in Epidemiology
Lyle D. Broemeling

Bayesian Methods in Health Economics
Gianluca Baio

**Bayesian Missing Data Problems: EM,
Data Augmentation and Noniterative
Computation**
Ming T. Tan, Guo-Liang Tian,
and Kai Wang Ng

Bayesian Modeling in Bioinformatics
Dipak K. Dey, Samiran Ghosh,
and Bani K. Mallick

**Benefit-Risk Assessment in
Pharmaceutical Research and
Development**
Andreas Sashegyi, James Felli, and
Rebecca Noel

**Biosimilars: Design and Analysis of
Follow-on Biologics**
Shein-Chung Chow

Biostatistics: A Computing Approach
Stewart J. Anderson

**Causal Analysis in Biomedicine and
Epidemiology: Based on Minimal
Sufficient Causation**
Mikel Aickin

**Clinical and Statistical Considerations
in Personalized Medicine**
Claudio Carini, Sandeep Menon,
and Mark Chang

Clinical Trial Data Analysis using R
Ding-Geng (Din) Chen and Karl E. Peace

Clinical Trial Methodology
Karl E. Peace and Ding-Geng (Din) Chen

Computational Methods in Biomedical Research
Ravindra Khattree and Dayanand N. Naik

Computational Pharmacokinetics
Anders Källén

Confidence Intervals for Proportions and Related Measures of Effect Size
Robert G. Newcombe

Controversial Statistical Issues in Clinical Trials
Shein-Chung Chow

Data and Safety Monitoring Committees in Clinical Trials
Jay Herson

Design and Analysis of Animal Studies in Pharmaceutical Development
Shein-Chung Chow and Jen-pei Liu

Design and Analysis of Bioavailability and Bioequivalence Studies, Third Edition
Shein-Chung Chow and Jen-pei Liu

Design and Analysis of Bridging Studies
Jen-pei Liu, Shein-Chung Chow, and Chin-Fu Hsiao

Design and Analysis of Clinical Trials with Time-to-Event Endpoints
Karl E. Peace

Design and Analysis of Non-Inferiority Trials
Mark D. Rothmann, Brian L. Wiens, and Ivan S. F. Chan

Difference Equations with Public Health Applications
Lemuel A. Moyé and Asha Seth Kapadia

DNA Methylation Microarrays: Experimental Design and Statistical Analysis
Sun-Chong Wang and Arturas Petronis

DNA Microarrays and Related Genomics Techniques: Design, Analysis, and Interpretation of Experiments
David B. Allison, Grier P. Page, T. Mark Beasley, and Jode W. Edwards

Dose Finding by the Continual Reassessment Method
Ying Kuen Cheung

Elementary Bayesian Biostatistics
Lemuel A. Moyé

Frailty Models in Survival Analysis
Andreas Wienke

Generalized Linear Models: A Bayesian Perspective
Dipak K. Dey, Sujit K. Ghosh, and Bani K. Mallick

Handbook of Regression and Modeling: Applications for the Clinical and Pharmaceutical Industries
Daryl S. Paulson

Inference Principles for Biostatisticians
Ian C. Marschner

Interval-Censored Time-to-Event Data: Methods and Applications
Ding-Geng (Din) Chen, Jianguo Sun, and Karl E. Peace

Joint Models for Longitudinal and Time-to-Event Data: With Applications in R
Dimitris Rizopoulos

Measures of Interobserver Agreement and Reliability, Second Edition
Mohamed M. Shoukri

Medical Biostatistics, Third Edition
A. Indrayan

Meta-Analysis in Medicine and Health Policy
Dalene Stangl and Donald A. Berry

Mixed Effects Models for the Population Approach: Models, Tasks, Methods and Tools
Marc Lavielle

Monte Carlo Simulation for the Pharmaceutical Industry: Concepts, Algorithms, and Case Studies
Mark Chang

Multiple Testing Problems in Pharmaceutical Statistics
Alex Dmitrienko, Ajit C. Tamhane, and Frank Bretz

Noninferiority Testing in Clinical Trials: Issues and Challenges
Tie-Hua Ng

Optimal Design for Nonlinear Response Models
Valerii V. Fedorov and Sergei L. Leonov

Patient-Reported Outcomes: Measurement, Implementation and Interpretation
Joseph C. Cappelleri, Kelly H. Zou, Andrew G. Bushmakin, Jose Ma. J. Alvir, Demissie Alemayehu, and Tara Symonds

Quantitative Evaluation of Safety in Drug Development: Design, Analysis and Reporting
Qi Jiang and H. Amy Xia

Randomized Clinical Trials of Nonpharmacological Treatments
Isabelle Boutron, Philippe Ravaud, and David Moher

Randomized Phase II Cancer Clinical Trials
Sin-Ho Jung

Sample Size Calculations for Clustered and Longitudinal Outcomes in Clinical Research
Chul Ahn, Moonseong Heo, and Song Zhang

Sample Size Calculations in Clinical Research, Second Edition
Shein-Chung Chow, Jun Shao and Hansheng Wang

Statistical Analysis of Human Growth and Development
Yin Bun Cheung

Statistical Design and Analysis of Stability Studies
Shein-Chung Chow

Statistical Evaluation of Diagnostic Performance: Topics in ROC Analysis
Kelly H. Zou, Aiyi Liu, Andriy Bandos, Lucila Ohno-Machado, and Howard Rockette

Statistical Methods for Clinical Trials
Mark X. Norleans

Statistical Methods in Drug Combination Studies
Wei Zhao and Harry Yang

Statistics in Drug Research: Methodologies and Recent Developments
Shein-Chung Chow and Jun Shao

Statistics in the Pharmaceutical Industry, Third Edition
Ralph Buncher and Jia-Yeong Tsay

Survival Analysis in Medicine and Genetics
Jialiang Li and Shuangge Ma

Theory of Drug Development
Eric B. Holmgren

Translational Medicine: Strategies and Statistical Methods
Dennis Cosmatos and Shein-Chung Chow

Chapman & Hall/CRC Biostatistics Series

Sample Size Calculations for Clustered and Longitudinal Outcomes in Clinical Research

Chul Ahn

University of Texas Southwestern Medical Center

Dallas, Texas, USA

Moonseong Heo

Albert Einstein College of Medicine

Bronx, New York, USA

Song Zhang

University of Texas Southwestern Medical Center

Dallas, Texas, USA

CRC Press

Taylor & Francis Group

Boca Raton London New York

CRC Press is an imprint of the
Taylor & Francis Group, an **informa** business

A CHAPMAN & HALL BOOK

CRC Press
Taylor & Francis Group
6000 Broken Sound Parkway NW, Suite 300
Boca Raton, FL 33487-2742

First issued in paperback 2020

© 2015 by Taylor & Francis Group, LLC
CRC Press is an imprint of Taylor & Francis Group, an Informa business

No claim to original U.S. Government works

Version Date: 20141029

ISBN 13: 978-0-367-57599-1 (pbk)
ISBN 13: 978-1-4665-5626-3 (hbk)

Visit the Taylor & Francis Web site at
http://www.taylorandfrancis.com

and the CRC Press Web site at
http://www.crcpress.com

Contents

Preface   ix

List of Figures   xi

List of Tables   xiii

1 Sample Size Determination for Independent Outcomes   **1**
1.1 Introduction   1
1.2 Precision Analysis   2
1.3 Power Analysis   5
1.4 Further Readings   18

2 Sample Size Determination for Clustered Outcomes   **23**
2.1 Introduction   23
2.2 One–Sample Clustered Continuous Outcomes   24
2.3 One–Sample Clustered Binary Outcomes   28
2.4 Two–Sample Clustered Continuous Outcomes   34
2.5 Two–Sample Clustered Binary Outcomes   38
2.6 Stratified Cluster Randomization for Binary Outcomes . . .   42
2.7 Nonparametric Approach for One–Sample Clustered Binary
     Outcomes   45
2.8 Further Readings   51

3 Sample Size Determination for Repeated Measurement
Outcomes Using Summary Statistics   **61**
3.1 Introduction   61
3.2 Information Needed for Sample Size Estimation   62
3.3 Summary Statistics   64
3.4 Further Readings   78

4 Sample Size Determination for Correlated Outcome
Measurements Using GEE   **83**
4.1 Motivation   83
4.2 Review of GEE   85
4.3 Compare the Slope for a Continuous Outcome   90
4.4 Test the TAD for a Continuous Outcome   110
4.5 Compare the Slope for a Binary Outcome   119

4.6　Test the TAD for a Binary Outcome 123
4.7　Compare the Slope for a Count Outcome 126
4.8　Test the TAD for a Count Outcome 130
4.9　Further Readings 134

5　Sample Size Determination for Correlated Outcomes from Two-Level Randomized Clinical Trials　　　　　　149
5.1　Introduction 149
5.2　Statistical Models for Continuous Outcomes 150
5.3　Testing Main Effects 151
5.4　Two-Level Longitudinal Designs: Testing Slope Differences . 158
5.5　Cross-Sectional Factorial Designs: Interactions between Treatments 167
5.6　Longitudinal Factorial Designs: Treatment Effects on Slopes　172
5.7　Sample Sizes for Binary Outcomes 176
5.8　Further Readings 181

6　Sample Size Determination for Correlated Outcomes from Three-Level Randomized Clinical Trials　　　　　187
6.1　Introduction 187
6.2　Statistical Model for Continuous Outcomes 187
6.3　Testing Main Effects 189
6.4　Testing Slope Differences 200
6.5　Cross-Sectional Factorial Designs: Interactions between Treatments 211
6.6　Longitudinal Factorial Designs: Treatment Effects on Slopes　218
6.7　Sample Sizes for Binary Outcomes 223
6.8　Further Readings 230

Index　　　　　　　　　　　　　　　　　　　　　　　　　　235

Preface

One of the most common questions statisticians encounter during interaction with clinical investigators is "How many subjects do I need for this study?" Clinicians are often surprised to find out that the required sample size depends on a number of factors. Obtaining such information for sample size calculation is not trivial, and often involves preliminary studies, literature review, and, more than occasionally, educated guess. The validity of clinical research is judged not by the results but by how it is designed and conducted. Accurate sample size calculation ensures that a study has adequate power to detect clinically meaningful effects and avoids the waste in resources and the risk of exposing excessive patients to experimental treatments caused by an overpowered study.

In this book we focus on sample size determination for studies with correlated outcomes, which are widely implemented in medical, epidemiological, and behavioral studies. Correlated outcomes are usually categorized into two types: clustered and longitudinal. The former arises from trials where randomization is performed at the level of some aggregates (e.g., clinics) of research subjects (e.g., patients). The latter arises when the outcome is measured at multiple time points during follow-up from each subject. A key difference between these two types is that for a clustered design, subjects within a cluster are considered exchangeable, while for a longitudinal design, the multiple measurements from the each subject are distinguished by their unique time stamps.

Designing a randomized trial with correlated outcomes poses special challenges and opportunities for researchers. Appropriately accounting for the correlation with different structures requires more sophisticated methodologies for analysis and sample size calculation. In practice it is also likely that researchers might encounter correlated outcomes with a hierarchical structure. For example, multiple levels of nested clustering (e.g., patients nested in clinics and clinics nested in hospital systems) can occur, and such designs can become more complicated if longitudinal measurements are obtained from each subject. Missing data leads to the challenge of "partially" observed data for clinical trials with correlated outcomes, and its impact on sample size requirement depends on many factors: the number of longitudinal measurements, the structure and strength of correlation, and the distribution of missing data. On the other hand, researchers enjoy some additional flexibility in designing randomized trials with correlated outcomes. When multiple levels of clustering are involved, the level at which to perform randomization actually becomes a design parameter, which can greatly impact trial administration, analysis, and sample size requirement. This issue is explored in Chapters 5 and 6. Another example is that in longitudinal studies, to certain extent, researchers can compensate the lack of unique subjects by increasing the number of measurements from each subject, and vice versa. This feature has profound implication for the design of clinical trials where the cost of recruiting an additional subject is drastically different from the cost of obtaining an additional measurement from an existing subject. It requires researchers to explore the trade-off between the number of subjects and the number of measurements per subject in order to achieve the optimal power under a given financial constraint. We explore this topic in Chapters 3 and 4.

The outline of this book is as follows. In Chapter 1 we review sample size determination for independent outcomes. Advanced readers who are already familiar with sample size problems can skip this chapter. In Chapter 2 we explore sample size determination for variants of clustered trials, including one- and two-sample trials, continuous and binary outcomes, stratified cluster design, and nonparametric approaches. In Chapter 3 we review sample size methods based on summary statistics (such as individually estimated means or slopes) obtained from longitudinal outcomes. In Chapter 4 we present sample size determination based on GEE approaches for various types of correlated outcomes, including continuous, binary, and count. The impact of missing data, correlation structures, and financial constraints is investigated. In Chapter 5 we present sample size determination based on mixed-effects model approaches for randomized clinical trials with two level data structure. Longitudinal and cross-sectional factorial designs are explored. In Chapter 6 we further extend the mixed-effects model sample size approaches to scenarios where three level data structures are involved in randomized trials.

We wish this book to serve as a useful resource for biostatisticians, clinical investigators, epidemiologists, and social scientists whose research involves randomized trials with correlated outcomes. While jointly addressing the overarching theme of sample size determination for correlated outcome under such settings, individual chapters are written in a self-contained manner so that readers can explore specific topics relevant to their research projects without having to refer to other chapters.

We give special thanks to Dr. Mimi Y. Kim for her enthusiastic support by providing critical reviews and suggestions, examples, edits, and corrections throughout the chapters. Without her input, this book would have not been in the present form. We also thank Acquisitions Editor David Grubbs for providing the opportunity to work on this book, and Production Manager Suzanne Lassandro for her outstanding support in publishing this book. In addition, we thank the support of the University of Texas Southwestern Medical Center and the Albert Einstein College of Medicine.

Chul Ahn, PhD
Moonseong Heo, PhD
Song Zhang, PhD

List of Figures

1.1 Sample size estimation for a one–sided test in a one–sample problem 12

4.1 Numerical study to explore the relationship between s_t^2 and ρ, under the scenario of complete data and various values of θ from the damped exponential family. $\theta = 1$ corresponds to AR(1) and $\theta = 0$ corresponds to CS. The measurement times are normalized such that $t_m - t_1 = 1$. Hence $\rho_{1m} = \rho$ under all values of θ. 95

4.2 Numerical study to explore the relationship between s_t^2 and ρ, under the scenario of incomplete data and various values of θ from the damped exponential family. $\theta = 1$ corresponds to AR(1) and $\theta = 0$ corresponds to CS. IM and MM represent the independent and monotone missing pattern, respectively. The measurement times are normalized such that $t_m - t_1 = 1$. Hence $\rho_{1m} = \rho$ under all values of θ. 97

4.3 A numerical study to explore $\frac{n\{m+1\}}{n\{m\}}$ under missing data and different correlation structures. The vertical axis is $\frac{n\{m+1\}}{n\{m\}}$. "Complete" indicates the scenario of complete data. "IM" and "'MM" indicate the independent and monotone missing patterns, respectively, with marginal observant probabilities computed by $\delta_j = 1 - 0.3 * (j - 1)/(m - 1)$. 101

4.4 Different trends in the marginal observant probabilities. δ_1 approximately follows a linear trend. δ_2 is relatively steady initially but drops quickly afterward. δ_3 drops quickly from the beginning but plateaus. 109

5.1 Geometrical representations of fixed parameters in model (5.12) for a parallel-arm longitudinal cluster randomized trial. 160

5.2 Geometrical representations of fixed parameters in model (5.31) for a 2-by-2 factorial longitudinal cluster randomized trial. 174

List of Tables

2.1   Proportion of infection (y_i/m_i) from $n = 29$ subjects (clusters)   33

2.2   Distribution of the number of infected sites (m_i)   33

2.3   Stepped wedge design, where C represents control and I represents intervention   53

4.1   Sample sizes under various scenarios   110

5.1   Sample size and power for detecting a main effect $\delta_{(2)}$ in model (5.3) when randomizations occur at the second level (two-sided significance level $\alpha = 0.05$)   154

5.2   Sample size and power for detecting a main effect $\delta_{(1)}$ in model (5.8) when randomizations occur at the first level (two-sided significance level $\alpha = 0.05$)   157

5.3   Sample size and power for detecting an effect $\delta_{(f)}$ on slope differences in a fixed-slope model (5.12) with $r_\tau = 0$ when randomizations occur at the second level (two-sided significance level $\alpha = 0.05$)   162

5.4   Sample size and power for detecting an effect $\delta_{(f)}$ on slope differences in a random-slope model (5.4.5) with $r_\tau = 0.1$ when randomizations occur at the second level (two-sided significance level $\alpha = 0.05$)   164

5.5   Sample size and power for detecting a main effect $\delta_{(e)}$ at the end of study in a fixed-slope model (5.22) when randomizations occur at the second level (two-sided significance level $\alpha = 0.05$)   167

5.6   Sample size and power for detecting a two-way interaction XZ effect $\delta_{XZ(2)}$ in model (5.25) for a 2-by-2 factorial design when randomizations occur at the second level (two-sided significance level $\alpha = 0.05$)   170

5.7   Sample size and power for detecting a two-way interaction XZ effect $\delta_{XZ(1)}$ in model (5.28) for a 2-by-2 factorial design when randomizations occur at the first level (two-sided significance level $\alpha = 0.05$)   173

5.8   Sample size and statistical power for detecting a three-way interaction XZT effect δ_{XZT} in model (5.31) for a 2-by-2 factorial design when randomizations occur at the second level (two-sided significance level $\alpha = 0.05$) 176

5.9   Sample size and statistical power for detecting a main effect $|p_1 - p_0|$ on binary outcome in model with $m = 2$ (5.34) when randomizations occur at the second level (two-sided significance level $\alpha = 0.05$) 179

5.10   Sample size and statistical power for detecting a main effect $|p_1 - p_0|$ on binary outcome in model with $m = 1$ (5.34) when randomizations occur at the first level (two-sided significance level $\alpha = 0.05$) 181

6.1   Sample size and power for detecting a main effect $\delta_{(3)}$ in model (6.4) when randomizations occur at the third level with $\rho_2 = 0.05$ (two-sided significance level $\alpha = 0.05$) . . . 192

6.2   Sample size and power for detecting a main effect $\delta_{(2)}$ in model (6.9) when randomizations occur at the second level with $\rho_2 = 0.05$ (two-sided significance level $\alpha = 0.05$) . . . 195

6.3   Sample size and power for detecting a main effect $\delta_{(1)}$ in model (6.13) when randomizations occur at the first level with $\rho_2 = 0.05$ (two-sided significance level $\alpha = 0.05$) . . . 198

6.4   Sample size and power for detecting an effect $\delta_{(f)}$ on slope differences in a three-level fixed-slope model (6.17) with $r_\tau = 0$ when randomizations occur at the third level (two-sided significance level $\alpha = 0.05$) 204

6.5   Sample size and power for detecting an effect $\delta_{(r)}$ on slope differences in a three-level random-slope model (6.22) with $r_\tau = 0.1$ when randomizations occur at the third level (two-sided significance level $\alpha = 0.05$) 207

6.6   Sample size and power for detecting a main effect $\delta_{(e)}$ at the end of study in a three-level fixed-slope model (6.28) when randomizations occur at the third level (two-sided significance level $\alpha = 0.05$) 211

6.7   Sample size and power for detecting a two-way interaction XZ effect $\delta_{XZ(3)}$ in model with $m = 3$ (6.31) for a 2-by-2 factorial design when randomizations occur at the third level with $\rho_2 = 0.05$ (two-sided significance level $\alpha = 0.05$) . . . 214

6.8   Sample size and power for detecting a two-way interaction XZ effect $\delta_{XZ(2)}$ in model with $m = 2$ (6.31) for a 2-by-2 factorial design when randomizations occur at the second level with $\rho_2 = 0.05$ (two-sided significance level $\alpha = 0.05$) . 216

6.9 Sample size and power for detecting a two-way interaction XZ effect $\delta_{XZ(1)}$ in model with $m = 1$ (6.31) for a 2-by-2 factorial design when randomizations occur at the first level with $\rho_2 = 0.05$ (two-sided significance level $\alpha = 0.05$) . . . 219

6.10 Sample size and power for detecting a three-way interaction XZT effect δ_{XZT} in model (6.38) for a 2-by-2 factorial design when randomizations occur at the third level (two-sided significance level $\alpha = 0.05$) 222

6.11 Sample size and statistical power for detecting a main effect $|p_1 - p_0|$ on binary outcome in model with $m = 3$ (6.41) when randomizations occur at third level (two-sided significance level $\alpha = 0.05$) 226

6.12 Sample size and statistical power for detecting a main effect $|p_1 - p_0|$ on binary outcome in model with $m = 2$ (6.41) when randomizations occur at second level (two-sided significance level $\alpha = 0.05$) 228

6.13 Sample size and statistical power for detecting a main effect $|p_1 - p_0|$ on binary outcome in model with $m = 1$ (6.41) when randomizations occur at first level (two-sided significance level $\alpha = 0.05$) 231

1

Sample Size Determination for Independent Outcomes

1.1 Introduction

One of the most common questions any statistician gets asked from clinical investigators is "How many subjects do I need?" Researchers are often surprised to find out that the required sample size depends on a number of factors and they have to provide information to a statistician before they can get an answer. Clinical research is judged to be valid not by the results but by how it is designed and conducted. The cliche "do it right or do it over" is particularly apt in clinical research.

One of the most important aspects in clinical research design is the sample size estimation. In planning a clinical trial, it is necessary to determine the number of subjects to be recruited for the clinical trial in order to achieve sufficient power to detect the hypothesized effect. The ICH E9 guidance [1] states: "The number of subjects in a clinical trial should always be large enough to provide a reliable answer to the questions addressed. This number is usually determined by the primary objective of the trial. If the sample size is determined on some other basis, then this should be made clear and justified. For example, a trial sized on the basis of safety questions or requirements or important secondary objectives may need larger or smaller numbers of subjects than a trial sized on the basis of the primary efficacy question." Sample size in clinical trials must be carefully estimated if the results are to be credible. If the number of subjects is too small, even a well–conducted trial will have little chance of detecting the hypothesized effect. Ideally, the sample size should be large enough to have a high probability of detecting a clinically important difference between treatment groups and to show it to be statistically significant if such a difference really exists. If the number of subjects is too large, the clinical trial will lead to statistical significance for an effect of little clinical importance. Conversely, the clinical trial may not lead to statistical significance despite a large difference that is clinically important if the number of subjects is too small.

When an investigator designs a study, an investigator should consider constraints such as time, cost, and the number of available subjects. However, these constraints should not dictate the sample size. There is no reason to

1

carry out a study that is too small, only to come up with results that are inconclusive, since an investigator will then need to carry out another study to confirm or refute the initial results. Selecting an appropriate sample size is a crucial step in the design of a study. A study with an insufficient sample size may not have sufficient statistical power to detect meaningful effects and may not produce reliable answers to important research questions. Krzywinski and Altman [2] say that the ability to detect experimental effect is weakened in studies that do not have sufficient power. Choosing the appropriate sample size increases the chance of detecting a clinically meaningful effect and ensures that the study is both ethical and cost-effective.

Sample size is usually estimated by precision analysis or power analysis. In precision analysis, sample size is determined by the standard error or the margin of error at a fixed significance level. The approach of precision analysis is simple and easy to estimate the sample size [3]. In power analysis, sample size is estimated to achieve a desired power for detecting a clinically or scientifically meaningful difference at a fixed type I error rate. Power analysis is the most commonly used method for sample size estimation in clinical research. The sample size calculation requires assumptions that typically cannot be tested until the data have been collected from the trial. Sample size calculations are thus inherently hypothetical.

1.2   Precision Analysis

Sample size estimation is needed for the study in which the goal is to estimate the unknown parameter with a certain degree of precision. Thus, some key decisions in planning a study are "How precise will the parameter estimate be if I select a particular sample size?" and "How large a sample size do I need to attain a desirable level of precision?" What we are essentially saying is that we want the confidence interval to be of a certain width, in which the $100(1-\alpha)\%$ confidence level reflects the probability of including the true (but unknown) value of the parameter. Since the precision is determined by the width of the confidence interval, the goal of precision analysis is to determine the sample size that allows the confidence interval to be within a pre-specified width. The narrower the confidence interval is, the more precise the parameter inference is. Confidence interval estimation provides a convenient alternative to significance testing in most situations. The confidence interval approach is equivalent to the method of hypothesis testing. That is, if the confidence interval does not include the parameter value under the null hypothesis, the null hypothesis is rejected at a two–sided significance level of α. For example, consider the hypothesis of no difference between means (μ_1 and μ_2). The method of hypothesis testing rejects the hypothesis $H_0 : \mu_1 - \mu_2 = 0$ at the two–sided significance level of α if and only if the $100(1-\alpha)\%$ confidence interval for the mean difference ($\mu_1 - \mu_2$) does not include the value zero. Thus, the significance test can be performed with the confidence interval approach.

1.2.1 Continuous Outcomes

Suppose that $Y_1, \ldots, Y_n$ are independent and identically distributed normal random variables with mean μ and variance σ^2. The parameter μ can be estimated by the sample mean $\bar{y} = \sum_{i=1}^{n} Y_i$. When σ^2 is known, the $100(1 - \alpha)\%$ confidence interval is

$$\bar{y} \pm z_{1-\alpha/2} \frac{\sigma}{\sqrt{n}},$$

where $z_{1-\alpha/2}$ is the $100(1 - \alpha/2)$th percentile of the standard normal distribution. Note that the sample size estimate based on precision analysis depends on the type I error rate, not on the type II error rate. The maximum half width of the confidence interval is called the maximum error of an estimate of the unknown parameter. Suppose that the maximum error of μ is δ. Then, the required minimum sample size is the smallest integer that is greater than or equal to n solved from the following equation:

$$z_{1-\alpha/2} \frac{\sigma}{\sqrt{n}} = \delta.$$

Thus, the required sample size is the smallest integer that is greater than or equal to n:

$$n = \frac{z_{1-\alpha/2}^2 \sigma^2}{\delta^2}. \tag{1.1}$$

From Equation (1.1), we can obtain the required sample size once the maximum error or the width of the $100(1 - \alpha)\%$ confidence interval of μ is specified.

1.2.1.1 Example

Suppose that a clinical investigator is interested in estimating how much reduction will be made on the fasting serum–cholesterol level with administration of a new cholesterol–lowering drug for 6 months among recent Hispanic immigrants with a given degree of precision. Suppose that the standard deviation (σ) for reduction in cholesterol level equals 40 mg/dl. We would like to estimate the minimum sample size needed to estimate the reduction in fasting serum–cholesterol level if we require that the 95% confidence interval for reduction in cholesterol level is no wider than 20 mg/dl. The $100(1 - \alpha)\%$ confidence interval for true reduction in fasting serum–cholesterol level is

$$\bar{y} \pm z_{1-\alpha/2} \frac{\sigma}{\sqrt{n}},$$

where $\bar{y}$ is the mean change in fasting serum–cholesterol level after administration of a drug, and $z_{1-\alpha/2}$ is the $100(1 - \alpha/2)$th percentile of the standard normal distribution. The width of a 95% confidence interval is

$$2 \cdot z_{1-\alpha/2} \frac{\sigma}{\sqrt{n}} = 2 \cdot 1.96 \cdot \frac{40}{\sqrt{n}}.$$

We want the width of the 95% confidence interval to be no wider than 20 mg/dl. The required sample size is the smallest integer satisfying $n \geq 4 \cdot (1.96)^2 (40)^2 / (20)^2 = 61.5$. In order for a 95% confidence interval of reduction in cholesterol level to be no wider than 20 mg/dl, we need at least 62 subjects when the standard deviation for reduction in cholesterol level equals to 40 mg/dl.

1.2.2   Binary Outcomes

The study goal may be based on finding a suitably narrow confidence interval for the statistics of interest at a given significance level (α), where the significance level is usually considered as the maximum probability of type I error that can be tolerated. We may want to know how many subjects are required for the $100(1 - \alpha)\%$ confidence interval to be a certain width.

Suppose that $Y_1, \ldots, Y_n$ are independent and identically distributed Bernoulli random variables with mean $p = E(Y_i), (i = 1, \ldots, n)$. The parameter p can be estimated by the sample mean $\hat{p} = \sum_{i=1}^{n} Y_i / n$. For large n, $\hat{p}$ is asymptotically normal with mean p and variance $p(1-p)/n$. The $100(1-\alpha)\%$ confidence interval for p is

$$\hat{p} \pm z_{1-\alpha/2} \sqrt{\frac{\hat{p}(1 - \hat{p})}{n}}.$$

Suppose that the maximum error of p is δ. Then, the sample size can be estimated by

$$z_{1-\alpha/2} \sqrt{\frac{\hat{p}(1 - \hat{p})}{n}} = \delta.$$

Thus, the required sample size is

$$n = \frac{z_{1-\alpha/2}^2 \hat{p}(1 - \hat{p})}{\delta^2}. \tag{1.2}$$

We can estimate the sample size from Equation (1.2) once the maximum error or the width of the $100(1 - \alpha)\%$ confidence interval for p is specified. There are a number of alternative ways to estimate the confidence interval for a binomial proportion [4].

1.2.2.1   Example

Suppose that a clinical investigator is interested in conducting a clinical trial with a new cancer drug to estimate the response rate with a maximum error of 20%. In oncology, the response rate (RR) is generally defined as the proportion of patients whose tumor completely disappears (termed a complete response, CR) or shrinks more than 50% after treatment (termed a partial response, PR). In simpler terms, RR = PR + CR. An investigator expects the response rate of a new cancer drug to be 30%. How many patients are needed to achieve a maximum error of 20%? Let $\hat{p}$ be the estimate of the response rate. The maximum error of the response rate is $z_{1-\alpha/2}\sqrt{\hat{p}(1-\hat{p})/n}$. With the guessed value of $\hat{p} = 0.3$, a maximum error of p is $z_{1-\alpha/2}\sqrt{0.3 \cdot 0.7/n}$. Thus, we need $z_{1-\alpha/2}\sqrt{0.3 \cdot 0.7/n} \leq 0.2$, or $n \geq 21$. That is, we need at least 21 subjects to obtain a maximum error $\leq 20\%$. When we do not know the value of p, a conservative approach is to use $\hat{p}$ that yields the maximum error. The maximum error of p occurs when $\hat{p} = 0.5$. So, a conservative maximum error of p is $z_{1-\alpha/2}\sqrt{0.5 \cdot 0.5/n} = z_{1-\alpha/2}0.5/\sqrt{n}$. Thus, $1.96 \cdot 0.5/\sqrt{n} \leq 0.2$ at a 5% significance level. Therefore, the required sample size is $n = 25$. An investigator should recruit at least 25 subjects to achieve a maximum error of 20% in the response rate estimation.

The larger the sample size, the more precise the estimate of the parameter will be if all the other factors are equal. An investigator should specify what degree of precision is aimed for the study. A trial will take more cost and time as the size of a trial increases. In order to estimate the sample size using precision analysis, we need to decide how large the maximum error of the unknown parameter is or how wide the confidence interval for the unknown parameter is, and we need to know the formula for the relevant maximum error.

1.3 Power Analysis

Power analysis uses two types of errors (type I and II errors) for sample size estimation while precision analysis uses only one type of error (type I error) for sample size estimation. Power analysis tests the null hypothesis at a predetermined level of significance with a desired power.

1.3.1 Information Needed for Power Analysis

A clinical trial that is conducted without attention to sample size or power information takes the risks of either failing to detect clinically meaningful differences (i.e., type II error) or using an unnecessarily excessive number of subjects for a study. Either case fails to adhere to the Ethical Guidelines of the American Statistical Association which says, "Avoid the use of excessive or inadequate number of research subjects by making informed recommendations for study size" [5]. The sample size estimate is important for economic and ethical reasons [6]. An oversized clinical trial exposes more than necessary number of subjects to a potentially harmful trial, and uses more resources than necessary. An undersized clinical trial exposes the subjects to a potentially harmful trial and leads to a waste of resources without producing useful results. The sample size estimate will allow the estimation of total cost of the proposed study. While the exact final number that will be used for analysis will be unknown due to missing information such as lack of demographic information and clinical information, it is still desirable to determine a target sample size based on the proposed study design. In this section, we describe the general information needed to estimate the sample size for the trial.

1. Choose the primary endpoint

The primary endpoint should be chosen so that the primary objective of the trial can be assessed, and the primary endpoint is generally used for sample size estimation. Primary endpoint measures the outcome that will answer the primary question being asked by a trial. Suppose that the primary hypothesis is to test whether the new cancer drug yields longer overall survival than the standard cancer drug. In this case, the primary endpoint is overall survival. The sample size for a trial is determined by the power needed to detect a clinically meaningful difference in overall survival at a given significance level. The secondary hypothesis is to investigate other relevant questions from the same trial. For example, the secondary hypothesis is to test whether the new cancer drug produces better quality of life than the standard cancer drug, or whether the new cancer drug yields longer progression–free survival than the standard cancer drug.

The sample size calculation depends on the type of primary endpoint. The variable type of the primary outcome must be defined before sample size and power calculations can be conducted. The variable type may be continuous, categorical, ordinal, or survival. Categorical variables may have only two categories or more than two categories.

- A quantitative (or continuous) outcome representing a specific measure (e.g., total cholesterol, quality of life, or blood pressure). Mean and median can be used to compare the primary endpoint between treatment groups.

- A binary outcome indicating occurrence of an event (e.g., the occurrence of myocardial infarction, or the occurrence of recurrent disease). Odds ratio, risk difference, and risk ratio can be used to compare the primary endpoint between treatment groups.

- Survival outcome for the time to occurrence of an event of interest (e.g., the time from study entry to death, or time to progression). A Kaplan–Meier survival curve is often used to graphically display the time to the event, and log–rank test or Cox regression analysis is frequently used to test if there is a significant difference in the treatment effect between treatment groups.

2. Determine the hypothesis of interest

The primary purpose of a clinical trial is to address a scientific hypothesis, which is usually related to the evaluation of the efficacy and safety of a drug product. To address a hypothesis, different statistical methods are used depending on the type of question to be answered. Most often the hypothesis is related to the effect of one treatment as compared to another. For example, one trial could compare the effectiveness of a new drug to that of a standard drug. Yet the specific comparison to be performed will depend on the hypothesis to be addressed. Let μ_1 and μ_2 be the mean responses of a new drug and a standard drug, respectively.

- A superiority test is designed to detect a meaningful difference in mean response between a standard drug and a new drug [7]. The primary objective is to show that the mean response of a new drug is different from that of a standard drug.

$$H_0 : \mu_1 = \mu_2 \; versus \; H_1 : \mu_1 \neq \mu_2$$

The null hypothesis (H_0) says that the two drugs are not different with respect to the mean response ($\mu_1 = \mu_2$). The alternative hypothesis (H_1) says that the two drugs are different with respect to the mean response ($\mu_1 \neq \mu_2$). The statistical test is a two–sided test since there are two chances of rejecting the null hypothesis ($\mu_1 > \mu_2$ or $\mu_1 < \mu_2$) with each side allocated an equal amount of the type I error of $\alpha/2$.

If the alternative hypothesis is $\mu_1 > \mu_2$ or $\mu_1 < \mu_2$ instead of $\mu_1 \neq \mu_2$, then the statistical test is referred to as a one–sided test since there is only one chance of rejecting the null hypothesis with one side allocated the type I error of α.

- An equivalence test is designed to confirm the absence of a meaningful difference between a standard drug and a new drug. The primary objective is to show that the mean responses to two drugs differ by an amount that is clinically unimportant. This is usually demonstrated by showing that the absolute difference in mean responses between drugs is likely to lie within an equivalence margin (Δ) of clinically acceptable differences.

$$H_0 : |\mu_1 - \mu_2| \geq \Delta \; versus \; H_1 : |\mu_1 - \mu_2| < \Delta$$

The null hypothesis (H_0) says that the two drugs are different with respect to the mean response ($|\mu_1 - \mu_2| \geq \Delta$). The alternative hypothesis (H_1) says that the two drugs are not different with respect to the mean response ($|\mu_1 - \mu_2| < \Delta$). In an equivalence test, an investigator wants to test if the difference between a new drug and a standard drug is of no clinical importance. This is to test for equivalence of two drugs.

The null hypothesis is expressed as a union ($\mu_1 - \mu_2 \geq \Delta$ or $\mu_1 - \mu_2 \leq -\Delta$) and the alternative hypothesis (H_1) as an intersection ($-\Delta < \mu_1 - \mu_2 < \Delta$). Each component of the null hypothesis needs to rejected to conclude equivalence.

- A non–inferiority test is designed to show that a new drug is not less effective than a standard drug by more than Δ, the margin of non–inferiority. The null and alternative hypotheses can be specified as:

$$H_0 : \mu_1 - \mu_2 \le -\Delta \ versus \ H_1 : \mu_1 - \mu_2 > -\Delta$$

The null hypothesis (H_0) says that a new drug is inferior to a standard drug with respect to the mean response. The alternative hypothesis (H_1) says that a new drug is non–inferior to a standard drug with respect to the mean response. That is, the alternative hypothesis of non–inferiority trial states that a standard drug may indeed be more effective than a new drug, but no more than Δ. In phase III clinical trials that compare a new drug with a standard drug, non–inferiority trials are more common than equivalence trials since it is only the non–inferiority limit that is usually of interest. This is to test for non–inferiority of the new drug.

Choice of hypothesis depends on which scientific question an investigator is trying to answer. All the above hypothesis tests are useful in the development of drugs. In comparison studies with a standard drug, a non–inferiority trial is used to demonstrate that a new drug provides at least the same benefit to the subject as a standard drug. Non–inferiority trials are commonly used when a new drug is easier to administer, less expensive, and less toxic than a standard drug. Equivalence trials are used to show that a new drug is identical (within an acceptable range) to a standard drug. This is used in the registration and approval of biosimilar drugs that are shown to be equivalent to their branded reference drugs [8]. Most equivalence trials are bioequivalence trials that aim to compare a generic drug with the original branded reference drug.

3. Determine Δ

Sample size calculation depends on the hypothesis of interest. For a superiority test, the necessary sample size depends on the clinically meaningful difference (Δ). In superiority trials, fewer subjects will be needed for a larger value of Δ while more subjects will be needed for a smaller value of Δ. For instance, we can detect a 40% difference in efficacy with a modest number of subjects. However, a larger number of subjects will be needed to reliably detect a 10% difference in efficacy. Because sample size is inversely related to the square of Δ, even the slightly misspecified difference can lead to a large change in the sample size. Clinically meaningful differences are commonly specified using one of two approaches. One is to select the drug effect deemed important to detect, and the other is to calculate the sample size according to the best guess concerning the true effect of drug [9].

For an equivalence test, the required sample size depends on the margin of clinical equivalence. In an equivalence test, the equivalence margin of clinically acceptable difference (Δ) depends on the disease being studied. For example, an absolute difference of 1% is often used as the clinically meaningful difference in thrombolytic trials while a 20% difference is considered as clinically meaningful in most other situations including migraine headache [10]. Bioequivalence trials aim to show the equivalent pharmacokinetic profile through the most commonly used pharmacokinetic variables such as area under the curve (AUC) and maximum concentration(C_{max}). Average bioequivalence is widely used for comparison of a generic drug with the original branded drug. The 80/125 rule is currently used as regulation for the assessment of average bioequivalence [11]. For average bioequivalence, the FDA [11] recommends that the geometric means ratio between the test drug and the reference drug is within 80% and 125% for the bioavailability measures $(AUC$ and $C_{max})$.

For a non–inferiority test, the necessary sample size depends on the upper bound for non–inferiority. Setting the non–inferiority margin is a major issue in designing a non–inferiority trial. The Food and Drug Administration [12] and the European Medicines Agency [13] issued guidances on the choice of non–inferiority margin. The choice of the non–inferiority margin needs to take account of both statistical reasoning and clinical judgement. An appropriate selection of non–inferiority margin should provide assurance that a new drug has a clinically relevant superiority over placebo, and a new drug is not substantially inferior to a standard drug, which results in a tighter margin.

The clinically or scientifically meaningful margin (Δ) needs to be specified to estimate the number of subjects for the trial since the purpose of the sample size estimation is to provide sufficient power to reject the null hypothesis when the alternative hypothesis is true.

In this book, we restrict the sample size estimation to a superiority test, which is most commonly used in clinical trials. Julious [7, 14, 15] and Chow *et al.* [3] provided general sample size formulas for equivalence trials and non–inferiority trials.

4. Determine the variance of the primary endpoint

The variance of the primary endpoint is usually unknown in advance. In cross-sectional studies, the variance or the standard deviation is generally obtained from either previous studies or pilot studies. However, for correlated outcomes such as clustered outcomes or repeated measurement outcomes, the variance of the primary endpoint generally needs to be estimated utilizing various sources of information such as missing proportion, correlation among measurements, and the number of measurements, etc. Detailed description of the estimation of the variance for correlated outcomes will be given in later chapters. A large variance will lead to a large sample size for a study. That is, as the variance increases, the sample size increases.

5. Choose type I error and power

Type I error (α) is the probability of rejecting the null hypothesis when the null hypothesis is actually true. Type II error (β) is the probability of not rejecting the null hypothesis when it is actually false. The aim of the sample size calculation is to estimate the minimal sample size required to meet the objectives of the study for a fixed probability of type I error to achieve a desired power, which is defined as $1 - \beta$. The power is the probability of rejecting the null hypothesis when it is actually false. A two–sided type I error of 5% is commonly used to reflect a 95% confidence interval for an unknown parameter, and this is familiar to most investigators as the conventional benchmark of 5%. As α decreases, the sample size increases. For example, a study with α level of 0.01 requires more sample size than a study with α level of 0.05.

Typically, the sample size is computed to provide a fixed level of power under a specified alternative hypothesis. The alternative hypothesis usually represents a minimal clinically or scientifically meaningful difference in efficacy between treatment groups. Power $(1 - \beta)$ is an important consideration in sample size determination. Low power can cause a true difference in a clinical outcome between study groups to go undetected. However, too much power may make results statistically significant when results do not show a clinically meaningful difference.

When there is a large difference such as a 100% real difference in therapeutic efficacy between a standard drug and a new drug, it is unlikely to be missed by most studies. That is, type II error (β) is small when there is a large difference in therapeutic efficacy. However, type II error is a common problem in studies that aim to distinguish between a standard drug and a new drug that may differ in therapeutic efficacy by only a small amount such as 1% or 5%. The number of subjects must be drastically increased to reduce type II error when the aim is to discriminate a small difference between a standard drug and a new drug. Otherwise, there is a high chance of incorrectly overlooking small differences in therapeutic efficacy with an insufficient number of subjects. Type II error (β) of 10% or 20% is commonly used for sample size estimation. That is, the power $(1 - \beta)$ of 80% or 90% is widely used for the design of the study. The higher the power, the less likely the risk of type II error. The power increases as the sample size increases. A sufficient sample size ensures that the study is able to reliably detect a true difference, and not underpowered.

6.   Select a statistical method for data analysis

A statistical method for sample size estimation should adequately align with the statistical method for data analysis [16]. For example, an investigator would like to test whether there is a significant difference in total cholesterol levels between those who take a new drug and who take a standard drug. The investigator plans to analyze the data using a two–sample t–test. In this case, a sample size calculation based on a two-group chi–square test with dichotomization of total cholesterol levels would be inappropriate since the statistical method used for power analysis is different from that to be used for data analysis. Discrepancy between the statistical method for sample size estimation and the statistical method for data analysis can lead to a sample size that is too large or too small. The statistical method used for sample size calculation should be the same as that used for data analysis.

1.3.2 One–Sample Test for Means

We illustrate the sample size calculation using a one–sided test through an example. Suppose that the total cholesterol levels for male college students are normally distributed with a mean (μ) of 180 mg/dl and a standard deviation (σ) of 80 mg/dl. Suppose that an investigator would like to examine whether the mean total cholesterol level of the physically inactive male college students is higher than 180 mg/dl using a one–sided 5% significance level (α). That is, an investigator would like to test the hypotheses: $H_0 : \mu = \mu_0 = 180$ mg/dl (or $\mu \leq 180$ mg/dl) versus $H_1 : \mu > 180$ mg/dl assuming that the standard deviation of the total cholesterol level is the same as that of male college students. An investigator wants to risk a 10% chance (90% power) of failing to reject the null hypothesis when the true mean (μ_1) of the total cholesterol level is as large as 210 mg/dl. How many subjects are needed to detect 30 mg/dl difference in total cholesterol level from the population mean of 180 mg/dl at a one–sided 5% significance level and a power of 90%?

For $\alpha = 0.05$, we would reject the null hypothesis (H_0) if the average total cholesterol level is greater than the critical value (C) in Figure 1.1, where $C = \mu_0 + z_{1-\alpha} \cdot \sigma/\sqrt{n} = 180 + 1.645 \cdot 80/\sqrt{n}$. If the true mean is 210 mg/dl with a power of 90% ($\beta = 0.1$), we would not reject the null hypothesis when the sample average is less than $C = \mu_1 + z_\beta \cdot \sigma/\sqrt{n} = 210 - 1.282 \cdot 80/\sqrt{n}$. The sample size ($n$) can be estimated by setting two equations equal to each other:

$$180 + 1.645 \cdot 80/\sqrt{n} = 210 - 1.282 \cdot 80/\sqrt{n}.$$

Therefore, the required number of subjects is

$$n = \frac{(1.645 + 1.282)^2 \cdot 80^2}{(180 - 210)^2} = 61.$$

In general, the estimated sample size for a one–sided test for testing $H_0 : \mu = \mu_0$ versus $H_1 : \mu > \mu_1$ with a significance level of α and a power of $1 - \beta$ is the smallest integer that is larger than or equal to n satisfying the following equation

$$n = \frac{(z_{1-\alpha} + z_{1-\beta})^2 \sigma^2}{(\mu_0 - \mu_1)^2}. \tag{1.3}$$

We will show how the sample size can be estimated for a two–sided one–sample test. Let n be the number of subjects. Let Y_i denote the response for subject i, $(i = 1, \ldots, n)$, and $\bar{y}$ be the sample mean. We assume that $Y_i's$ are independent and normally distributed random variables with mean μ_0 and variance σ^2. Suppose that we want to test the hypotheses $H_0 : \mu = \mu_0$ versus $H_1 : \mu = \mu_1 \neq \mu_0$.

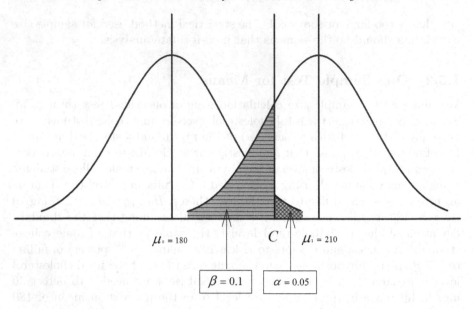

FIGURE 1.1
Sample size estimation for a one–sided test in a one–sample problem

When σ^2 is known, we reject the null hypothesis at the significance level α if

$$\left| \frac{\bar{y} - \mu_0}{\sigma/\sqrt{n}} \right| > z_{1-\alpha/2},$$

where $z_{1-\alpha/2}$ is the $100(1 - \alpha/2)$th percentile of the standard normal distribution. Under the alternative hypothesis ($H_1 : \mu = \mu_1$), the power is given by

$$\Phi\left(\frac{\sqrt{n}(\mu_1 - \mu_0)}{\sigma} - z_{1-\alpha/2} \right) + \Phi\left(-\frac{\sqrt{n}(\mu_1 - \mu_0)}{\sigma} - z_{1-\alpha/2} \right),$$

where Φ is the cumulative standard normal distribution function. By ignoring the small value of the second term in the above equation, the power is approximated by the first term. Thus, the sample size required to achieve the power of $1 - \beta$ can be obtained by solving the following equation

$$\frac{\sqrt{n}(\mu_1 - \mu_0)}{\sigma} - z_{1-\alpha/2} - z_{1-\beta}.$$

The required sample size is the smallest integer that is larger than or equal to n satisfying the following equation

$$n = \frac{(z_{1-\alpha/2} + z_{1-\beta})^2 \sigma^2}{(\mu_1 - \mu_0)^2}. \tag{1.4}$$

If the population variance σ^2 is unknown, σ^2 can be estimated by the sample variance $s^2 = \sum_{i=1}^{n}(y_i - \bar{y})^2/(n-1)$, which is an unbiased estimator of σ^2. For large n, we reject the null hypothesis $H_0 : \mu = \mu_0$ at the significance level α if

$$\left|\frac{\bar{y} - \mu_0}{s/\sqrt{n}}\right| > z_{1-\alpha/2}.$$

Therefore, the sample size estimates for a one–sided test and a two–sided test can be obtained by replacing σ^2 by s^2 in Equations (1.3) and (1.4).

1.3.2.1 Example

Consider the design of a single-arm psychiatric study that evaluates the effect of a test drug on cognitive functioning of children with mental retardation before and after administration of a test drug. A pilot study shows that the mean difference in cognitive functioning before and after taking a test drug was 6 with a standard deviation equal to 9. We would like to estimate the sample size needed to detect the mean difference of 6 in cognitive functioning to achieve 80% power at a two–sided 5% significance level assuming a standard deviation of 9. Let μ denote the mean difference in cognitive functioning between pre- and post-drug administration. The null hypothesis $H_0 : \mu = 0$ is to be tested against the alternative hypothesis $H_1 : \mu = 6$. From Equation (1.4), $n = (1.960 + 0.842)^2 \cdot 9^2/6^2 = 17.7$. Therefore, a sample size of 18 subjects is needed to detect a change in mean difference of 6 in cognitive functioning, assuming a standard deviation of 9 using a normal approximation with a two–sided significance level of 5% and a power of 80%.

1.3.2.2 Example

Concerning the effect of a test drug on systolic blood pressure before and after the treatment, a pilot study shows that the mean systolic blood pressure changes after a 4–month administration of a test drug was 15 mm Hg with a standard deviation of 40 mm Hg. We would like to estimate the sample size needed to detect 15 mm Hg in systolic blood pressure to achieve 80% power at a two–sided 5% significance level assuming the standard deviation of 40 mm Hg. From Equation (1.4), $n = (1.960 + 0.842)^2 \cdot 40^2/15^2 = 55.8$. Therefore, a sample size of 56 subjects will have 80% power to detect a change in mean of 15 mm Hg in systolic blood pressure, assuming a standard deviation of 40 mm Hg at a two–sided 5% significance level.

1.3.3 One–Sample Test for Proportions

Let Y_i denote a binary response variable of the ith subject with $p = E(Y_i)$, $(i = 1, \ldots, n)$, where n is the number of subjects in the trial. For example, Y_i can denote the response or non–response in cancer clinical trials, where $Y_i = 0$ denotes non–response, and $Y_i = 1$ denotes response, which includes either complete response or partial response. The response rate can be estimated by the observed proportion $\hat{p} = \sum_{i=1}^{n} Y_i/n$, where n is the number of subjects. We illustrate the sample size calculation using the one–sided test. Suppose we wish to test the null hypothesis $H_0 : p = p_0$ versus the alternative hypothesis $H_1 : p = p_1 > p_0$ at the one–sided significance level of α. Under the null hypothesis, the test statistic

$$Z = \frac{\hat{p} - p_0}{\sqrt{\hat{p}(1 - \hat{p})/n}}$$

approximately has a standard normal distribution for large n. We reject the null hypothesis at a significance level α if the test statistic Z is greater than $z_{1-\alpha}$.

For $\alpha = 0.05$, we would reject the null hypothesis (H_0) if the average response rate is greater than the critical value (C), where $C = p_0 + z_{1-\alpha}\sqrt{p_0(1 - p_0)/n}$. If the alternative hypothesis is true, that is, if the true response rate is p_1, we would not reject the null hypothesis if the response rate is less than $C = p_1 + z_\beta\sqrt{p_1(1 - p_1)/n}$.

By setting the two equations equal, we get

$$p_0 + z_{1-\alpha}\sqrt{p_0(1 - p_0)/n} = p_1 + z_\beta\sqrt{p_1(1 - p_1)/n}.$$

The required sample size to test $H_0 : p = p_0$ versus $H_1 : p = p_1 > p_0$ at a one–sided significance level of α and a power of $1 - \beta$ is

$$n = \frac{(z_{1-\alpha}\sqrt{p_0(1 - p_0)} + z_{1-\beta}\sqrt{p_1(1 - p_1)})^2}{(p_1 - p_0)^2}.$$

The sample size for a two–sided test $H_0 : p = p_0$ versus $H_1 : p = p_1$ for $p_1 \neq p_0$ can be obtained by replacing $z_{1-\alpha}$ by $z_{1-\alpha/2}$ as shown in a one–sample test for means:

$$n = \frac{(z_{1-\alpha/2}\sqrt{p_0(1 - p_0)} + z_{1-\beta}\sqrt{p_1(1 - p_1)})^2}{(p_1 - p_0)^2}. \tag{1.5}$$

1.3.3.1   Example

Consider the design of a single-arm oncology clinical trial that evaluates if a new molecular therapy has at least a 40% response rate. Let p be the response rate of a new molecular therapy. We would like to estimate the sample size needed to test the null hypothesis $H_0 : p = p_0 = 0.20$ against the alternative hypothesis $H_1 : p = p_1 \neq p_0$. The trial is designed based on a two–sided test that achieves 80% power at $p = p_1 = 0.40$ with a two–sided 5% significance level. From Equation (1.5),

$$n = \frac{(1.96\sqrt{0.2(1 - 0.2)} + 0.842\sqrt{0.4(1 - 0.4)})^2}{(0.4 - 0.2)^2} = 35.8.$$

The required number of subjects is 36 to detect the difference between the null hypothesis proportion of 0.2 and the alternative proportion of 0.4 at a two–sided significance level of 5% and a power of 80%.

1.3.4   Two–Sample Test for Means

Suppose that $Y_{1i}, (i = 1, ..., n_1)$ and $Y_{2i}, (i = 1, ..., n_2)$ represent observations from groups 1 and 2, and Y_{1i} and Y_{2i} are independent and normally distributed with means μ_1 and μ_2 and variances σ_1^2 and σ_2^2, respectively. Let's consider a one–sided test. Suppose that we want to test the hypotheses $H_0 : \mu_1 = \mu_2$ versus $H_1 : \mu_1 > \mu_2$.

Let $\bar{y}_1$ and $\bar{y}_2$ be the sample means of Y_{1i} and Y_{2i}. Assume that the variances σ_1^2 and σ_2^2 are known, and $n_1 = n_2 = n$. Then, the Z–test statistic can be written as

$$Z = \frac{\bar{y}_1 - \bar{y}_2}{\sqrt{\sigma_1^2/n + \sigma_2^2/n}}.$$

Under the null hypothesis (H_0), the test statistic Z is normally distributed with mean 0 and variance 1. Thus, we reject the null hypothesis if $Z > z_{1-\alpha}$. Under the alternative hypothesis (H_1), let $\mu_1 - \mu_2 = \Delta$, which is the clinically meaningful difference to be detected. Then, under the alternative hypothesis (H_1), the expected value of $(\bar{y}_1 - \bar{y}_2)$ is Δ, and Z follows the normal distribution with mean μ^* and variance 1, where $\mu^* = \Delta/\sqrt{\sigma_1^2/n + \sigma_2^2/n}$.

Under the null hypothesis (H_0),

$$P\{Z > z_{1-\alpha}|H_0\} < \alpha.$$

Similarly, under the alternative hypothesis (H_1),

$$P\{Z > z_{1-\alpha}|H_1\} > 1 - \beta.$$

That is,

$$P\{\frac{\bar{y}_1 - \bar{y}_2}{\sqrt{\sigma_1^2/n + \sigma_2^2/n}} > z_{1-\alpha}|H_1\} > 1 - \beta.$$

Under the alternative hypothesis, the expected value of $(\bar{y}_1 - \bar{y}_2)$ is Δ. Thus,

$$P\{\frac{(\bar{y}_1 - \bar{y}_2) - \Delta}{\sqrt{\sigma_1^2/n + \sigma_2^2/n}} > z_{1-\alpha} - \frac{\Delta}{\sqrt{\sigma_1^2/n + \sigma_2^2/n}}|H_1\} > 1 - \beta.$$

The above equation can be written as follows due to the symmetry of the normal distribution:

$$z_{1-\alpha} - \frac{\Delta}{\sqrt{\sigma_1^2/n + \sigma_2^2/n}} = z_\beta = -z_{1-\beta}.$$

The simple manipulation yields the required sample size per group assuming equal allocation of subjects in each group,

$$n = \frac{(\sigma_1^2 + \sigma_2^2)(z_{1-\alpha} + z_{1-\beta})^2}{\Delta^2}.$$

If $\sigma_1^2 = \sigma_2^2 = \sigma^2$, then the required sample size per group is

$$n = \frac{2\sigma^2(z_{1-\alpha} + z_{1-\beta})^2}{\Delta^2}. \tag{1.6}$$

In some randomized clinical trials, more subjects are assigned to the treatment group than to the control group to encourage participation of subjects in a trial due to their higher chance of being randomized to the treatment group than the control group. Let $n_1 = n$ be the number of subjects in the control group and $n_2 = kn$ be the number of subjects in the treatment group. Then, the sample size for the study will be

$$n_1 = n = (1 + 1/k)\sigma^2 \frac{(z_{1-\alpha} + z_{1-\beta})^2}{\Delta^2}. \tag{1.7}$$

The total sample size for the trial is $n_1 + n_2$. The relative sample size required to maintain the power and type I error rate of a trial against the trial with an equal number of subjects in each group is $(2 + k + 1/k)/4$. For example, in a trial that randomizes subjects in a 2:1 ratio requires a 12.5% larger sample size in order to maintain the same power as a trial with a 1:1 randomization.

The sample size needed to detect the difference in means between two groups with a two–sided test can be obtained by replacing $z_{1-\alpha}$ by $z_{1-\alpha/2}$ as shown in a one–sample test for means:

$$n_1 = n = (1 + 1/k)\sigma^2 \frac{(z_{1-\alpha/2} + z_{1-\beta})^2}{\Delta^2}. \tag{1.8}$$

If the population variance σ^2 is unknown, σ^2 can be estimated by the sample pooled variance $s^2 = \{\sum_{i=1}^{n_1}(y_{1i}-\bar{y}_1)^2 + \sum_{i=1}^{n_2}(y_{2i}-\bar{y}_2)^2\}/(n_1+n_2-2)$, which is an unbiased estimator of σ^2. For large n_1 and n_2, we reject the null hypothesis $H_0 : \mu_1 = \mu_2$ against the alternative hypothesis $H_1 : \mu_1 \neq \mu_2$ at the significance level α if the absolute value of the test statistic Z is greater than $z_{1-\alpha/2}$.

$$Z = \frac{\bar{y}_1 - \bar{y}_2}{s\sqrt{\frac{1}{n_1} + \frac{1}{n_2}}}.$$

If $n_1 = n$ and $n_2 = kn$, the Z test statistic becomes

$$Z = \frac{\bar{y}_1 - \bar{y}_2}{s\sqrt{\frac{k+1}{kn}}}.$$

Therefore, the sample size estimates for a one–sided test and a two–sided test can be obtained by replacing σ^2 by s^2 in Equations (1.7) and (1.8).

1.3.4.1 Example

In a prior randomized clinical trial [17] investigating the effect of propranolol versus no propranolol in geriatric patients with New York Heart Association functional class II or III congestive heart failure (CHF), the changes in mean left ventricular ejection fraction (LVEF) from baseline to 1 year after treatment were 6% and 2% for propranolol and no propranolol groups, respectively. We will conduct a two–arm randomized clinical trial with a placebo and a new beta blocker drug to investigate if patients taking propranolol significantly improve LVEF after 1 year compared with patients taking placebo. We assume the similar increase in LVEF as in the prior study and a common standard deviation of 8% in changes in LVEF from baseline to 1 year after treatment. How many subjects are needed to test the superiority of a new drug in improving LVEF over placebo with a two–sided 5% significance level and 80% power? The required sample size is

$$n = \frac{2\sigma^2(z_{1-\alpha/2} + z_{1-\beta})^2}{\Delta^2} = 2 \cdot 8^2 \cdot (1.960 + 0.842)^2/4^2 = 62.8.$$

The required sample size is 63 subjects per group.

1.3.5 Two–Sample Test for Proportions

In a randomized clinical trial subjects are randomly assigned to one of two treatment groups. Let Y_{ij} be the binary random variable ($Y_{ij} = 1$ for response, 0 for no response) of the jth subject in the ith treatment, $j = 1, \ldots, n_i$, and $i = 1, 2$. We assume that $Y_{ij}'s$ are independent and identically distributed with $E(Y_{ij}) = p_i$ for a fixed i. The response rate p_i is usually estimated by the observed proportion in the ith treatment group:

$$\hat{p}_i = \sum_{j=1}^{n_i} Y_{ij}/n_i.$$

Let p_1 and p_2 be the response rates of control and treatment arms, respectively. The sample sizes are n_1 and n_2 in each treatment group, respectively. Suppose that an investigator wants to test whether there is a difference in the response rates between control and treatment arms. The null (H_0) and alternative (H_1) hypotheses are:

H_0: The response rates are equal ($p_1 = p_2$).

H_1: The response rates are different ($p_1 \neq p_2$).

We reject the null hypothesis $H_0 : p_1 = p_2$ at the significance level of α if

$$\left| \frac{\hat{p}_1 - \hat{p}_2}{\sqrt{\hat{p}_1(1 - \hat{p}_1)/n_1 + \hat{p}_2(1 - \hat{p}_2)/n_2}} \right| > z_{1-\alpha/2}.$$

Under the alternative hypothesis, the power of the test is approximated by

$$\Phi\left(\frac{|p_1 - p_2|}{\sqrt{p_1(1 - p_1)/n_1 + p_2(1 - p_2)/n_2}} - z_{1-\alpha/2} \right).$$

The sample size estimate needed to achieve a power of $1 - \beta$ can be obtained by solving the following equation:

$$\frac{|p_1 - p_2|}{\sqrt{p_1(1 - p_1)/n_1 + p_2(1 - p_2)/n_2}} - z_{1-\alpha/2} = z_{1-\beta}.$$

When $n_2 = k \cdot n_1$, n_1 can be written as

$$n_1 = \frac{(z_{1-\alpha/2} + z_{1-\beta})^2}{(p_1 - p_2)^2} \left[p_1(1 - p_1) + p_2(1 - p_2)/k \right].$$

Under equal allocation, $n_1 = n_2 = n$, the required sample size per group is

$$n_1 = n_2 = n = \frac{(z_{1-\alpha/2} + z_{1-\beta})^2}{(p_1 - p_2)^2} \left[p_1(1 - p_1) + p_2(1 - p_2) \right].$$

1.4   Further Readings

Sample size calculation is an important issue in the experimental design of biomedical research. The sample size formulas presented in this chapter are based on asymptotic approximation and superiority trials. Closed–form sample size estimates for independent outcomes can be obtained using normal approximation for equivalence trials, cross–over trials, non–inferiority trials, and bioequivalence trials [14]. In some clinical trials such as phase II cancer clinical trials [18], sample sizes are usually small. Therefore, the sample size calculation based on asymptotic approximation would not be appropriate for clinical trials with a small number of subjects. The small sample sizes for typical phase II clinical trials imply the need for the use of exact statistical methods in sample size estimation [19]. Chow *et al.* [3] provided procedures for sample size estimation for proportions based on exact tests for small samples. Even though the closed–form formulas cannot be obtained for sample size estimates based on exact tests, the sample size estimates can be obtained numerically.

The tests for proportions using normal approximation to the binomial outcome are equivalent to the usual chi–square tests since $Z^2 = \chi^2$. The p–values for the two tests are equal. For example, the critical value of the chi–square with 1 degree of freedom is $\chi^2_{0.05} = 3.841$ at the $\alpha = 0.05$ level, which is equal to the square of two–sided $Z_{\alpha/2} = Z_{0.025} = 1.96$. If one wishes to use a two–sided chi–square test, one should use a two–sided sample size or power determination by using $Z_{\alpha/2}$ instead of Z_α [20]. Others [21, 22, 23] have used arcsine transformation of proportions, $A(p) = 2 \ arcsin \ (\sqrt{p})$, to stabilize variance in the sense that the variance formula of $A(p)$ is free of the proportion p. Given a proportion $\hat{p}$ with $E(\hat{p}) = p$, $A(\hat{p})$ is asymptotically normal with mean $A(p)$ and variance $1/n$, where n is the sample size. Since the variance of $A(p)$ does not depend on the expectation, the sample size and power calculation becomes simplified.

Pre– and post–intervention studies have been widely used in medical and social behavioral studies [24, 25, 26, 27, 28]. In pre–post studies, each subject contributes a pair of dependent observations: one observation at pre–intervention and the other observation at post–intervention. Paired t–test has been used to detect the intervention effect on a continuous outcome while McNemar's test [29] has been the most widely used approach to detect the intervention effect on a binary outcome in pre–post studies. Paired t–test can be conducted by applying the one–sample t–test on the difference between pre–test and post–test observations. Sample size needed to detect a difference between a pair of continuous outcomes from pre–post tests can be estimated by using the sample size formula for a one–sample test for means in Equation (1.4). However, unlike paired continuous outcomes from pre–post tests, sample size formulas for independent outcomes presented in this chapter cannot be used to estimate the sample size needed to detect a difference between a pair of binary observations from pre–post studies. Sample size determination for studies involving a pair of binary observations from pre–post studies will be discussed in Chapter 4.

Clustered data often arise in medical and behavioral studies such as dental, ophthalmologic, radiologic, and community intervention studies in which data are obtained from multiple units of each cluster. In radiologic studies, as many as 60 lesions may be observed through positron emission tomography (PET) in one patient since PET offers the possibility of imaging the whole body [30]. Sample size estimation for clustered outcomes should be done incorporating the dependence of within–cluster observations. Here, the unit of data collection is a cluster (subject), and the unit of data analysis is a lesion within a cluster. Two major problems arise in a sample size calculation for clustered data. One is that the number of units in each cluster, called cluster size, tends to vary cluster by cluster with a certain distribution. The other is that observations within each cluster are correlated. The sample size estimate needs to incorporate the variable cluster size and the correlation among observations within a cluster.

Controlled clinical trials often employ a parallel–groups repeated measures design in which subjects are randomly assigned between treatment groups, evaluated at baseline, and then evaluated at intervals across a treatment period of fixed total duration. The repeated measurements are usually equally spaced, although not necessarily so. The hypothesis of primary interest in short–term efficacy trials concerns the difference in the rates of changes or the time–averaged responses between treatment groups [31]. Major problems in the sample size estimation of repeated measurement data are missing data and the correlation among repeated observations within a subject. As in the sample size estimate of clustered outcomes, sample size should be estimated incorporating the correlation among repeated measurements within each subject and the missing data mechanisms for studies with repeated measurements. Here, a sample size means the number of subjects.

In the subsequent chapters, sample size estimates will be provided using large sample approximation for correlated outcomes such as clustered outcomes and repeated measurement outcomes. There are many complexities in estimating sample size. For example, different sample size formulas are appropriate for different types of study designs, with computations more complex for studies that recruit study subjects at multiple centers. Sample size determinations also have to take into account that some subjects will be lost to follow-up or otherwise drop out of a study. Certain manipulations, such as increased precision of measurements or repeating measurements at various time points, can be used to maximize power for a given sample size.

Bibliography

[1] ICH. *Statistical Principles for Clinical Trials.* Tripartite International Conference on Harmonized Guidelines, E9, 1998.

[2] M. Krzywinski and N. Altman. Points of significance: Power and sample size. *Nature Methods*, 10:1139–1140, 2013.

[3] S. C. Chow, J. Shao, and H. Wang. *Sample Size Calculations in Clinical Research.* Chapman & Hall/CRC, 2008.

[4] R. G. Newcombe. Two sided confidence intervals for the single proportion: Comparison of seven methods. *Statistics in Medicine*, 17:857–872, 1998.

[5] ASA. Ethical guidelines for statistical practice: Executive summary. *Amstat News*, April:12–15, 1999.

[6] R. V. Lenth. Some practical guidelines for effective sample size determination. *American Statistician*, 55(3):187–193, 2001.

[7] S. A. Julious. Tutorial in biostatistics: Sample size for clinical trials. *Statistics in Medicine*, 23:1921–1986, 2004.

[8] S. C. Chow. *Biosimilars: Design and Analysis of Follow-on Biologics.* Chapman & Hall/CRC, 2013.

[9] J. Wittes. Sample size calculations for randomized clinical trials. *Epidemiologic Reviews*, 24(1):39–53, 1984.

[10] J. S. Lee. Understanding equivalence trials (and why we should care). *Canadian Association of Emergency Physicians*, 2(3):194–196, 2000.

[11] FDA. *Guidance for Industry Bioavailability and Bioequivalence Studies for Orally Administered Drug Products General Considerations*. Center for Drug Evaluation and Research, the U.S. Food and Drug Administration, Rockville, MD., 2003.

[12] FDA. *Guideline for Industry on Non-Inferiority Clinical Trials*. Center for Drug Evaluation and Research and Center for Biologics Evaluation and Research, Food and Drug Administration, Rockville, MD, 2010.

[13] EMEA. *Guidelines on the Choice of the Non-Inferiority Margin*. European Medicines Agency CHMP/EWP/2158/99, London, UK, 2005.

[14] S. A. Julious. *Sample Sizes for Clinical Trials*. Chapman & Hall/CRC, 2009.

[15] S. A. Julious and M. J. Campbell. Tutorial in biostatistics: Sample size for parallel group clinical trials with binary data. *Statistics in Medicine*, 31:2904–2936, 2010.

[16] K. E. Muller, L. M. Lavange, S. L. Ramey, and C. T. Ramey. Power calculations for general linear multivariate models including repeated measures applications. *Journal of American Statistical Association*, 87(420):1209–1226, 1992.

[17] W. S. Aronow and C. Ahn. Postprandial hypotension in 499 elderly persons in a long-term health care facility. *Journal of the American Geriatrics Society*, 42(9):930–932, 1994.

[18] S. Piantadosi. *Clinical Trials: A Methodologic Perspective, (2nd ed.)*. John Wiley & Sons, Inc, 2005.

[19] R. P. Hern. Sample size tables for exact single–stage phase II designs. *Statistics in Medicine*, 20:859–866, 2001.

[20] J. M. Lachin. Introduction to sample size determination and power analysis for clinical trials. *Controlled Clinical Trials*, 2:93–113, 1981.

[21] R. D. Sokal and F. J. Rohlf. *Biometry: The Principles and Practice of Statistics in Biometric Research*. San Francisco: Freeman, 1969.

[22] S. H. Jung and C. Ahn. Estimation of response probability in correlated binary data: A new approach. *Drug Information Journal*, 34:599–604, 2000.

[23] S. H. Jung, S. H. Kang, and C. Ahn. Sample size calculations for clustered binary data. *Statistics in Medicine*, 20:1971–1982, 2001.

[24] M. C. Rossi, C. Perozzi, C. Consorti, T. Almonti, P. Foglini, N. Giostra, P. Nanni, S. Talevi, D. Bartolomei, and G. Vespasiani. An interactive diary for diet management (DAI): A new telemedicine system able to

promote body weight reduction, nutritional education, and consumption of fresh local produce. *Diabetes Technology and Therapeutics*, 12(8):641–647, 2010.

[25] A. Wajnberg, K. H. Wang, M. Aniff, and H. V. Kunins. Hospitalizations and skilled nursing facility admissions before and after the implementation of a home-based primary care program. *Journal of the American Geriatric Society*, 58(6):1144–1147, 2010.

[26] E. J. Knudtson, L. B. Lorenz, V. J. Skaggs, J. D. Peck, J. R. Goodman, and A. A. Elimian. The effect of digital cervical examination on group b streptococcal culture. *Journal of the American Geriatric Society*, 202(1):58.e1–4, 2010.

[27] T. Zieschang, I. Dutzi, E. Müller, U. Hestermann, K. Grunendahl, A. K. Braun, D. Huger, D. Kopf, N. Specht-Leible, and P. Oster. Improving care for patients with dementia hospitalized for acute somatic illness in a specialized care unit: a feasibility study. *International Psychogeriatrics*, 22(1):139–146, 2010.

[28] A. M. Spleen, B. C. Kluhsman, A. D. Clark, M. B. Dignan, E. J. Lengerich, and The ACTION Health Cancer Task Force. An increase in HPV–related knowledge and vaccination intent among parental and non–parental caregivers of adolescent girls, age 9–17 years, in Appalachian Pennsylvania. *Journal of Cancer Education*, 27(2):312–319, 2012.

[29] Q. McNemar. Note on the sampling error of the difference between correlated proportions or percentages. *Psychometrika*, 12(2):153–157, 1947.

[30] M. Gonen, K. S. Panageas, and S. M. Larson. Statistical issues in analysis of diagnostic imaging experiments with multiple observations per patient. *Radiology*, 221:763–767, 2001.

[31] P. J. Diggle, P. Heagerty, K. Y. Liang, and S. L. Zeger. *Analysis of longitudinal data (2nd ed.)*. Oxford University Press, 2002.

2

Sample Size Determination for Clustered Outcomes

2.1 Introduction

Clustered data frequently arise in many fields of applications. We frequently make observations from multiple sites of each subject (called a cluster). For example, observations from the same subject are correlated although those from different subjects are independent. In periodontal studies that observe each tooth, each patient usually contributes data from more than one tooth to the studies. In this case, a patient corresponds to a cluster, and a tooth corresponds to a site.

The degree of similarity or correlation is typically measured by intracluster correlation coefficient (ρ). If one simply ignores the clustering effect and analyzes clustered data using standard statistical methods developed for the analysis of independent observations, one may underestimate the true p-value and inflate the type I error rate of such tests since the correlation among observations within a cluster tends to be positive [1, 2]. Therefore, clustered data should be analyzed using statistical methods that take into account of the dependence of within–cluster observations. If one fails to take into account the clustered nature of the study design during the planning stage of the study, one will obtain smaller sample size estimate and statistical power than planned. However, one will obtain larger sample size estimate and statistical power than planned in some studies such as split–mouth trials [3, 4, 5] in which each of two treatments is randomly assigned to two segments of a subject's mouth. In split–mouth trials, both intervention and control treatments are applied in each subject.

Intracluster correlation coefficient (ρ) is defined by $\rho = \sigma_B^2/(\sigma_B^2 + \sigma_W^2)$, where σ_B^2 is the between–cluster variance, and σ_W^2 is the within–cluster variance. As the within–cluster variance (σ_W^2) approaches to 0, ρ approaches to 1. Let n be the number of clusters and m be the number of observations in each cluster. When $\rho = 1$, all responses within a cluster are identical. The effective sample size (ESS) is reduced to the number of clusters (n) when $\rho = 1$ since all responses within a cluster are identical. A very small value of ρ implies that the within–cluster variance (σ_W^2) is much larger than the between–cluster variance (σ_B^2). When $\rho = 0$, there is no correlation among observations within a cluster. The effective sample size is the total number of observations across all clusters (nm) when $\rho = 0$. To get the effective sample size, the total number of observations (the number of observations per cluster (m) times the number of clusters (n)) is divided by a correction factor $[1 + (m-1)\rho]$ that includes ρ and the number of observations per cluster (m). That is, the effective sample size is $nm/[1 + (m-1)\rho]$. The correction factor, $[1 + (m-1)\rho]$, is called the design effect or the variance inflation factor [6].

In the TOSS (trial of cilostazol in symptomatic intracranial arterial stenosis) clinical trial [7], investigators examined the effect of cilostazol on the progression of intracranial arterial stenosis, which narrows an artery inside the brain that can lead to stroke. Cilostazol is a medication for the treatment of intermittent claudication, a condition caused by narrowing of the arteries that supply blood to the legs. One hundred thirty–six subjects were randomly allocated to receive either cilostazol or placebo with an equal probability. Three arteries (two middle cerebral arteries and one basilar artery) were evaluated for the progression of intracranial stenosis in both cilostazol and placebo groups.

The number of arteries evaluated in each treatment group is 204 (=3 arteries/subject x 68 subjects). If observations in three arteries are independent $(\rho = 0)$, then the effective number of observations is 204. If the observations in three arteries are completely dependent $(\rho = 1)$, then the effective number of observations is 68. If ρ takes the value between 0 and 1, the effective number of observations is $204/[1 + (m-1)\rho]$, where $m = 3$. The effective number of observations in each treatment group is $nm/[1 + (m-1)\rho]$ when $0 \le \rho \le 1$. As a special case, the effective number of observations is nm when $\rho = 0$, and n when $\rho = 1$.

2.2   One–Sample Clustered Continuous Outcomes

Clustered continuous outcomes occur frequently in biomedical studies. Examples include size of tumors in cancer patients, and pocket probing depth and clinical attachment level in teeth of subjects undergoing root planning under local anesthetic.

2.2.1   Equal Cluster Size

We assume that the number of observations in each cluster (m) is small compared to the number of clusters (n) so that asymptotic theories can be applied to n for sample size estimation. Let Y_{ij} be a random variable of the jth $(j = 1, \ldots, m)$ observation in the ith $(i = 1, \ldots, n)$ cluster, where Y_{ij} is assumed to be normally distributed with mean $E(Y_{ij}) = \mu$ and common variance $V(Y_{ij}) = \sigma^2$. We assume a pairwise common intracluster correlation coefficient, $\rho = corr(Y_{ij}, Y_{ij'})$ for $j \neq j'$.

Let $y_i = \sum_{j=1}^{m} Y_{ij}$ denote the sum of responses in the ith cluster, and $\bar{y}_i$ be the mean response computed over m observations in the ith cluster. The total number of observations is nm. The mean of Y_{ij} computed over all observations is written as

$$\bar{y} = \frac{\sum_{i=1}^{n} \sum_{j=1}^{m} Y_{ij}}{nm},$$

where $\bar{y}$ estimates the population mean μ.

The degree of dependence within clusters is measured by the intracluster correlation coefficient (ρ), which can be estimated by analysis of variance (ANOVA) estimate [8] as

$$\hat{\rho} = \frac{MSC - MSW}{MSC + (m-1)MSW},$$

where

$$MSC = m \sum_{i=1}^{n} \frac{(\bar{y}_i - \bar{y})^2}{n-1},$$

$$MSW = \sum_{i=1}^{n} \sum_{j=1}^{m} \frac{(y_{ij} - \bar{y}_i)^2}{n(m-1)}.$$

The overall mean $\bar{y}$ has a normal distribution with mean μ and variance V, where

$$V = \frac{\sum_{i=1}^{n} m\{1 + (m-1)\hat{\rho}\}\sigma^2}{(nm)^2} = \frac{\{1 + (m-1)\hat{\rho}\}\sigma^2}{nm}.$$

We test the null hypothesis $H_0 : \mu = \mu_0$ versus the alternative hypothesis $H_1 : \mu = \mu_1$ for $\mu_0 \neq \mu_1$. The test statistic $Z = (\bar{y} - \mu_0)/\sqrt{V}$ is asymptotically normal with mean 0 and variance 1. We reject $H_0 : \mu = \mu_0$ if the absolute value of Z is larger than $z_{1-\alpha/2}$, the $100(1-\alpha/2)$th percentile of the standard normal distribution.

We are interested in estimating the sample size n with a power of $1 - \beta$ for the projected alternative hypothesis $H_1 : \mu = \mu_1$. The sample size (n) needed to achieve a power of $1 - \beta$ can be obtained by solving the following equation:

$$\frac{|\mu_1 - \mu_0|}{\sqrt{V}} = z_{1-\alpha/2} + z_{1-\beta}.$$

The required number of clusters is

$$n = \frac{(z_{1-\alpha/2} + z_{1-\beta})^2}{(\mu_1 - \mu_0)^2} \frac{\{1 + (m-1)\hat{\rho}\}}{m} \sigma^2. \tag{2.1}$$

The total number of observations is

$$n \cdot m = \frac{(z_{1-\alpha/2} + z_{1-\beta})^2 \{1 + (m-1)\hat{\rho}\}\sigma^2}{(\mu_1 - \mu_0)^2}.$$

When the cluster size is 1 ($m = 1$), the required number of observations is

$$n_1 = \frac{(z_{1-\alpha/2} + z_{1-\beta})^2 \sigma^2}{(\mu_1 - \mu_0)^2}.$$

When cluster size is $m(m > 1)$, the variance is inflated by a factor of $\{1 + (m-1)\hat{\rho}\}$ compared with the variance under $m = 1$. The factor $\{1+(m-1)\hat{\rho}\}$ is called variance inflation factor or design effect. That is, the total number of observations can be computed by multiplying n_1 by the design effect $\{1 + (m-1)\hat{\rho}\}$.

2.2.2   Unequal Cluster Size

Cluster sizes are often unequal in cluster randomized studies. When the cluster sizes are not constant, one approach is to replace the cluster size (m) by an advance estimate of the average cluster sizes, which was referred to as the average cluster size method [9, 10]. The average cluster size method is likely to underestimate the actual required sample size [11]. Another approach is to replace the cluster size (m) by the largest expected cluster size in the sample, which was called as the maximum cluster size method [10]. Here, we provide the sample size estimate under variable cluster size.

Let n be the number of clusters in a clinical trial, and m_i be the cluster size in the ith cluster ($i = 1, \ldots, n$). The number of observations in the ith cluster, m_i, may vary at random with a certain distribution. Here, we estimate the sample size using the information on varying cluster sizes. We assume that the cluster sizes (m_i, $i = 1, \ldots n$) are independent and identically distributed, and the cluster sizes (m_i's) are small compared to n so that asymptotic theories can be applied to n for sample size estimation. Let Y_{ij} be a random variable of the jth observation ($j = 1, \ldots, m_i$) in the ith cluster, where Y_{ij} is assumed to be normally distributed with mean μ and variance σ^2. We assume a pairwise common intracluster correlation coefficient, $\rho = corr(Y_{ij}, Y_{ij'})$ for $j \neq j'$. The correlation is assumed not to vary with the number of observations per cluster.

Let $y_i = \sum_{j=1}^{m_i} Y_{ij}$ denote the sum of responses in the ith cluster, and $\bar{y}_i = \sum_{j=1}^{m_i} Y_{ij}/m_i$ be the mean response computed over m_i responses in the ith cluster. Then, the mean of y_{ij} computed over all clusters is written as

$$\bar{y} = \frac{\sum_{i=1}^n m_i \bar{y}_i}{\sum_{i=1}^n m_i},$$

where $\bar{y}$ estimates the population mean μ. The mean cluster size is $\bar{m} = \sum_{i=1}^n m_i/n$.

The degree of dependence within clusters is measured by the intracluster correlation coefficient (ρ), which can be estimated by analysis of variance (ANOVA) estimate [8].

It can be shown that conditional on the empirical distribution of m_i's, the overall mean ($\bar{y}$) has a normal distribution with mean μ and variance V, where

$$V = \frac{\sum_{i=1}^n m_i\{1 + (m_i - 1)\hat{\rho}\}\sigma^2}{(\sum_{i=1}^n m_i)^2}.$$

Based on the asymptotic result, we can reject $H_0 : \mu = \mu_0$ if the absolute value of the test statistic $Z = (\bar{y} - \mu_0)/\sqrt{V}$ is larger than $z_{1-\alpha/2}$, the $100(1-\alpha/2)$th percentile of the standard normal distribution.

We are interested in estimating the sample size n with a power of $1 - \beta$ for the projected alternative hypothesis $H_1 : \mu = \mu_1$. Since m_i's are independent and identically distributed random variables, by the law of large numbers, as $n \to \infty$,

$$nV \to \frac{E[m\{1 + (m - 1)\hat{\rho}\}]\sigma^2}{E(m)^2},$$

where m is the random variable associated with the cluster size and $E(\cdot)$ is the expectation with respect to the distribution of the cluster size.

The sample size needed to achieve a power of $1 - \beta$ can be obtained by solving the following equation:

$$\frac{|\mu_1 - \mu_0|}{\sqrt{V}} = z_{1-\alpha/2} + z_{1-\beta}.$$

This leads to

$$n = \frac{(z_{1-\alpha/2} + z_{1-\beta})^2\sigma^2}{(\mu_1 - \mu_0)^2} \frac{E[m\{1 + (m - 1)\hat{\rho}\}]}{E(m)^2}.$$

Let $E(m) = \theta$, $V(m) = \tau^2$, and $\gamma = \tau/\theta$, where γ is the coefficient of variation of the cluster size. Then, we can write

$$n = \frac{(z_{1-\alpha/2} + z_{1-\beta})^2\sigma^2}{(\mu_1 - \mu_0)^2}\{(1 - \hat{\rho})\frac{1}{\theta} + \hat{\rho} + \hat{\rho}\gamma^2\}. \tag{2.2}$$

The sample size formula (2.2) provides the sample size estimate by accounting for variability in cluster size. When cluster sizes are equal across all clusters, then the sample size formula (2.2) is the same as the sample size formula (2.1) with $\gamma = 0$.

Let $(w_1, \ldots, w_n)$ be a set of weights assigned to clusters with $w_i \geq 0$ and $\sum_{i=1}^n w_i = 1$. The overall mean can be expressed as $\bar{y} = \sum_{i=1}^n w_i\bar{y}_i$. The overall mean ($\bar{y}$) is an unbiased estimate of μ. The above sample size estimate is based on equal weights to observations by letting $w_i = m_i/\sum_{i=1}^n m_i$. Sample size can be also estimated by an estimator that assigns equal weights ($w_i = 1/n$) to each cluster or an estimator that minimizes the variance of an overall mean ($\bar{y}$). These weighting schemes will be described in detail for clustered binary outcomes.

2.2.2.1   Example

Reports have established the effectiveness of minimally invasive periodontal surgery (MIPS) in treating osseous defects [12, 13]. Since these papers were published, new devices (including a videoscope and ultrasonic tips) have been incorporated to enhance the effectiveness of the procedure. Haffajee *et al.* [14] computed the intracluster correlation coefficients of periodontal measurements for five groups of treated periodontal disease subjects and one group of untreated subjects with periodontal disease. The median intracluster correlation coefficient (ρ) is 0.067 for clinical attachment level change. Harrel *et al.* [12] showed clinical attachment loss (CAL) gains of 4.05 mm following application of minimally invasive periodontal surgery (MIPS) in 16 subjects presenting multiple sites with deep pockets associated with different morphologies, including furcation involvements.

An investigator is proposing a prospective cohort study to evaluate the effectiveness of the MIPS using these new devices. He expects CAL gains of 3.0 mm with a standard deviation of 3.5 mm over the 1–year study period. An investigator will evaluate three sites in each subject and would like to estimate the sample size to detect the mean difference of 1.05 mm in clinical attachment loss (CAL) gains over the 1–year study period to achieve 80% power at a two–sided 5% significance level. We estimate the sample size (n) to test the null hypothesis of $H_0 : \mu = 4.05$ versus the alternative hypothesis $H_1 : \mu = 3.0$ with a two–sided 5% significance level and 80% power assuming three sites per subjects ($m = 3$) and $\rho = 0.067$. From Equation (2.1) with the fixed number of sites per subject ($m = 3$), the required sample size for testing $H_0 : \mu = 4.05$ versus $H_1 : \mu = 3.0$ is

$$n = \frac{(1.96 + 0.842)^2 \{1 + (3 - 1)0.067\}}{(4.05 - 3.0)^2} \cdot \frac{}{3} 3.5^2 = 33.$$

Suppose that the number of sites examined per subject varies among subjects with a mean of 3 and a standard deviation of 2. Then, from Equation (2.2) with a variable number of sites per subject ($\theta = 3$ and $\gamma = 2/3$), the required sample size is

$$n = \frac{(1.96 + 0.842)^2 3.5^2}{(4.05 - 3.0)^2} \{(1 - 0.067)/3 + 0.067 + 0.067(2/3)^2\} = 36.$$

2.3   One–Sample Clustered Binary Outcomes

Clustered binary outcomes occur frequently in medical and behavioral studies. Examples include the presence of cavities in one or more teeth, the presence of arthritic pain in one or more joints, the presence of infection in one or two eyes, and the occurrence of lymph node metastases in cancer patients.

2.3.1 Equal Cluster Size

We assume that cluster sizes are equal across clusters. Let n be the total number of clusters in an experiment and m be the number of observations in each cluster. Let Y_{ij} be the binary random variable of the jth ($j = 1, \ldots, m$) observation in the ith ($i = 1, \ldots, n$) cluster, which is coded as 1 for response and 0 for non–response.

We assume that observations within a cluster are exchangeable in the sense that, given m, $Y_{i1}, \ldots, Y_{im}$ have a common marginal response probability $P(Y_{ij} = 1) = p(0 < p < 1)$ and a common pairwise intracluster correlation coefficient $\rho = \text{corr}(Y_{ij}, Y_{ij'})$ for $j \neq j'$.

Let $y_i = \sum_{j=1}^{m} Y_{ij}$ denote the total number of responses in the ith cluster. Under the exchangeability assumption, we have $E(y_i) = mp$ and $var(y_i) = mp(1 - p)\{1 + (m - 1)\rho\}$. The proportion of responses in the ith cluster is estimated by $\hat{p}_i = y_i/m$ with $E(\hat{p}_i) = p$. An unbiased estimate of p is $\hat{p} = \sum_{i=1}^{n} \hat{p}_i/n$.

For large n, $\sqrt{n}(\hat{p} - p)$ is approximately normal with mean 0 and variance

$$\hat{\sigma}^2 = \hat{p}(1 - \hat{p})\frac{\{1 + (m - 1)\hat{\rho}\}}{m},$$

where $\hat{\rho}$ can be obtained by ANOVA method. The ANOVA method suitable for continuous variables can be used to estimate the intracluster correlation coefficient for binary outcomes. Ridout *et al.* [15] conducted simulation studies to investigate the performance of various estimators of intracluster correlation coefficient for clustered binary data under the common intracluster correlation, $\rho = corr(Y_{ij}, Y_{ij'})$ for $j \neq j'$. Their simulation studies showed that the ANOVA estimator performed well for clustered binary data. The ANOVA estimator of intracluster correlation coefficient can be written as

$$\hat{\rho} = \frac{MSC - MSW}{MSC + (m - 1)MSW},$$

where $MSC = \sum m(\hat{p}_i - \hat{p})^2/(n - 1)$, and $MSW = \sum y_i(1 - \hat{p}_i)/\{n(m - 1)\}$.

Suppose that we wish to test the null hypothesis $H_0 : p = p_0$ versus $H_1 : p = p_1$ for $p_0 \neq p_1$ at a two–sided significance level of α. Under the null hypothesis, the test statistic

$$Z = \frac{\sqrt{n}(\hat{p} - p_0)}{\hat{\sigma}}$$

is asymptotically normal with mean 0 and variance 1. We reject $H_0 : p = p_0$ if the absolute value of the test statistic Z is larger than $z_{1-\alpha/2}$, the $100(1 - \alpha/2)$th percentile of the standard normal distribution. We are interested in calculating the sample size n against the alternative hypothesis $H_1 : p = p_1$ with a two–sided significance level of α and power of $1 - \beta$. The required sample size can be obtained by solving $\sqrt{n}|p_0 - p_1|/\hat{\sigma} = z_{1-\alpha/2} + z_{1-\beta}$. The required number of clusters is

$$n = \frac{\hat{\sigma}^2 (z_{1-\alpha/2} + z_{1-\beta})^2}{(p_0 - p_1)^2} = \frac{p_1(1 - p_1)(z_{1-\alpha/2} + z_{1-\beta})^2}{(p_0 - p_1)^2} \cdot \frac{\{1 + (m-1)\hat{\rho}\}}{m}.$$

When the cluster size is 1 ($m = 1$), the required sample size becomes

$$n_1 = \frac{p_1(1 - p_1)(z_{1-\alpha/2} + z_{1-\beta})^2}{(p_0 - p_1)^2}.$$

When cluster size is $m(m > 1)$, the total number of observations (nm) is $\{1 + (m-1)\hat{\rho}\}$ times the required number of observations under $m = 1$. The factor $\{1 + (m-1)\hat{\rho}\}$ is called variance inflation factor or design effect.

2.3.2 Unequal Cluster Size

Let n be the total number of clusters in an experiment and m_i be the number of observations in the ith ($i = 1, \ldots, n$) cluster. The number of observations per cluster may vary at random with a certain distribution. Let Y_{ij} be the binary random variable of the jth ($j = 1, \ldots, m_i$) observation in the ith cluster, which is coded as 1 for response and 0 for non–response.

We assume that observations within a cluster are exchangeable with $P(Y_{ij}) = p$ ($0 < p < 1$) and $Corr(Y_{ij}, Y_{ij'}) = \rho$ for $j \neq j'$ as in equal cluster size. The intracluster correlation is assumed not to vary with the number of observations per cluster.

Let $y_i = \sum_{j=1}^{m_i} Y_{ij}$ denote the total number of responses in the ith cluster. The proportion of responses in the ith cluster is estimated by $\hat{p}_i = y_i/m_i$ with $E(\hat{p}_i) = p$. Under the exchangeability assumption, we have $E(y_i) = m_i p$ and $var(y_i) = m_i p(1 - p)\{1 + (m_i - 1)\rho\}$. Let $(w_1, \ldots, w_n)$ be a set of weights assigned to clusters with $w_i \geq 0$ and $\sum_{i=1}^{n} w_i = 1$. An unbiased estimate of p is $\hat{p} = \sum_{i=1}^{n} w_i \hat{p}_i$. Three weighting schemes have been proposed for parametric and nonparametric sample size estimation for one–sample clustered binary data [16, 17]. Three weighting schemes are equal weights to observations, equal weights to clusters, and minimum variance weights that minimize the variance of the weighted estimator.

Cochran [18] and Donner and Klar [11] used the estimator $\hat{p}_u = \sum y_i / \sum m_i$ that assigns equal weights to observations with $w_i = m_i / \sum_{i'=1}^{n} m_{i'}$. Lee [19] and Lee and Dubin [20] used the estimator $\hat{p}_c = \sum \hat{p}_i/n$ that assigns equal weights to clusters with $w_i = 1/n$. Ahn [21] showed that the method of assigning equal weights to clusters is preferred to the method of assigning equal weights to observations when the intracluster correlation is 0.6 or greater in a simulation study. Jung *et al.* [16] also showed that the sample size under equal weights to observations (n_u) is usually smaller than that under equal weights to clusters (n_c) for small ρ while n_c is generally smaller than n_u for large ρ. If observations within a cluster are highly dependent, then making another observation from the same cluster will not add much information. In this case, the method assigning equal weights to clusters is preferred to the method assigning equal weights to observations. If all clusters have an equal number of observations, then these two weighting methods are identical.

Jung and Ahn [22] proposed a minimum variance estimator, $\hat{p}_m$, that minimizes the variance of $\hat{p} = \sum_{i=1}^{n} w_i \hat{p}_i$. The variance of the estimator $(\hat{p}_m)$ is minimized with weights

$$w_i = \frac{m_i\{1 + (m_i - 1)\hat{\rho}\}^{-1}}{\sum_{i=1}^{n} m_i\{1 + (m_i - 1)\hat{\rho}\}^{-1}},$$

where $\hat{\rho}$ can be obtained by the ANOVA method. The ANOVA estimator of intracluster correlation coefficient can be written as

$$\hat{\rho} = \frac{MSC - MSW}{MSC + (m_A - 1)MSW},$$

where $MSC = \sum_{i=1}^{n} m_i(\hat{p}_i - \hat{p})^2/(n-1)$, $MSW = \sum_{i=1}^{n} y_i(1 - \hat{p}_i)/(M - n)$, $m_A = (M - \sum_{i=1}^{n} m_i^2/M)/(n - 1)$, and $M = \sum_{i=1}^{n} m_i$. Note that $p_m = p_u$ if $\rho = 0$ and $p_m = p_c$ if $\rho = 1$. If cluster sizes are equal across all clusters $(m_i = m)$, then $p_m = p_u = p_c$.

We would like to test the null hypothesis $H_0 : p = p_0$ versus the alternative hypothesis $H_1 : p = p_1$ for $p_0 \neq p_1$. The test statistic

$$Z_w = \frac{\sqrt{n}(\hat{p}_w - p_0)}{\hat{\sigma}_w}$$

is asymptotically normal with mean 0 and variance 1, where $w = u, c$, and m. Hence, we reject H_0 if the absolute value of Z_w is larger than $z_{1-\alpha/2}$, which is the $100(1 - \alpha/2)$th percentile of the standard normal distribution.

Jung *et al.* [16] provided the sample size formulas needed to test the null hypothesis $H_0 : p = p_0$ versus the alternative hypothesis $H_1 : p = p_1$ with a power of $1 - \beta$ using three weighting schemes of equal weights to observations, equal weights to clusters, and minimum variance weights.

2.3.2.1 Equal Weights to Observations

Under equal weights to observations with $w_i = m_i/\sum_{i=1}^{n} m_i$, the variance of $\sqrt{n}(\hat{p}_u - p_0)$ is

$$\hat{\sigma}_u^2 = V\{\sqrt{n}(\hat{p}_u - p_0)\} = \hat{p}_u(1 - \hat{p}_u)\frac{n\sum_i m_i\{1 + (m_i - 1)\hat{\rho}\}}{(\sum_i m_i)^2}.$$

The test statistic

$$Z_u = \frac{\sqrt{n}(\hat{p}_u - p_0)}{\hat{\sigma}_u}$$

has a standard normal distribution with mean 0 and variance 1 for large n. Under the alternative hypothesis $(H_1 : p = p_1)$, $\hat{\sigma}_u^2$ converges to σ_u^2, where

$$\sigma_u^2 = p_1(1 - p_1)\frac{E(m) + [E(m^2) - E(m)]\hat{\rho}}{[E(m)]^2},$$

and $E(m)$ and $E(m^2)$ are computed using the probability distribution of cluster sizes. The required sample size to test $H_0 : p = p_0$ versus $H_1 : p = p_1$ at a two–sided significance level of α and a power of $1 - \beta$ is

$$n_u = \frac{p_1(1 - p_1)(z_{1-\alpha/2} + z_{1-\beta})^2}{(p_0 - p_1)^2} \frac{\{E(m) + [E(m^2) - E(m)]\hat{\rho}\}}{[E(m)]^2}.$$

2.3.2.2   Equal Weights to Clusters

Under equal weights to clusters with $w_i = 1/n$, the variance of $\sqrt{n}(\hat{p}_c - p_0)$ is

$$\hat{\sigma}_c^2 = V\{\sqrt{n}(\hat{p}_c - p_0)\} = \hat{p}_c(1 - \hat{p}_c)\frac{1}{n}\sum_i \frac{1 + (m_i - 1)\hat{\rho}}{m_i}.$$

The test statistic

$$Z_c = \frac{\sqrt{n}(\hat{p}_c - p_0)}{\hat{\sigma}_c}$$

is asymptotically normal with mean 0 and variance 1. Under the alternative hypothesis ($H_1 : p = p_1$), $\hat{\sigma}_c^2$ converges to σ_c^2, where

$$\sigma_c^2 = p_1(1 - p_1)\{E(1/m) + \{1 - E(1/m)\}\hat{\rho}\},$$

and $E(1/m)$ is computed using the probability distribution of cluster sizes. The required sample size with the power of $1 - \beta$ for the alternative hypothesis $H_1 : p = p_1$ is

$$n_c = \frac{p_1(1 - p_1)(z_{1-\alpha/2} + z_{1-\beta})^2}{(p_0 - p_1)^2}\{E(1/m) + \{1 - E(1/m)\}\hat{\rho}\}.$$

2.3.2.3   Minimum Variance Weights

The variance of the estimator $\hat{p} = \sum_{i=1}^{n} w_i\hat{p}_i$ is minimized when the weight, w_i, is inversely proportional to the variance of $\hat{p}_i$, $V(\hat{p}_i) = V(y_i)/m_i^2$ [22]. The weight that minimizes the variance of the estimator is

$$w_i = \frac{m_i\{1 + (m_i - 1)\hat{\rho}\}^{-1}}{\sum_{i=1}^{n} m_i\{1 + (m_i - 1)\hat{\rho}\}^{-1}},$$

where $\hat{\rho}$ can be obtained by the ANOVA method. The variance of $\hat{p}_m$ is consistently estimated by

$$\hat{\sigma}_m^2 = \frac{\hat{p}_m(1 - \hat{p}_m)}{n^{-1}\sum_i m_i\{1 + (m_i - 1)\hat{\rho}\}^{-1}}.$$

The test statistic

$$Z_m = \frac{\sqrt{n}(\hat{p}_m - p_0)}{\hat{\sigma}_m}$$

has a standard normal distribution with mean 0 and variance 1 for large n. Under the alternative hypothesis ($H_1 : p = p_1$), $\hat{\sigma}_m^2$ converges to σ_m^2, where

$$\sigma_m^2 = p_1(1 - p_1) \frac{1}{E[m + \{1 + (m-1)\hat{\rho}\}^{-1}]}.$$

The required sample size against the alternative hypothesis $H_1 : p = p_1$ for a two–sided significance level of α and power of $1 - \beta$ is

$$n_m = \frac{p_1(1 - p_1)(z_{1-\alpha/2} + z_{1-\beta})^2}{(p_0 - p_1)^2} \frac{1}{E[m + \{1 + (m-1)\hat{\rho}\}^{-1}]}.$$

The sample size (n_m) under minimum variance estimate is always smaller than or equal to n_u and n_c.

2.3.2.4 Example

We use the data of Hujoel *et al.* [23] as a pilot data to illustrate sample size calculation for clustered binary outcomes. An enzymatic diagnostic test was used to determine whether a site was infected by two specific organisms, *treponema denticola* and *bacteroides gingivalis*. Each subject had a different number of infected sites, as determined by the gold standard (an antibody assay against the two organisms).

In a sample of 29 subjects, the number of true positive test results (y_i) and the number of infected sites (m_i) are given in Table 2.1.

In the example of an enzymatic diagnostic test in Table 2.1, the ANOVA estimate ($\hat{\rho}$) of the intracluster correlation coefficient is 0.20.

Suppose that we would like to estimate the sample size based on the hypothesis $H_0 : p_0 = .6$ versus $H_1 : p_1 = .7$ using a two–sided significance level of 5% and a power of 80%. Table 2.2 shows the distribution of the number of infected sites (m_i).

Using the observed relative frequency from Table 2.2, $E(m) = 4.897, E(1/m) = 0.224, E(m^2) = 25.379$, and $E[m\{1 + (m-1)\hat{\rho}\}^{-1}] = 2.704$. Therefore, the required sample sizes are $n_u = 62, n_c = 63$, and $n_m = 61$.

TABLE 2.1
Proportion of infection (y_i/m_i) from $n = 29$ subjects (clusters)

3/6, 2/6, 2/4, 5/6, 4/5, 5/5, 4/6, 3/4, 2/4, 3/4, 5/5, 4/4, 6/6, 3/3, 5/6, 1/2, 4/6, 0/4, 5/6, 4/5, 4/6, 0/6, 4/5, 3/5, 0/2, 2/6, 2/4, 5/5, 4/6.

TABLE 2.2
Distribution of the number of infected sites (m_i)

	m			
	2	3	4	5	6
Relative frequency, f(m)	2/29	1/29	7/29	7/29	12/29

2.4   Two–Sample Clustered Continuous Outcomes

Cluster randomization trials have become increasingly popular among health researchers over the last three decades. In healthcare research using cluster randomization trials, clusters (for example, physicians, hospitals, villages) of subjects are randomized into two or more intervention groups and all subjects within a cluster receive the same intervention. In cluster randomization, the unit of randomization is the cluster while the unit of analysis is the subject. That is, the unit of analysis is different from the unit of randomization in cluster randomization. The GoodNEWS (Genes, Nutrition, Exercise, Wellness, and Spiritual Growth) trial is a community-based participatory research (CBPR) study that aims to reduce risk factors for cardiovascular disease among African–American individuals, using a Lay Health Promoter (LHP) approach in the faith–based organization (FBO) setting [24]. The GoodNEWS trial assessed the effect of a health promotion program on increasing levels of physical activity and dietary change using a cluster randomization design, in which the church congregation is the unit of randomization and the individual participant is the unit of analysis. Both the sample size calculation (the number of required clusters) and the subsequent statistical analysis must take into account the intracluster correlation induced by the cluster randomization since the responses of observations within a cluster are usually correlated.

The setting described here is the same as that of a two-level randomized trial described in Chapter 5. Chapter 5 provides description on two-level randomized clinical trials, where subjects are considered the first level units and clusters such as hospitals or primary care clinics are considered the second level units. Chapter 5 considers the cases where randomization to interventions can occur at either the first or second level. In this chapter, we focus our discussion on randomization at the cluster level (second level).

2.4.1   Equal Cluster Size

We assume cluster sizes (m) are constant across all clusters. For simplicity, we assume clusters are randomly allocated to one of two intervention groups with an equal proportion, $n_1 = n_2 = n$. The total number of clusters in two intervention groups is $2n$. Let Y_{ijk} be the continuous random variable denoting the response of the jth $(j = 1, \ldots, m)$ observation in the ith $(i = 1, \ldots, n)$ cluster of the kth treatment $(k - 1, 2)$. Let $Y_{ik} = (Y_{i1k}, \ldots, Y_{imk})'$ follow multivariate normal distribution with mean vector $\boldsymbol{\mu_k} = (\mu_k, \ldots, \mu_k)'$ of length m, and variance-covariance matrix Σ of size $m \times m$. We also assume that observations in a cluster are exchangeable since all the observations receive the same intervention within each cluster. That is, the diagonal elements of Σ are σ^2 and the off-diagonal elements of Σ are $\rho\sigma^2$, where ρ is the intracluster correlation. Let $\bar{y}_{ik}$ denote the mean response computed over all m observations in the ith cluster of the kth treatment. The mean of Y_{ijk} computed over all observations in group k is written as

$$\bar{y}_k = \frac{\sum_{i=1}^{n} \sum_{j=1}^{m} Y_{ijk}}{nm}.$$

The overall mean at intervention group k, $\bar{y}_k$, will have a normal distribution with mean μ_k and variance V_k,

$$V_k = \frac{\{1 + (m-1)\hat{\rho}\}\sigma^2}{nm}.$$

The degree of dependence within clusters is measured by the intracluster correlation coefficient (ρ), which can be estimated by the ANOVA estimate [8] as

$$\hat{\rho} = \frac{MSC - MSW}{MSC + (m-1)MSW},$$

where

$$MSC = m \sum_{k=1}^{2} \sum_{i=1}^{n} \frac{(\bar{y}_{ik} - \bar{y}_k)^2}{2(n-1)},$$

$$MSW = \sum_{k=1}^{2} \sum_{i=1}^{n} \sum_{j=1}^{m} \frac{(Y_{ijk} - \bar{y}_{ik})^2}{2n(m-1)}.$$

We test the null hypothesis $H_0 : \mu_1 = \mu_2$ versus the alternative hypothesis $H_1 : \mu_1 \neq \mu_2$ with the test statistic $Z = (\bar{y}_1 - \bar{y}_2)/\sqrt{V_1 + V_2}$, which is asymptotically normal with mean 0 and variance 1. We reject $H_0 : \mu_1 = \mu_2$ if the absolute value of Z is larger than $z_{1-\alpha/2}$, the $100(1 - \alpha/2)$th percentile of the standard normal distribution.

The required sample size for testing the hypothesis $H_0 : \mu_1 = \mu_2$ versus $H_1 : \mu_1 \neq \mu_2$ at a two–sided significance level of α and a power of $1 - \beta$ is

$$n = \frac{2\sigma^2(z_{1-\alpha/2} + z_{1-\beta})^2\{1 + (m-1)\hat{\rho}\}}{m(\mu_1 - \mu_2)^2}. \tag{2.3}$$

When the cluster size is 1 ($m = 1$), the required number of observations is

$$n_1 = \frac{2\sigma^2(z_{1-\alpha/2} + z_{1-\beta})^2}{(\mu_1 - \mu_2)^2}.$$

When cluster size is $m (m > 1)$, the total number of observations (nm) is equal to $n_1 \cdot \{1 + (m-1)\hat{\rho}\}$.

2.4.2 Unequal Cluster Size

Cluster sizes are not constant in many cluster randomization trials. Manatunga *et al.* [10] derived the sample size estimate under varying cluster sizes.

For simplicity, we assume an equal allocation of clusters between two intervention groups, $n_1 = n_2 = n$. Let m_{ik} be the cluster size of the ith cluster of the kth intervention, and be independently and identically distributed. Let Y_{ijk} be the continuous random variable denoting the response of the jth ($j = 1, \ldots, m_{ik}$) observation in the ith ($i = 1, \ldots, n$) cluster of the kth treatment ($k = 1, 2$). We assume that the vector $Y_{ik} = (Y_{i1k}, \ldots, Y_{im_{ik}k})'$ is multivariate normal with mean vector $\mu_k = (\mu_k, \ldots, \mu_k)'$ of length m_{ik}, and variance-covariance matrix Σ of size $m_{ik} \times m_{ik}$. We also assume that observations in a cluster are exchangeable with $\Sigma_{ii} = \sigma^2$ and $\Sigma_{ij} = \rho\sigma^2$ for $i \neq j$, where ρ is the intracluster correlation.

When the cluster size varies across clusters, we derive the sample size formula using the test statistic based on the unbiased estimator $\hat{\mu}_k = \bar{y}_k = \sum_{i=1}^{n} \sum_{j=1}^{m_{ik}} Y_{ijk} / \sum_{i=1}^{n} m_{ik}$. Conditional on the cluster size m_{ik}, $\bar{y}_k$ follows a normal distribution with mean μ_k and variance V_k, where

$$V_k = \frac{\sum_{i=1}^{n} m_{ik}\{1 + (m_{ik} - 1)\hat{\rho}\}\sigma^2}{(\sum_{i=1}^{n} m_{ik})^2}.$$

A test statistic for testing $H_0 : \mu_1 = \mu_2$ against $H_1 : \mu_1 \neq \mu_2$ is

$$Z = \frac{\hat{\mu}_1 - \hat{\mu}_2}{\sqrt{\hat{V}_1 + \hat{V}_2}},$$

which is asymptotically normal with mean 0 and variance 1.

Based on the asymptotic result, we reject $H_0 : \mu_1 = \mu_2$ if the absolute value of the test statistic Z is larger than $z_{1-\alpha/2}$, the $100(1 - \alpha/2)$th percentile of the standard normal distribution.

We are interested in estimating the sample size n with a two–sided significance level of α and power of $1 - \beta$ for the alternative hypothesis $H_1 : \mu_1 \neq \mu_2$. Since m_{ik}'s are independent and identically distributed random variables, by the law of large numbers, as $n \to \infty$,

$$nV_k \to \frac{E[m\{1 + (m - 1)\hat{\rho}\}]\sigma^2}{E(m)^2},$$

where m is the random variable associated with the cluster size and $E(\cdot)$ is the expectation with respect to the distribution of the cluster size.

The required sample size for testing the hypothesis $H_0 : \mu_1 = \mu_2$ versus $H_1 : \mu_1 \neq \mu_2$ at a two–sided significance level of α and a power of $1 - \beta$ is

$$n = \frac{2(z_{1-\alpha/2} + z_{1-\beta})^2 \sigma^2}{(\mu_1 - \mu_2)^2} \frac{E[m\{1 + (m - 1)\hat{\rho}\}]}{E(m)^2}.$$

Let $E(m) = \theta$, $V(m) = \tau^2$, and $\gamma = \tau/\theta$, where γ is the coefficient of variation of the cluster size. Then, the sample size can be written accounting for the mean and the coefficient of variation as follows:

$$n = \frac{2(z_{1-\alpha/2} + z_{1-\beta})^2 \sigma^2}{(\mu_1 - \mu_2)^2} \left\{ (1 - \hat{\rho})\frac{1}{\theta} + \hat{\rho} + \hat{\rho}\gamma^2 \right\}. \tag{2.4}$$

The sample size formula (2.4) provides the sample size estimate by accounting for variability in cluster size. When cluster sizes are equal across all clusters, then the sample size formula (2.4) is the same as the sample size formula (2.3) with $\gamma = 0$.

2.4.2.1 Example

An investigator would like to assess the effect of a health promotion program on increasing level of physical activity using a cluster randomization design, in which the church congregation is the unit of randomization and the individual participant is the unit of analysis. The primary outcome for sample size and power estimates is physical activity response in kcal/kg/day, based on the 7–day physical activity recall (PAR). Churches will be randomly allocated to receive either a health maintenance intervention or a control condition with an equal proportion. Let μ_1 and μ_2 be the physical activity responses of subjects in the intervention and the control groups, respectively. The investigator aims to recruit 20 church members ($m = 20$) from each participating church. A net difference ($\mu_1 - \mu_2$) of 1.1 kcal/kg/day is expected between intervention groups based on the results observed in Project PRIME [25]. We expect a standard deviation of $\sigma = 3.67$ kcal/kg/day, and a standardized difference $\Delta^* = (\mu_1 - \mu_2)/\sigma = 0.3$. Adams *et al.* [26] conducted reanalysis of data from 31 cluster-based studies in primary care to estimate intracluster correlation coefficients. The median intracluster correlation (ρ) was 0.010 (interquartile range [IQR] of 0 to 0.032). Here, we estimated the sample sizes using $\rho = 0.025$. The null hypothesis is $H_0 : \mu_1 = \mu_2$ and the alternative hypothesis is $H_1 : \mu_1 \neq \mu_2$.

How many churches are needed to achieve 80% power at a two–sided 5% significance level? When we recruit a fixed number of subjects ($m = 20$) from each participating church, the required number of churches for each intervention group is the smallest integer greater than or equal to n from Equation (2.3), which is 13 churches:

$$n = \frac{2(3.67)^2(1.96 + 0.842)^2[1 + (20 - 1)0.025]}{20(1.1)^2} = 12.9.$$

Suppose that the number of participating subjects differs across churches with a mean of 20 and a standard deviation of 4. Then, the required number of churches for each intervention group is the smallest integer greater than or equal to n from Equation (2.4), which is 14 churches:

$$n = \frac{2(z_{1-\alpha/2} + z_{1-\beta})^2\sigma^2}{(\mu_1 - \mu_2)^2}\left\{(1 - \hat{\rho})\frac{1}{\theta} + \hat{\rho} + \hat{\rho}\gamma^2\right\} = 13.1.$$

2.5   Two–Sample Clustered Binary Outcomes

In an innovative cancer risk intake system (CRIS) trial, physicians are randomly allocated to either a comparison group or a risk–based innovative cancer risk intake system (CRIS) group that delivers patient-tailored print–outs based on personal risk factors and perceived barriers to colon cancer testing [27]. In this CRIS trial, a physician corresponds to a cluster. Based on assignment of his or her physician, each patient will be assigned either to the CRIS intervention or the comparison group, in which patients and physicians will receive non–tailored print–outs that are simple reminders about testing, but are not risk–based; nor will they list or address patient barriers to testing.

2.5.1   Equal Cluster Size

For simplicity, we assume an equal allocation of clusters between two intervention groups, $n_1 = n_2 = n$, and an equal cluster size (m) across all clusters. Let Y_{ijk} be the binary random variable of the jth $(j = 1, \ldots, m)$ observation in the ith $(i = 1, \ldots, n)$ cluster of the kth treatment $(k = 1, 2)$, which is coded as 1 for response and 0 for non–response. We also assume that observations in a cluster are exchangeable in the sense that, given m, $Y_{i1k}, \ldots, Y_{imk}$ have a common marginal response probability $P(Y_{ijk} = 1) = p_k (0 < p_k < 1)$ and a common intracluster correlation coefficient, $\rho = corr(Y_{ijk}, Y_{ij'k})$ for $j \neq j'$. We wish to test the null hypothesis $H_0 : p_1 = p_2$ versus the alternative hypothesis $H_1 : p_1 \neq p_2$.

Here, we consider the estimator of the response probability based upon equal weight to clusters. Specifically, the variance estimate of $\hat{p}_k$ is

$$V(\hat{p}_k) = \hat{p}_k(1 - \hat{p}_k)\frac{1 + (m-1)\hat{\rho}}{nm}, k = 1, 2.$$

The variance of $(\hat{p}_1 - \hat{p}_2)$ is estimated by

$$V(\hat{p}_1 - \hat{p}_2) = [\hat{p}_1(1 - \hat{p}_1) + \hat{p}_2(1 - \hat{p}_2)]\frac{1 + (m-1)\hat{\rho}}{nm}.$$

Since ρ is unknown, the analysis of variance estimator $\hat{\rho}$ will be used. When the sample size (n) is sufficiently large, the null hypothesis is rejected if

$$|(\hat{p}_1 - \hat{p}_2)/\sqrt{V(\hat{p}_1 - \hat{p}_2)}| > z_{1-\alpha/2},$$

where $z_{1-\alpha/2}$ is the $100(1 - \alpha/2)$th percentile of the standard normal distribution. By the law of large numbers, as $n \to \infty$,

$$n \cdot V(\hat{p}_k) \to p_k(1 - p_k)\frac{1 + (m-1)\hat{\rho}}{m}, k = 1, 2.$$

With the two–sided significance level of α and a power of $1 - \beta$, the sample size for comparing two proportions has the form

$$n = (z_{1-\alpha/2} + z_{1-\beta})^2 \frac{p_1(1 - p_1) + p_2(1 - p_2)}{(p_1 - p_2)^2} \frac{1 + (m - 1)\hat{\rho}}{m}. \qquad (2.5)$$

2.5.2 Unequal Cluster Size

It is common to assume a constant cluster size in the estimation of sample size for cluster randomization trials. Kang *et al.* [9] derived the sample size formula for binary outcomes using a large sample approximation in cases of varying cluster sizes. Let m_{ik} denote the cluster size of the ith cluster of the kth treatment, $i = 1, \ldots, n$ and $k = 1, 2$. We assume that the cluster sizes are independent and identically distributed. Based on equal weights to clusters, the response probability in group k is estimated by

$$\hat{p}_k = \frac{\sum_{i=1}^{n} \sum_{j=1}^{m_{ik}} Y_{ijk}}{\sum_{i=1}^{n} m_{ik}}, k = 1, 2.$$

It can be shown that conditional on the empirical distribution of m_{ik}, the variance of $\hat{p}_k$ is

$$V(\hat{p}_k) = \hat{p}_k(1 - \hat{p}_k) \frac{\sum_{i=1}^{n} m_{ik}[1 + (m_{ik} - 1)\hat{\rho}]}{(\sum_{i=1}^{n} m_{ik})^2}.$$

Since m_{ik}'s are independent and identically distributed random variables, by the law of large numbers, as $n \to \infty$,

$$n \cdot V(\hat{p}_k) \to p_k(1 - p_k) \frac{E(m) + \hat{\rho}E(m^2) - \hat{\rho}E(m)}{[E(m)]^2},$$

where $E(\cdot)$ is the expectation with respect to the distribution of the cluster size, and m is the random variable denoting the size of the clusters with $\theta = E(m)$, $\tau^2 = Var(m)$, and $\gamma = \tau/\theta$ (the coefficient of variation of the cluster size). Then, we get

$$n \cdot V(\hat{p}_k) \to p_k(1 - p_k) \left\{ (1 - \hat{\rho})\frac{1}{\theta} + \hat{\rho} + \hat{\rho}\gamma^2 \right\}.$$

The variance of $(\hat{p}_1 - \hat{p}_2)$ is

$$V(\hat{p}_1 - \hat{p}_2) = \sum_{k=1}^{2} \hat{p}_k(1 - \hat{p}_k) \frac{\sum_{i=1}^{n} m_{ik}[1 + (m_{ik} - 1)\hat{\rho}]}{(\sum_{i=1}^{n} m_{ik})^2},$$

where ρ can be estimated by the ANOVA estimator of intracluster correlation coefficient

$$\hat{\rho} = \frac{MSG - MSE}{MSG + (m_0 - 1)MSE},$$

$$MSG = \sum_{k=1}^{2}\sum_{i=1}^{n} m_{ik}(\bar{Y}_{i.k} - \bar{Y}_{...})^2/(2n-1)$$

$$MSE = \sum_{k=1}^{2}\sum_{i=1}^{n}\sum_{j=1}^{m_{ik}}(Y_{ijk} - \bar{Y}_{i.k})^2/(M_T - 2n)$$

$$M_T = \sum_{k=1}^{2}\sum_{i=1}^{n} m_{ik}$$

$$\bar{Y}_{i.k} = \sum_{j=1}^{m_{ik}} Y_{ijk}/m_{ik}$$

$$\bar{Y}_{...} = \sum_{k=1}^{2}\sum_{i=1}^{n}\sum_{j=1}^{m_{ik}} Y_{ijk}/M_T$$

$$m_0 = (M_T - \sum_{k=1}^{2}\sum_{i=1}^{n} m_{ik}^2/M_T)/(2n-1).$$

When the sample size n is sufficiently large, the null hypothesis is rejected if the absolute value of $(\hat{p}_1 - \hat{p}_2)/\sqrt{V(\hat{p}_1 - \hat{p}_2)}$ is larger than $z_{1-\alpha/2}$.

The sample size for comparing two proportions with the power of $1 - \beta$ and the two–sided significance level of α is

$$n = (z_{1-\alpha/2} + z_{1-\beta})^2 \frac{[p_1(1 - p_1) + p_2(1 - p_2)]}{(p_1 - p_2)^2}\left\{\frac{1 - \hat{\rho}}{\theta} + \hat{\rho} + \hat{\rho}\gamma^2\right\}. \quad (2.6)$$

Note that the above sample size depends on proportions (p_1 and p_2), ρ, and the first two moments of cluster size. If cluster sizes are equal across clusters, then the coefficient of variation, γ, is equal to 0. Thus, the above sample size estimate (2.6) is equal to the sample size estimate under equal cluster size (2.5). The method that uses the average cluster size for unequal cluster sizes underestimates (overestimates) the sample size when ρ is positive (negative) [9]. In most clustered outcomes, ρ is positive.

2.5.2.1 Example

We illustrate the sample size calculation using a periodontal study [28], which followed 40 chronically ill subjects to examine the development of root lesions over a one–year period. The data of the trial is provided in Rosner [29], which shows that the number of investigated surfaces had a mean of 3.15 and a standard deviation of 1.46. The ANOVA estimate of the intracluster correlation ($\hat{\rho}$) was 0.354. Investigators are proposing a double–blinded randomized clinical trial to investigate if a new drug decreases the incidence rate of surfaces with root lesions over a one–year period than placebo. Subjects will be randomly allocated to receive either a new drug or a placebo with an equal probability.

Investigators assumed that similar cluster sizes and an intracluster correlation coefficient will be obtained from the proposed clinical trial. Let p_1 and p_2 be the probabilities of incidence of surfaces with root lesions in a placebo and a new drug, respectively. How many subjects are needed to detect the difference in the incidence rates of $p_1 = 0.3$ and $p_2 = 0.1$ with a two–sided 5% significance level and a 90% power? The required sample size is 40 subjects in each treatment group from Equation (2.6). The average size method that uses the average cluster size requires 35 subjects in each treatment group from Equation (2.5).

2.5.3 Estimation of Cluster Size for a Fixed Number of Clusters

We occasionally encounter situations in which the number of clusters cannot exceed a specified maximum value due to financial constraints or other practical reasons. In those cases, we need to specify the sample size requirement in terms of the number of observations per cluster instead of the number of clusters. Donner and Klar [11] and Taljaard *et al.* [30] provided the sample size formula for the number of observation per cluster when there is an upper limit in the number of clusters to be studied in a cluster randomization trial. We assume constant cluster size (m) across all clusters. The required sample size for testing the hypothesis $H_0 : \mu_1 = \mu_2$ versus $H_1 : \mu_1 \neq \mu_2$ at a two–sided significance level of α and a power of $1 - \beta$ is

$$n = \frac{2\sigma^2(z_{1-\alpha/2} + z_{1-\beta})^2\{1 + (m - 1)\rho\}}{m(\mu_1 - \mu_2)^2}. \tag{2.7}$$

Let n_1 be the sample size estimate from a simple randomization trial with the cluster size of 1. Then, n_1 is given by

$$n_1 = \frac{2\sigma^2(z_{1-\alpha/2} + z_{1-\beta})^2}{(\mu_1 - \mu_2)^2}. \tag{2.8}$$

In cluster randomization trials with an equal cluster size (m), the total number of observations per intervention group can be obtained by multiplying n_1 by the variance inflation factor $[1 + (m - 1)\rho]$, where ρ is the intracluster correlation coefficient. Thus, the number of clusters required per group is computed by $k = n_1[1 + (m - 1)\rho]/m$ and the total number of observations required per group is $km = n_1[1 + (m - 1)\rho]$.

When the number of clusters per intervention group is fixed by k^*, the number of observations per cluster is given by

$$m = \frac{n_1(1 - \rho)}{k^* - n_1\rho}. \tag{2.9}$$

When the number of clusters is small, the use of critical values $z_{1-\alpha/2}$ and $z_{1-\beta}$ in Equation (2.8) instead of critical values $t_{1-\alpha/2}$ and $t_{1-\beta}$ corresponding to the t-distribution underestimates the required sample size. To adjust for underestimation, one cluster per group may be added when the sample size is determined with a 5% level of significance, and two clusters per group with a 1% level of significance [11]. For example, at a 5% significance level, the number of clusters required per group is computed by $k = 1 + n_1[1 + (m - 1)\rho]/m$. When the number of clusters per intervention group is fixed by k^*, the number of observations per cluster needed at a 5% significance level is given by

$$m = \frac{n_1(1 - \rho)}{k^* - 1 - n_1\rho}.$$ (2.10)

Equations (2.9) and (2.10) can be also used to obtain the cluster size for clustered binary outcomes under the fixed number of clusters. When the cluster size (m) is 1, the required sample size for testing the hypothesis $H_0 : p_1 = p_2$ versus $H_1 : p_1 \neq p_2$ at a two-sided significance level of α and a power of $1 - \beta$ is

$$n_1 = \frac{(z_{1-\alpha/2} + z_{1-\beta})^2 \{p_1(1 - p_1) + p_2(1 - p_2)\}}{(p_1 - p_2)^2}.$$ (2.11)

The number of observations per cluster can be obtained by plugging the above n_1 into Equations (2.9) or (2.10).

In practical situations, the number of observations may be different among clusters. Ahn *et al.* [31] conducted simulation studies to investigate the effect of the cluster size variability and the intracluster correlation coefficient (ρ) on the power of the study in which the number of available clusters is fixed in advance. The simulation study showed that the performance of the sample size formula depends on the number of clusters per group and the cluster size variability.

2.6 Stratified Cluster Randomization for Binary Outcomes

Woolson *et al.* [32] provided a sample size formula for the Cochran-Mantel-Haenszel statistic involving a set of 2×2 contingency tables in which only one of the margins in each table is fixed. They used Cochran's statistics [33] to test the null hypothesis that an assumed common odds ratio is equal to one assuming that subjects in each stratum (table) are independent and the total number of subjects in each stratum is large. Their sample size formula can be also used to estimate the power of the Mantel-Haenszel statistic since the two test procedures are essentially equivalent in this case. Donner [34] provided the sample size formula for stratified cluster randomized trials by generalizing the sample size formula of Woolson *et al.* [32] that was derived for case-control studies. Donner's sample size formula accounts for clustering in a stratified cluster randomization design.

Suppose that clusters are randomly assigned to the intervention and control groups within each stratum, and n_i is the number of clusters in each intervention group in stratum i ($i = 1, \ldots, k$), where k is the number of strata. The cluster sizes are assumed to be equal within each stratum, but may vary across strata. Let m_i be the cluster size in stratum i. Let P_{iC} denote the probability of response among subjects in stratum i of the control group, and P_{iT} denote the probability of response among subjects in stratum i of the intervention group. Let Ψ be the common odds ratio that characterizes the effect of intervention. Then, we may write

$$P_{iT} = \frac{P_{iC}\Psi}{1 - P_{iC} + P_{iC}\Psi}.$$

We would like to estimate the sample size $n_i (i = 1, \ldots, k)$ for testing $H_0 : \Psi = 1$ versus $H_1 : \Psi \neq 1$ with a power of $(1 - \beta)$ and a two–sided significance level of α. Let $\bar{P}_i = (P_{iT} + P_{iC})/2$.

Using the approach of Donald and Donner [35], the variance of the observed difference in the ith stratum, $\hat{P}_{iT} - \hat{P}_{iC}$, can be written as

$$V(\hat{P}_{iT} - \hat{P}_{iC}) = \bar{P}_i(l - \bar{P}_i) \left\{ \frac{1}{n_i m_i} + \frac{1}{n_i m_i} \right\} [1 + (m_i - 1)\rho] \text{ under } H_0,$$

and

$$V(\hat{P}_{iT} - \hat{P}_{iC}) = \left\{ \frac{P_{iT}(1 - P_{iT})}{n_i m_i} + \frac{P_{iC}(1 - P_{iC})}{n_i m_i} \right\} [l + (m_i - 1)\rho] \text{ under } H_1.$$

Let N be the total number of observations needed for this trial, and let t_i denote the proportion of subjects belonging to stratum i, where $\sum_{i=1}^{k} t_i = 1$. Donner [34] showed that the sample size can be obtained from Equations (2.12) and (2.13):

$$z_{\alpha/2} = \frac{\sum_{i=1}^{k} t_i N(\hat{P}_{iT} - \hat{P}_{iC})/4}{\sqrt{\{\sum_{i=1}^{k} t_i N \bar{P}_i (1 - \bar{P}_i)[1 + (m_i - 1)\rho]\}/4}}, \tag{2.12}$$

and

$$z_{1-\beta} = \frac{\sum_{i=1}^{k} t_i N(\hat{P}_{iT} - \hat{P}_{iC})/4 - \sum_{i=1}^{k} t_i N(P_{iT} - P_{iC})/4}{\sqrt{\{\sum_{i=1}^{k} t_i N\{P_{iT}(1 - P_{iT}) + P_{iC}(1 - P_{iC})\}[1 + (m_i - 1)\rho]\}/8}}. \tag{2.13}$$

From Equations (2.12) and (2.13),

$$\sqrt{N}(-z_{\alpha/2}T + z_{1-\beta}U) = NV, \tag{2.14}$$

where

$$T = \frac{1}{2}\sqrt{\sum_{i=1}^{k} t_i[1 + (m_i - 1)\rho]\bar{P}_i(1 - \bar{P}_i)},$$

$$U = \sqrt{\frac{1}{8}\sum_{i=1}^{k} t_i[1 + (m_i - 1)\rho][P_{iT}(1 - P_{iT}) + P_{iC}(1 - P_{iC})]},$$

and

$$V = \frac{1}{4}\sum_{i=1}^{k} t_i(P_{iT} - P_{iC}).$$

Therefore, the required total number of observations (N) is

$$N = [z_{1-\alpha/2}T + z_{1-\beta}U]^2/V^2,$$

The number of clusters needed in each intervention group within stratum i is given by $n_i = Nt_i/(2m_i)$.

2.6.1   Example

Investigators will conduct a randomized pragmatic clinical trial of management of patients with CKD, diabetes, and hypertension with a clinician support model enhanced by technology support compared with the standard of care. The trial will investigate if patients in the intervention group have a lower hospitalization rate than those in the standard medical care group. The study will employ a prospective stratified cluster randomization design. Physicians will be stratified by healthcare systems, and randomly allocated to either an intervention group or a standard medical care group using a randomized permutation block within stratum. Based on assignment of his or her physician, each patient will be assigned either to an intervention group or a standard medical care group. All eligible patients of physicians who are randomized to the study will be included in the comparison of two intervention groups, regardless of intervention compliance (intent-to-treat analysis).

In studies of adults with diabetes in primary care practices, intracluster correlation coefficient (ρ) had median value of 0.0185 with an interquartile range of 0.006–0.037 [36]. Adams *et al.* [26] conducted reanalysis of data from 31 cluster-based studies in primary care to estimate intracluster correlation coefficients. The median intracluster correlation coefficient was 0.010 with interquartile range of 0 to 0.032. Here, we estimate the sample sizes using $\rho = 0.025$.

Let P_{iC} denote the probability of hospitalization among subjects during a 2–year follow-up period in stratum i of standard medical care group, and P_{iT} denote the probability of hospitalization among subjects in stratum i of the intervention group. We would like to estimate the number of primary care providers to be assigned to each intervention group within each hospital system in order to have 80% power at a two–sided 0.05 significance level for detecting a clinically meaningful difference in the proportions of hospitalizations. We assume the proportion of hospitalization during the 2–year follow-up period to be 60% across all four large health care systems in standard medical care group based on the studies of Daratha *et al.*[37] and Go *et al.*[38]. We expect that the hospitalization rate in the intervention group will be 6% lower than that in the standard medical care group. We illustrate the sample size estimate using the null and alternative hypotheses of $H_0 : P_{iC} = P_{iT} = 0.60$ and $H_1 : P_{1C} \neq P_{iT}$ with $P_{iC} = 0.60$ and $P_{iT} = 0.54$ for $i = 1, \ldots, 4$. Because patients are clustered within primary care providers, the sample size estimate needs to incorporate the clustering effect.

Electronic health records show that the number of patients with coexistent CKD, hypertension, and diabetes are 2,002, 6,931, 6,813, and 5,478 in four large healthcare systems. The fractions of patients belonging to each healthcare system are $t_1 = 0.094$, $t_2 = 0.327$, $t_3 = 0.321$, and $t_4 = 0.258$ with the total number of patients equal to 21,224. The numbers of primary care providers are 248, 242, 239, and 89 in four healthcare systems. The average numbers of patients seen by primary care providers $(m_i's)$ are 8.2, 28.6, 28.5, and 61.6 in each of four hospital systems.

The primary care providers will be equally allocated to either an intervention group or a standard medical care group within each hospital system. The numbers of primary care providers $(n_i's)$ needed for the trial are 23, 23, 23, and 9 in each intervention group of four healthcare systems, respectively.

Based on the assignment of his or her physician, each patient will be assigned either to an intervention group or a standard medical care group. The total numbers of patients needed for the trial are 187, 647, 636, and 512 in each treatment group of four healthcare systems, respectively, using a two–sided 5% significance level and 80% power. The total number of patients needed for the trial is 1,982 in each treatment group across all four hospital systems.

2.7 Nonparametric Approach for One–Sample Clustered Binary Outcomes

Clustered data has been recently analyzed by nonparametric statistical methods [39, 40, 41, 42, 43, 44] while it has been extensively analyzed using parametric methods [1, 20, 22, 45, 46] for the last four decades. Larocque [39] used a signed-rank test by incorporating clustering effect in a variance estimate that is based on the sums of squares over independent clusters. Rosner *et al.* [40] used a signed-rank test by incorporating clustering with variance estimate from the ranks of absolute observations with a common intracluster correlation. Datta and Satten [44] used a signed-rank test for clustered data with the distribution of pairwise differences within a cluster depending on cluster size. Sample size formulas were provided for one–sample and two–sample tests using a parametric approach in previous sections. In this section, we provide sample size formulas for one–sample clustered binary outcomes using a nonparametric approach.

2.7.1   Equal Cluster Size

Noether [47] provided sample size formulas for some nonparametric tests for mutually independent observations. Hu *et al.* [48] extended Noether's nonparametric sample size formula to obtain sample size formula for clustered binary outcomes. We assume that cluster sizes m_i's are small relative to the number of clusters (n) so that asymptotic results can be obtained with respect to n.

Hu *et al.* [48] provided the sample size formula for testing a proportion in clustered binary data using a sign test, which incorporates the intracluster correlation coefficient assuming an equal number of observations in each cluster.

The number of observations (m) per cluster is assumed to be equal across all clusters. Let Y_{ij} be the binary random variable of the jth $(j = 1, \ldots, m)$ observation in the ith cluster, which is coded as 1 for success (response) and -1 for failure (non–response). This coding scheme has some desirable properties. With this coding we can express the total as the difference between the total number of successes and the total number of failures in each cluster [49, 50]. We assume $P(Y_{ij} = 1) = p(0 < p < 1)$ and a common intracluster correlation coefficient $\rho = \mathrm{corr}(Y_{ij}, Y_{ij'})$ for $j \neq j'$.

We are interested in testing the null hypothesis $H_0 : p = p_0$ versus $H_1 : p = p_1$ for $p_0 \neq p_1$. Let m_i^+ and m_i^- denote the total numbers of successes and failures in the ith cluster, respectively. Then, $\sum_{i=1}^{n} \sum_{j=1}^{m} Y_{ij} = m_i^+ - m_i^-$. The test statistic we use in the sign test is $T = \sum_{i=1}^{n} \sum_{j=1}^{m} Y_{ij}/(mn)$.

The expected value of T under the null hypothesis is given by

$$E(T|H_0) = \mu_0(T) = \frac{\sum_{i=1}^{n} \sum_{j=1}^{m} E(Y_{ij}|H_0)}{mn} = 2p_0 - 1.$$

The variance of T under the null hypothesis can be estimated by

$$V(T|H_0) = \sigma_0(T)^2 = \frac{\sum_{i=1}^{n} \sum_{j=1}^{m} Var(Y_{ij}|H_0)}{n^2 m^2}$$

$$= 4p_0(1 - p_0) \frac{\sum_{i=1}^{n} m\{1 + (m-1)\hat{\rho}\}}{n^2 m^2},$$

where $\hat{\rho}$ can be obtained by the ANOVA method. The test statistic

$$Z = \frac{\sum_{i=1}^{n} \sum_{j=1}^{m} Y_{ij} - nm(2p_0 - 1)}{\sqrt{4p_0(1 - p_0)nm\{1 + (m-1)\hat{\rho}\}}}$$

is asymptotically normal with mean 0 and variance 1. Hence, we reject H_0 if the absolute value of Z is larger than $z_{1-\alpha/2}$, which is the $100(1 - \alpha/2)$th percentile of the standard normal distribution.

We are interested in estimating the sample size based on testing the null hypothesis $H_0 : p = p_0$ versus the alternative hypothesis $H_1 : p = p_1$ for $p_0 \neq p_1$.

Noether [47] proposed a sample size determination for some common non-parametric tests such as a sign test, and showed that the sample size or the power of the test can be estimated by solving the following equation:

$$\left\{ \frac{\mu_1(T) - \mu_0(T)}{\sigma_0(T)} \right\}^2 = (z_{1-\alpha/2} + r z_{1-\beta})^2,$$

where $\mu_0(T) = n(2p_0 - 1)$ and $\mu_1(T) = n(2p_1 - 1)$ are the expected values of T under null and alternative hypotheses, respectively, and $r = \sigma_1(T)/\sigma_0(T)$. The required sample size (n) to test $H_0 : p = p_0$ against $H_1 : p = p_1$ for a two–sided significance level of α and a power of $1 - \beta$ is given by

$$n = \frac{(z_{1-\alpha/2} + r z_{1-\beta})^2}{(p_1 - p_0)^2} \left\{ \frac{1 + (m-1)\hat{\rho}}{m} \right\} p_0(1 - p_0),$$

where $r = \sigma_1(T)/\sigma_0(T) = \sqrt{\frac{p_1(1-p_1)}{p_0(1-p_0)}}$.

2.7.2 Unequal Cluster Sizes

Hu *et al.* [48] provided the sample size formula for testing a proportion in clustered binary data using a sign test, which incorporates the intracluster correlation coefficient, and varying number of observations per cluster by assigning equal weights to observations. The sample size formula was extended by incorporating equal weights to clusters and weights minimizing variance estimate in a signed test [17].

Let Y_{ij} be the binary random variable taking values $Y_{ij} = 1$ and $Y_{ij} = -1$ for success and failure, respectively for the jth ($j = 1, \ldots, m_i$) observation in the ith ($i = 1, \ldots, n$) cluster. The sum of Y_{ij} can be expressed as the difference between the total number of successes and the total number of failures for each cluster by coding $Y_{ij} = 1$ and -1 for success and failure, respectively [51, 52]. Cluster size (m_i) may vary at random with probability mass function $f(\cdot)$. We assume $P(Y_{ij} = 1) = p(0 < p < 1)$ and a common intracluster correlation coefficient $\rho = corr(Y_{ij}, Y_{ij'})$ for $j \neq j'$. We test the null hypothesis $H_0 : p = p_0$, versus $H_1 : p = p_1$ for $p_0 \neq p_1$.

Let S_i be the difference between the total number of successes and the total number of failures in the ith cluster ($S_i = \sum_{j=1}^{n_i} Y_{ij}$), and $(w_1, w_2, \ldots, w_n)$ be a sequence of cluster weights with $w_i \geq 0$ and $\sum_{i=1}^{n} w_i = 1$. Then we have a class of weighted test statistics used in this sign test

$$T_w = \sum_{i=1}^{n} w_i \frac{\sum_{j=1}^{m_i} Y_{ij}}{m_i} = \sum_{i=1}^{n} w_i \frac{(m_i^+ - m_i^-)}{m_i},$$

where m_i^+ and m_i^- are the total numbers of successes and failures in the ith cluster, respectively.

The expected value of T_w under the null hypothesis is given by

$$\mu_0(T_w) = \sum_{i=1}^n w_i \frac{E(\sum_{j=1}^{m_i} Y_{ij}|H_0)}{m_i} = 2p_0 - 1.$$

Under the null hypothesis, the variance of T_w can be estimated by

$$\sigma_0(T_w)^2 = \sum_{i=1}^n w_i^2 \frac{Var(\sum_{j=1}^{m_i} Y_{ij}|H_0)}{m_i^2}$$

$$= 4p_0(1-p_0) \sum_{i=1}^n w_i^2 \frac{\{1 + (m_i - 1)\hat{\rho}\}}{m_i},$$

where $\hat{\rho}$ can be estimated by the ANOVA method [53].

The standardized weighted sign test statistic is defined by

$$Z_w = \frac{\sum_{i=1}^n w_i(m_i^+ - m_i^-)/m_i - (2p_0 - 1)}{\sqrt{4p_0(1-p_0)\sum_{i=1}^n w_i^2\{1 + (m_i - 1)\hat{\rho}\}/m_i}},$$

which is asymptotically normal with mean 0 and variance 1. Hence, we reject H_0 if the absolute value of Z_w is larger than $z_{1-\alpha/2}$, which is the $100(1-\alpha/2)$th percentile of the standard normal distribution.

2.7.2.1   Equal Weights to Observations

The choice of $w_i = m_i/\sum_{i=1}^n m_i$ provides equal weights to observations, and the test statistic T_u under equal weights to observations can be expressed as

$$T_u = \sum_{i=1}^n w_i \frac{(m_i^+ - m_i^-)}{m_i} = \frac{\sum_{i=1}^n(m_i^+ - m_i^-)}{\sum_{i=1}^n m_i}. \tag{2.15}$$

The expected values of T_u under null and alternative hypotheses are $\mu_0(T_u) = E(T_u|H_0) = 2p_0 - 1$ and $\mu_1(T_u) = E(T_u|H_1) = 2p_1 - 1$, respectively. The variance of test statistics under equal weights to observations, $\sigma_0(T_u)^2$ is

$$\sigma_0(T_u)^2 = 4p_0(1-p_0)\frac{\sum_{i=1}^n m_i\{1 + (m_i - 1)\hat{\rho}\}}{(\sum_{i=1}^n m_i)^2}. \tag{2.16}$$

The standardized test statistic

$$Z_u = \frac{\sum_{i=1}^n(m_i^+ - m_i^-) - \sum_{i=1}^n m_i(2p_0 - 1)}{\sqrt{4p_0(1-p_0)\sum_{i=1}^n m_i\{1 + (m_i - 1)\hat{\rho}\}}} \tag{2.17}$$

is asymptotically normal with mean 0 and variance 1.

As n gets large, $V(\sqrt{n}T_u)$ converges to $\sigma_0(T_u)^2$, where

$$\sigma_0(T_u)^2 = 4p_0(1-p_0)\{(1-\hat{\rho})E(M) + E(M^2)\hat{\rho}\}/E(M)^2,$$

where M is the random variable denoting the size of the clusters with $\theta = E(M)$ and $\tau^2 = V(M)$.

The sample size for a two–sided significance level of α and a power of $1 - \beta$ can be obtained by solving the following equation:

$$\left\{ \frac{\mu_1(T_u) - \mu_0(T_u)}{\sigma_0(T_u)} \right\}^2 = (z_{1-\alpha/2} + rz_{1-\beta})^2.$$

The required sample size is

$$n_u = \frac{(z_{1-\alpha/2} + rz_{1-\beta})^2}{(p_1 - p_0)^2} \left\{ \frac{1 - \hat{\rho}}{\theta} + \hat{\rho} + \frac{\tau^2}{\theta^2}\hat{\rho} \right\} p_0(1 - p_0),$$

where $r = \sigma_1(T_u)/\sigma_0(T_u) = \sqrt{\frac{p_1(1-p_1)}{p_0(1-p_0)}}$.

2.7.2.2 Equal Weights to Clusters

Datta and Satten [44] used the nonparametric statistic that assigns equal weights to clusters with $w_i = 1/n$:

$$T_c = \frac{1}{n} \sum_{i=1}^{n} \left(\frac{m_i^+ - m_i^-}{m_i} \right).$$

The expected values of T_c under the null and alternative hypotheses are $\mu_0(T_c) = E(T_c|H_0) = 2p_0 - 1$ and $\mu_1(T_c) = E(T_c|H_1) = 2p_1 - 1$, respectively. The variance of T_c under the null hypothesis is

$$\sigma_0(T_c)^2 = \frac{4p_0(1 - p_0)}{n^2} \sum_{i=1}^{n} \{1 + (m_i - 1)\hat{\rho}\}/m_i.$$

The standardized test statistic

$$Z_c = \frac{\sum_{i=1}^{n} \left(\frac{m_i^+ - m_i^-}{m_i} \right) - n(2p_0 - 1)}{\sqrt{4p_0(1 - p_0) \sum_{i=1}^{n} \{1 + (m_i - 1)\hat{\rho}\}/m_i}}$$

is asymptotically normal with mean 0 and variance 1.

As n gets large, $V(\sqrt{n}T_c)$ converges to $\sigma_0(T_c)^2$, where

$$\sigma_0(T_c)^2 = 4p_0(1 - p_0)\{(1 - \hat{\rho})E(1/M) + \hat{\rho}\}.$$

The sample size for a two–sided significance level of α and power of $1 - \beta$ can be obtained by solving the following equation:

$$\left\{ \frac{\mu_1(T_c) - \mu_0(T_c)}{\sigma_0(T_c)} \right\}^2 = (z_{1-\alpha/2} + rz_{1-\beta})^2.$$

The required sample size is

$$n_c = \frac{(z_{1-\alpha/2} + rz_{1-\beta})^2}{(p_1 - p_0)^2} \{(1 - \hat{\rho})E(1/M) + \hat{\rho}\} p_0(1 - p_0),$$

where $r = \sigma_1(T_c)/\sigma_0(T_c) = \sqrt{\frac{p_1(1-p_1)}{p_0(1-p_0)}}$.

2.7.2.3 Minimum Variance Weights

The variance of T_w is minimized under the constraint of $\sum_{i=1}^{n} w_i = 1$ when

$$w_i = \frac{m_i\{1 + (m_i - 1)\hat{\rho}\}^{-1}}{\sum_{i=1}^{n} m_i\{1 + (m_i - 1)\hat{\rho}\}^{-1}}.$$

That is, the minimum variance, $\sigma(T_m)^2$, is obtained with the above w_i. We refer the test statistic and the standardized test statistic using minimum variance weights as T_m and Z_m, respectively. The standardized test statistic Z_m is asymptotically normal with mean 0 and variance 1.

As n gets large, $V(\sqrt{n}T_m)$ converges to $\sigma_0(T_m)^2$, where

$$\sigma_0(T_m)^2 = 4p_0(1 - p_0)\frac{1}{E[M\{1 + (M - 1)\hat{\rho}\}^{-1}]}.$$

The sample size for a two–sided significance level of α and power of $1 - \beta$ can be obtained by solving the following equation:

$$\left\{\frac{\mu_1(T_m) - \mu_0(T_m)}{\sigma_0(T_m)}\right\}^2 = (z_{1-\alpha/2} + r z_{1-\beta})^2.$$

The required sample size is

$$n_m = \frac{(z_{1-\alpha/2} + r z_{1-\beta})^2}{(p_1 - p_0)^2}\frac{1}{E[M\{1 + (M - 1)\hat{\rho}\}^{-1}]}p_0(1 - p_0),$$

where $r = \sigma_1(T_m)/\sigma_0(T_m) = \sqrt{\frac{p_1(1-p_1)}{p_0(1-p_0)}}$.

When cluster size is constant, $n_u = n_c = n_m$. If all responses are independent ($\rho = 0$) and all cluster sizes are equal to 1, then all the sample size formulas reduce to Noether's sample size formula [47].

2.7.2.4 Example

We use the data of Hujoel et al. [23] for the illustration of sample size calculation, which was used in the parametric approach in Section 2.3.2.4. We would like to estimate the sample size based on the hypotheses $H_0 : p_0 = .6$ versus $H_1 : p_1 = .7$ using a two–sided significance level of 5% and a power of 80%. From the observed relative frequency distribution, $E(m) = 4.897$ and $V(m) = 1.403$. Therefore, the required sample size estimates are $n_u = 69$, $n_c = 69$, and $n_m = 68$. The sample size (n_m) under minimum variance estimate is always smaller than or equal to n_u and n_c. In this example, the sample size estimates under a nonparametric approach are larger than those under a parametric approach. Noether [47] stated that it will be often appropriate to assume that $\sigma_1(t)$ is close to $\sigma_0(T)$ for alternatives that do not differ too much from the alternative hypothesis. That is, we may set $r = 1$ for sample size calculation. Hu *et al.* [48] estimated the sample size using $r = 1$. The required sample size estimates are $n_u = 71$, $n_c = 71$, and $n_m = 70$ under $r = 1$.

2.7.3 Relative Efficiency of Weighted Estimators

The distribution of cluster sizes may not be exactly specified before the trial starts. Therefore, the sample size is usually estimated using an average cluster size without taking into account any potential imbalance in cluster size, even though cluster size usually varies among clusters. Because the average cluster size method yields the minimum required sample size (number of clusters), a good question is how many more clusters are needed due to variation of cluster size. Ahn *et al.* [54] investigated the relative efficiency (RE) of unequal versus equal cluster sizes for clustered binary data using the weighted sign test estimators. They investigated the RE numerically for a range of cluster size distributions with weighted sign test estimators. They presented formulas for computing the RE as a function of the cluster size and the intracluster correlation, which can be used to adjust the sample size for varying cluster sizes. The required sample size for unequal cluster sizes will not exceed the sample size for an equal cluster size multiplied by the maximum RE. Ahn *et al.* [54] showed that the maximum RE for various cluster size distributions does not exceed 1.50, 1.61, and 1.12 for equal weights to observations, equal weights to clusters, and minimum variance weights, respectively. It suggests sampling 50%, 61%, and 12% more clusters depending on the intracluster correlation and the weighting schemes than the number of clusters computed using an average cluster size.

2.8 Further Readings

Sample size formulas were presented for varying cluster sizes using weighting schemes of equal weights to clusters, equal weights to observations, and minimum variance weights for one–sample clustered outcomes. For two–sample clustered outcomes, sample size formulas were presented for varying cluster sizes using equal weights to observations. Sample size formulas need to be developed using different weighting schemes such as equal weights to clusters and minimum variance weights for two–sample clustered outcomes.

The Wilcoxon rank sum test has been frequently used to compare the location parameter of two distributions. Rosner and Glynn [55] generalized the ordinary Wilcoxon rank sum test for clustered outcomes and provided a sample size formula for comparison of clustered continuous outcomes between two groups accounting for intracluster correlation.

The split–mouth trial is a popular trial in oral health research [3, 4, 5]. In the split–mouth trial study, each of two treatments is randomly assigned to two segments of a subject's mouth. That is, both intervention and control treatments are applied in each subject. The advantage of the split–mouth design is to remove a large portion of the inter–subject variation from the estimates of the treatment effect. Thus, the split–mouth trial requires less sample size because of the reduced variability.

Let n_s be the number of subjects involved in the split–mouth study, n_p be the number of subjects for a whole–mouth study, and ρ be the within–subject correlation. Hujoel and Loesche [3] showed that the efficiency of a split–mouth study and a whole–mouth study is equal when $2n_s = n_p(1 - \rho)$. When $\rho = 0$, a split–mouth trial requires half of subjects compared with a whole–mouth trial. A split–mouth study becomes more useful as ρ increases. Sample size and power estimates are needed for a split–mouth study with unequal number of sites between treatment groups within a subject. Split–mouth trials are not recommended when contamination between sites is suspected and when finding matching sites is not possible in subjects [5].

Since the cost of the epidemiologic studies or clinical trials depends not only on the total number of observations, but also on the number of clusters, it is important to consider designing the study under financial constraints. Sample size formulas for computing optimal sample sizes that are based on the effectiveness of an intervention in cluster randomization trials under a given budget and statistical power are available in the literature [56, 57, 58]. For example, Tokola *et al.* [58] provided the explicit variance formulas that allow for cost minimization with a fixed power and power maximization with a given budget.

The most commonly used randomized cluster design is a parallel cluster randomized trial in which clusters are assigned to either an intervention or a control group in a random fashion. The advantages of a parallel design are simplicity and universal acceptance. The disadvantages are time and effort involved in their effective implementation, and large sample size for comparison, especially with low incidence outcomes.

In crossover trials, every cluster will receive both the intervention and the control treatment with the order of receiving interventions in a random way. Crossover trials have the advantage of potentially reducing variability because each cluster acts as its own control. A crossover trial will require less sample size than a comparable parallel design because the within–cluster variability is usually smaller than the between–cluster variability, and within–cluster responses to interventions are usually positively correlated. Carryover effects is a potential problem since carryover effects can cause intervention by period interactions, which means that the intervention effect is not constant over time. Thus, a washout period is required so that the effect of the earlier intervention does not influence the results for the next intervention. Sometimes, this constraint can make recruiting difficult, and there can be other logistic issues that can make it infeasible. Giraudeau *et al.* [59] provided the sample size formula for continuous outcomes that incorporates the intracluster correlation and interperiod correlation arising from crossover trials.

A stepped wedge design has received increasing attention in cluster randomized trials. A stepped wedge design is a type of crossover design [60] in which different clusters switch interventions at different time points, and

TABLE 2.3

Stepped wedge design, where C represents control and I represents intervention

Cluster Number	Intervention						
1	C	I	I	I	I	I	I
2	C	C	I	I	I	I	I
3	C	C	C	I	I	I	I
4	C	C	C	C	I	I	I
5	C	C	C	C	C	I	I
6	C	C	C	C	C	C	I

clusters cross over in one direction only, for example from control to intervention. Table 2.3 shows the intervention sequence of the stepped wedge design.

Woertman *et al.* [61] derived the design effect (also called sample size correction factor) from the formulas provided by Hussey and Hughes [60] assuming that cluster size (m) is equal across all clusters. The design effect can be used to estimate the sample size needed for stepped wedge cluster randomization design. The stepped wedge design effect is given by

$$DE_{sw} = \frac{1 + \rho(stm + bm - 1)}{1 + \rho(\frac{1}{2}stm + bm - 1)} \cdot \frac{3(1 - \rho)}{2t(s - \frac{1}{s})},$$

where ρ is the intracluster correlation coefficient, m is the cluster size, s is the number of steps, b is the number of baseline measurements, and t is the number of measurements after each step. That is, the clusters will be measured $b + s \cdot t$ times each. The above design effect depends on m, s, b, and t, and incorporates clustering effect and the stepped wedge design.

Woertman *et al.* [61] illustrated that stepped wedge designs could reduce the required sample size over a parallel design in cluster randomized trials. Their design effect formula incorporates the intracluster correlation that characterizes the correlation between subjects within the same cluster. The total number of subjects needed for the stepped wedge design is obtained by multiplying the design effect by the unadjusted sample size (N_u) which is obtained by not correcting for clustering and repeated measurements. That is, $N_{sw} = N_u \cdot DE_{sw}$. The number of clusters can be obtained by dividing the required sample size (N_{sw}) by the cluster size (m). The number of clusters switching intervention at each step is computed by dividing the number of clusters (c) by the number of steps (s).

Sample size estimate for clustered randomization trials can be derived using the sample size estimation method for time-averaged difference [62, 63]. Zhang and Ahn [62] derived the sample size formula for comparing the time-averaged responses of repeated measurement outcomes under compound symmetry correlation structure, which yielded the identical sample size formula in Equation (2.3). The sample size formula for time-averaged difference in Zhang and Ahn [63] can be used to estimate the sample size for cluster randomization trials accommodating various correlation structures and varying cluster sizes.

Clustered count data occurs frequently in biomedical studies. Amatya *et al.* [64] derived the sample size estimate for the required number of clusters with clustered count outcomes using GEE, and compared with that of the Hayes-Donner method [65]. Amatya *et al.* [64] showed that the difference in sample size estimates between two methods was negligible for very small values of intracluster correlation coefficient (ρ). However, they showed that the Hayes-Donner method performed better than GEE for larger values of ρ.

Cluster randomization trials often employ three–level hierarchical structures. Three–level hierarchical study designs are increasingly common. In three–level hierarchical studies, clusters such as hospitals (level 3 unit) are randomly assigned to one of two intervention groups. Health care professionals such as primary care physicians (level 2 unit) within the same cluster are trained with the randomly assigned intervention to provide care to subjects (level 1 unit). Heo and Leon [66] derived closed–form sample size and power formulas required to detect an intervention effect on outcomes at the subject's level using a three–level linear mixed effects model. Fazzari *et al.* [67] provided the explicit closed–form sample size and power formulas for three–level cluster randomized trials with randomization at the first or second level. They derived the sample size formulas with the maximum likelihood estimates from mixed effects linear models. Sample size determination for studies involving three–level hierarchical structures will be discussed in Chapter 6.

McNemar's test is used to compare two paired proportions. Gonen [68] presented a method to estimate the sample size and power for McNemar's test for clustered paired binary outcomes, using the adjustment presented by Eliasziw and Donner [69].

In this chapter, we restrict our attention to two–level data structures. Chapters 4 and 5 discuss sample size estimation for clustered outcomes from two-level randomized clinical trials using GEE and the mixed effects linear model. Chapter 6 extends the approach of Chapter 5 to deal with sample size determination for clustered outcomes from three–level randomized clinical trials using the mixed effects linear model.

Bibliography

[1] J.N.K. Rao and A.J. Scott. A simple method for the analysis of clustered binary data. *Biometrics*, 48:577–585, 1992.

[2] J.K. Haseman and L.L. Kupper. Analysis of dichotomous response data from certain toxicological experiments. *Biometrics*, 34:69–76, 1979.

[3] P. Hujoel and W. Loesche. Efficiency of split–mouth designs. *Journal of Clinical Peridontology*, 17:722–728, 1990.

[4] E. Lesaffre, B. Philstrom, I. Needleman, and H. Worthington. The design and analysis of split–mouth studies: What statisticians and clinicians should know. *Statistics in Medicine*, 28:3470–3842, 2009.

[5] N. Pandis. Sample calculation for split–mouth designs. *American Journal of Orthodontics and Dentofacial Orthopedics*, 141(6):818–819, 2012.

[6] A. Donner and N. Klar. Pitfalls of and controversies in cluster randomization trials. *American Journal of Public Health*, 94(3):416–422, 2004.

[7] S.U. Kwon, Y.J. Cho, J.S. Koo, H.J. Bae, Y.S. Lee, K.S. Hong, J.H. Lee, and J.S. Kim. Cilostazol prevents the progression of the symptomatic intracranial arterial stenosis: The multicenter double–blind placebo-controlled trial of cilostazol in symptomatic intracranial arterial stenosis. *Stroke*, 36:782–786, 2005.

[8] A. Donner and J. Koval. A procedure for generating group sizes from a one–way classification with a specified degree of imbalance. *Biometrical Journal*, 29(2):181–187, 1987.

[9] S.H. Kang, C. Ahn, and S.H. Jung. Sample size calculations for dichotomous outcomes in cluster randomization trials with varying cluster size. *Drug Information Journal*, 37:109–114, 2003.

[10] A. K. Manatunga, M. G. Hudgens, and S. Chen. Sample size estimation in cluster randomized studies with varying cluster size. *Biometrical Journal*, 43(1):75–86, 2001.

[11] A. Donner and N. Klar. *Design and Analysis of Cluster Randomization Trials in Health Research*. Oxford University Press, 2000.

[12] S.K. Harrel, T.G. Wilson, and M.E. Nunn. Prospective assessment of the use of enamel matrix proteins with minimally invasive surgery. *Journal of Periodontology*, 76:380–384, 2005.

[13] P. Cortellini and M.S. Tonetti. A minimally invasive surgical technique with an enamel matrix derivative in the regenerative treatment of intra-bony defects: A novel approach to limit morbidity. *Journal of Clinical Periodontology*, 34:87–93, 2007.

[14] A.D. Haffajee, S.S. Socransky, J.M. Goodson, and J. Lindhe. Intraclass correlations of periodontal measurements. *Journal of Clinical Periodontology*, 12(3):216–224, 1985.

[15] M.S. Ridout, C.G.B. DemUetrio, and D. Firth. Estimating intraclass correlation for binary data. *Biometrics*, 55:137–148, 1999.

[16] S.H. Jung, S.H. Kang, and C. Ahn. Sample size calculations for clustered binary data. *Statistics in Medicine*, 20:1971–1982, 2001.

[17] C. Ahn, F. Hu, and W.R. Schucany. Sample size calculation for clustered binary data with sign tests using different weighting schemes. *Statistics in Biopharmaceutical Research*, 3(1):65–72, 2011.

[18] W.G. Cochran. *Sampling Techniques (3rd ed.)*. Wiley: New York, 1977.

[19] E.W. Lee. Two sample comparison for large groups of correlated binary responses. *Statistics in Medicine*, 15:1187–1197, 1996.

[20] E. Lee and N. Dubin. Estimation and sample size considerations for clustered binary responses. *Statistics in Medicine*, 13:1241–1252, 1994.

[21] C. Ahn. An evaluation of methods for the estimation of sensitivity and specificity of site–specific diagnostic tests. *Biometrical Journal*, 39:793–807, 1997.

[22] S.H. Jung and C. Ahn. Estimation of response probability in correlated binary data: a new approach. *Drug Information Journal*, 34:599–604, 2000.

[23] P. Hujoel, L. Moulton, and W. Loesche. Estimation of sensitivity and specificity of site-specific diagnostic tests. *Journal of Periodontal Research*, 25:193–196, 1990.

[24] M.J. DeHaven, M.A. Ramos-Roman, N. Gimpel, J. Carson, J. DeLemos, S. Pickens, C. Simmons, T. Powell-Wiley, K. Banks-Richard, K. Shuval, J. Duval, L. Tong, N. Hsieh, and J.J. Lee. The GoodNEWS (genes, nutrition, exercise, wellness, and spiritual growth) trial: A community–based participatory research (CBPR) trial with African–American church congregations for reducing cardiovascular disease risk factors – recruitment, measurement, and randomization. *Contemporary Clinical Trials*, 32:630–640, 2011.

[25] A.L. Dunn, B.H. Marcus, J.B. Kampert, M.E. Garcia, H.W. Kohl, and S.N. Blair. Comparison of lifestyle and structured interventions to increase physical activity and cardiorespiratory fitness: A randomized trial. *JAMA*, 281(4):327–334, 1999.

[26] G. Adams, M.C. Gulliford, O.C. Ukoumunne, S. Eldridge, S. Chinn, and M.J. Campbell. Patterns of intra-cluster correlation from primary care research to inform study design and analysis. *Journal of Clinical Epidemiology*, 57(8):785–794, 2006.

[27] C. Ahn, F. Hu, and C.S. Skinner. Effect of imbalance and intracluster correlation coefficient in cluster randomized trials with binary outcomes. *Computational Statistics and Data Analysis*, 53:596–602, 2009.

[28] D.W. Banting, R.P. Ellen, and E.D. Fillery. A longitudinal study of root caries: baseline and incidence data. *Journal of Dental Research*, 64:1141–1144, 1985.

[29] B. Rosner. *Fundamentals of Biostatistics (7th ed.)*. Brooks/Cole, 2011.

[30] M. Taljaard, A. Donner, and N. Klar. Accounting for expected attrition in the planning of community intervention trials. *Statistics in Medicine*, 26:2615–2628, 2007.

[31] C. Ahn, F. Hu, C.S. Skinner, and D. Ahn. Effect of imbalance and intracluster correlation coefficient in cluster randomized trials with binary outcomes when the available number of clusters is fixed in advance. *Contemporary Clinical Trials*, 30:317–320, 2009.

[32] R.F. Woolson, J.A. Bean, and P.B. Rojas. Sample size for case-control studies using cochrans statistic. *Biometrics*, 42:927–932, 1986.

[33] W.G. Cochran. Some methods of strengthening the common chi-square tests. *Biometrics*, 10:417–451, 1954.

[34] A. Donner. Sample size requirements for stratified cluster randomization designs. *Statistics in Medicine*, 11(6):743–750, 1992.

[35] A. Donald and A. Donner. Adjustments to the Mantel–Haenszel chi–square statistic and odds ratio variance estimator when the data are clustered. *Statistics in Medicine*, 6:43–52, 1987.

[36] B. Littenberg and C.D. MacLean. Intra–cluster correlation coefficients in adults with diabetes in primary care practices: The vermont diabetes information system field survey. *BMC Medical Research Methodology*, 6(20):1–11, 2006.

[37] K.B. Daratha, R.A. Short, C.F. Corbett, M.E. Ring, R. Alicic, R. Choka, and K.R. Tuttle. Risks of subsequent hospitalization and death in patients with kidney disease. *Clinical Journal of the American Society of Nephrology*, 7, 2012.

[38] A.S. Go, G.M. Chertow, D. Fan, C.E. McCulloch, and C.Y. Hsu. Chronic kidney disease and the risks of death, cardiovascular events, and hospitalization. *New England Journal of Medicine*, 351(13):1296–1305, 2004.

[39] D. Larocque. The Wilcoxon signed–rank test for cluster correlated data. *Statistical Modeling and Analysis for Complex Data Problems*, pages 309–323, 2005.

[40] B. Rosner, R.J. Glynn, and M.L.T. Lee. The Wilcoxon signed–rank test for paired comparisons of clustered data. *Biometrics*, 62:185–192, 2006.

[41] B. Rosner, R.J. Glynn, and M.L.T. Lee. Extension of rank sum test for clustered data: Two–group comparison with group membership defined at the subunit level. *Biometrics*, 62:1251–1259, 2006.

[42] P.D. Gerard and W.R. Schucany. An enhanced sign test for dependent binary data with small numbers of clusters. *Computational Statistics and Data Analysis*, 51:4622–4632, 2007.

[43] D. Larocque, J. Nevalainen, and H. Oja. A weighted multivariate sign test for cluster–correlated data. *Biometrika*, 94:267–283, 2007.

[44] S. Datta and G.A. Satten. A signed-rank test for clustered data. *Biometrics*, 64:501–507, 2008.

[45] K. Liang and S. L. Zeger. Longitudinal data analysis for discrete and continuous outcomes using generalized linear models. *Biometrika*, 84:3–32, 1986.

[46] A. Donner, N. Birkett, and C. Buck. Randomization by cluster: Sample size requirements and analysis. *American Journal of Epidemiology*, 114:906–914, 1981.

[47] G.E. Noether. Sample size determination for some common nonparametric tests. *Journal of the American Statistical Association*, 82:645–647, 1987.

[48] F. Hu, W.R. Schucany, and C. Ahn. Nonparametric sample size estimation for sensitivity and specificity with multiple observations per subject. *Drug Information Journal*, 44:609–616, 2010.

[49] G. Molenbergh and L.M. Ryan. An exponential family model for clustered multivariate binary data. *Environmetrics*, 10:279–300, 1999.

[50] D.R. Cox and N. Wermuth. A note on the quadratic exponential binary distribution. *Biometrika*, 81:403–408, 1994.

[51] D.R. Cox and N Wermuth. A note on the quadratic exponential binary distribution. *Biometrika*, 81:403–408, 1994.

[52] G. Molenberghs and L.M. Ryan. An exponential family model for clustered multivariate binary data. *Environmetrics*, 10:279–300, 1999.

[53] A. Donald and A. Donner. A simulation study of the analysis of sets of 2 x 2 contingency tables under cluster sampling: Estimation of a common odds ratio. *Journal of the American Statistical Association*, 85:537–543, 1990.

[54] C. Ahn, F. Hu, and S.C. Lee. Relative efficiency of unequal versus equal cluster sizes for the nonparametric weighted sign test estimators in clustered binary data. *Drug Information Journal*, 46(4):428–433, 2013.

[55] B. Rosner and R.J. Glynn. Power and sample size estimation for the clustered Wilcoxon test. *Biometrics*, 67:646–653, 2011.

[56] M. Moerbeek, G.J.P. Van Breukelen, and M.P.F. Berger. Design issues for experiments in multilevel population. *Journal of Educational and Behavioral Statistics*, 25:271–284, 2000.

[57] S.W. Raudenbush. Statistical analysis and optimal design for cluster randomized trials. *Psychological Methods*, 2:173–185, 1997.

[58] K. Tokola, D. Larocque, J. Nevalainen, and H. Oja. Power, sample size and sampling costs for clustered data. *Statistics and Probability Letters*, 81(7):852–860, 2011.

[59] B. Giraudeau, P. Ravaud, and A. Donner. Sample size estimation for cluster randomized cross–over trials. *Statistics in Medicine*, 27(27):5578–5585, 2008.

[60] M.A. Hussey and J.P. Hughes. Design and analysis of stepped wedge cluster randomized trials. *Contemporary Clinical Trials*, 28:182–191, 2007.

[61] W. Woertman, E. de Hoop, M. Moerbeek, S.U. Zuidema, D.L. Gerritsen, and S. Teerenstra. Stepped wedge designs could reduce the required sample size in cluster randomized trials. *Journal of Clinical Epidemiology*, 66:752–758, 2013.

[62] S. Zhang and C. Ahn. How many measurements for time–averaged differences in repeated measurement studies. *Contemporary Clinical Trials*, 32:412–417, 2011.

[63] S. Zhang and C. Ahn. Sample size calculation for time–averaged differences in the presence of missing data. *Contemporary Clinical Trials*, 33:550–556, 2012.

[64] A. Amatya, D. Bhaumik, and R. Gibbons. Sample size estimation for cluster count data. *Statistics in Medicine*, 32:4162–4179, 2013.

[65] R.J. Hayes and S. Bennett. Sample size calculation for cluster randomized trials. *International Journal of Epidemiology*, 28:319–326, 1999.

[66] M. Heo and A.C. Leon. Statistical power and sample size requirements for three-level hierarchical cluster randomized trials. *Biometrics*, 64(4):1256–1262, 2008.

[67] M.J. Fazzari, M.Y. Kim, and M. Heo. Sample size determination for three–level randomized clinical trials with randomization at the first or second level. *Journal of Biopharmaceutical Statistics*, 24(3):579–599, 2014.

[68] M. Gonen. Sample size and power for McNemar's test with clustered data. *Statistics in Medicine*, 23(14):2283–2294, 2004.

[69] M. Eliasziw and A. Donner. Application of McNemar test to non-independent matched paired data. *Statistics in Medicine*, 10:1981–1991, 1991.

3

Sample Size Determination for Repeated Measurement Outcomes Using Summary Statistics

3.1 Introduction

A repeated measures design deals with response outcomes measured on the same experimental unit at different times or under different conditions. Repeated measurements can refer to multiple measurements on an experimental unit, such as the measurements of thickness of skin in arms, hands, fingers, face, neck, and trunk of scleroderma patients. Repeated measurements can be taken over a period of time, such as daily measurements of blood glucose levels or weekly measurements of systolic blood pressures or weights. The repeated assessments might be measured under different experimental conditions. Collecting repeated measurements can provide a more definitive evaluation of within–subject change across time. Moreover, collecting repeated measurements can simultaneously increase statistical power for detecting changes while reducing the costs of conducting a study. The repeated measurement design is generally more efficient for determining a treatment effect than completely randomized designs [1].

Controlled clinical trials tend to employ a parallel-groups repeated measurements design in which study subjects are randomly assigned between treatment groups, evaluated at baseline and intervals across a treatment period of fixed duration. The repeated measurements are usually equally spaced, although not necessarily so. The hypothesis of primary interest in short–term efficacy trials concerns the difference in patterns and magnitudes of change from baseline between treatment groups.

There are a number of complications in determining the sample size in repeated measures designs since the profusion of inputs needed for repeated measures design can make estimation of the sample size difficult. Unlike studies with independent observations, repeated measurements taken from the same subject are correlated, and the correlations must be accounted for in calculating the appropriate sample size. Failure to include correlation between the repeated measurement outcomes can result in inappropriate sample size estimates and a more costly experiment. Missing data occurs more frequently due to missed visits in longitudinal studies than cross–sectional studies. Missing data may make it difficult to write down the simple expression for the test statistic, and derive a sample size formula.

3.2   Information Needed for Sample Size Estimation

Estimation of a sample size for repeated measurement studies can be daunting due to complicated nature of repeated measurement data. We will provide description on what kind of information is needed to estimate the sample size for the repeated measurement studies. As we need some information to estimate the sample size for cross–sectional studies, we need the following information to obtain sample size estimate for repeated measurement studies: (1) the type I error (α), (2) the power ($1 - \beta$), (3) the primary hypothesis of interest, (4) the clinically or scientifically meaningful difference such as the difference in the rates of changes over time between two treatment groups, (5) the statistical method to be used for data analysis, (6) the variance of repeated measurement outcome, (7) the correlation among repeated measurements, and (8) the missing data patterns.

The statistical concepts of type I error and power are used to estimate the number of subjects needed for trials. The type I error (α) of 0.01 and 0.05 and the power ($1 - \beta$) of 0.80 and 0.90 have been widely used for the sample size estimation in clinical trials. Investigators designing the study need to provide the information on the primary hypothesis of interest and the clinically or scientifically meaningful difference. Sample size estimate must be conducted based on the specific hypothesis using a pre–specified data analysis method.

Sample size estimate for repeated measurement studies is complicated since it involves specifying the missing data patterns, and the values of variance and correlation among the repeated measurements. Failing to specify appropriate variance and correlation patterns can lead to incorrect sample size estimates since the sample size estimate depends on the variances of each repeated measurements and the correlations among repeated measurements. The variance and correlation among repeated measurements within a subject can be estimated from previous studies, a pilot study, or an educated speculation based on investigator's experience.

Popular correlation structures used in repeated measurement studies are compound symmetry (CS) correlation structure, autocorrelation structure with order 1, AR(1), a damped exponential family of correlation structures [2], and a linear exponent AR(1) family of correlation structures [3]. The AR(1) correlation structure assumes an exponentially decaying pattern of correlation according to the temporal distance between the repeated measurements while the CS correlation structure assumes a constant correlation between two distinct measurements regardless of distance. In a damped exponential family of correlation structure, the correlation between observations at two time points (j and j') is modeled by $\rho_{jj'} = \rho^c$, where $c = |j - j'|^\theta$. Here, ρ is the correlation between observations separated by one time unit and θ is a damping parameter. The damped exponential correlation structure provides a rich family of correlation structures. CS and AR(1) correlation structures are special cases of a damped exponential family of correlation structure with $\theta = 0$ and $\theta = 1$, respectively. The dampened exponential model has difficulties with convergence due to its parameterization [4]. A linear exponent autoregressive (LEAR) correlation structure has superior convergence properties as a two–parameter generalization of the AR(1) model. AR(1), CS, and the first–order moving average correlation structures are special cases of the LEAR family of correlation structures.

Missing data occurs frequently due to missed visits in longitudinal clinical trials. In order to estimate the sample size, we should project the missing patterns. Let p_j be the estimated proportion of subjects with assessment at measurement time j (t_j), and $p_{jj'}$ be the estimated proportion of subjects with assessments at both measurement times of t_j and $t_{j'}$. In order to specify $p_{jj'}$, we need to estimate the missing pattern more specifically. If missing at time t_j is independent of missing at time t'_j for each subject, then we have $p_{jj'} = p_j p_{j'}$ and call this independent missing (IM) or random missing (RM). In some studies, subjects missing at a measurement time may be missing at all following measurement times. This type of missing is often called monotone missing (MM). In this case, we have $p_{jj'} = p_{j'}$ for $j < j'$ ($p_1 \geq \cdots \geq p_K$), where K is the number of repeated measurements per subject.

One simple strategy to account for missing data is to estimate an attrition rate, and then recruit more subjects incorporating an attrition rate. To compensate for the loss of information due to premature withdrawals, Patel and Rowe [5] suggest dividing the sample size estimate under no missing data by the proportions of subjects without dropouts. For example, if a 20% dropout rate is expected, then they suggest the use of $N/(1 - 0.2)$ as the total sample size when N is the estimated total sample size under no missing data. This approach uses no information other than the final observation. Thus, this approach is too conservative if the analysis actually uses available information from incomplete cases. Ahn and Jung [6] showed that correcting the sample size estimate for dropouts should not be obtained by dividing the sample size estimate obtained under no missing data by the proportions of completers, which yields empirical powers larger than nominal powers. Overestimation of the sample size has disadvantage on cost and feasibility even though it may seem preferable to overestimate rather than to underestimate the sample size in a trial. Incorporation of missing patterns into sample size estimation will be described in detail in the next chapter.

Investigators may want to test more than one hypothesis in one study due to financial and ethical reasons. If more than one primary hypothesis is used for a trial, a simple, but conservative, Bonferroni correction is typically used to help control the type I error rate. For example, with 3 primary hypotheses, a type I error rate of $\alpha = 0.05/3 = 0.017$ would be used. An investigator may choose the largest sample size to guarantee power for all 3 tests in the absence of time, cost, and ethical issues [7].

3.3   Summary Statistics

The analysis of repeated measurement data has stimulated a wide range of methodological developments including generalized estimating equations (GEE) and generalized linear mixed models (GLMM). Frison and Pocock [8] stated that "However, we feel that a considerable gap has opened up between quite complex statistical theory and the day–to–day reality of statistical reporting in medical journals." Summary statistics have been widely used for statistical reporting in medical journals.

Many authors suggested using summary statistics to capture the clinically relevant information from the repeated measurements study since quite often the data for each subject may be effectively summarized by summary measures such as post–treatment means and the mean difference between pre– and post–treatment [8, 9, 10, 11, 12, 13, 14, 15]. Obvious summary measures include post–treatment mean, the mean of the repeated measurements for each subject, baseline to end–of–trial (endpoint) difference scores, linear trend scores calculated by applying linear orthogonal polynomial coefficients to the original measurements for each subject, and the area under the curve.

When designing repeated measurement trials, investigators need to decide which method to choose among repeated measurement analysis methods that incorporate all data measured over time, an endpoint analysis based on outcomes measured at the endpoint, or a difference score analysis based on change from the baseline to the endpoint. The endpoint analysis is a two–sample test that compares the group means of the final endpoint between two treatment groups. The difference score analysis is a two–sample test that compares the difference scores between baseline and endpoint measures between treatment groups. Rate of changes such as slope uses all data measured over time, and has been widely used as the summary statistics that compares the rates of changes between treatment groups.

In this chapter, we provide the methods to estimate the sample size for repeated measurements using summary statistics. Sample sizes can be estimated using a simple normal distribution model for continuous outcomes by reducing the repeated measurements of each individual to a single summary statistic. That is, the approach is to decide an appropriate summary of each subject's outcome, and then to estimate the sample size using simple statistical comparison methods such as Student's t-test, analysis of variance (ANOVA), and analysis of covariance (ANCOVA), to evaluate group differences in summary measures.

Vossoughi *et al.* [16] stated that the summary statistics can be used to confidently analyze linear trend data with a moderate to large number of measurements and/or small to moderate sample sizes (subjects) through their simulation study. Their simulation study showed that summary statistics approach performed close to linear mixed model (LMM) in terms of empirical type I errors and empirical powers. Summary statistic is a simple and powerful method to test main and interaction effects of repeated measurement outcomes.

In a randomized controlled clinical trial, pre–treatment means are expected to be equal. Thus, a preliminary estimate of the final endpoint is occasionally used to compare the treatment effect of two treatment groups. One can use a simple analysis using the mean for each subject's post–treatment measurement as the summary measures. Let μ_1 and μ_2 be the post–treatment means of treatment groups 1 and 2, respectively. The sample size to test $H_0 : \mu_1 = \mu_2$ versus $H_1 : \mu_1 \neq \mu_2$ in a repeated measurement study can be estimated from the following equation for a two–sided significance level of α and power of $1 - \beta$ using summary statistics. Let σ_1^2 and σ_2^2 be the variances of measurement in treatment groups 1 and 2, respectively. If $\sigma_1^2 = \sigma_2^2 = \sigma^2$, then the required sample size per group is

$$n = \frac{2\sigma^2(z_{1-\alpha/2} + z_{1-\beta})^2}{(\mu_1 - \mu_2)^2}, \tag{3.1}$$

where $z_{1-\alpha/2}$ is the $100 \cdot (1 - \alpha/2)$th percentile of the standard normal distribution.

To estimate the sample size for repeated measurement outcomes using summary statistics, Overall and Doyle [10] and Ahn *et al.* [15] considered a two–stage analysis of the repeated measurements of continuous outcomes in which an index or coefficient of change is calculated for each subject in stage 1, and the significance of the difference between group means on the derived measure of change is evaluated against the within–group variability of that measure in stage 2 using Student's t-tests, analysis of variance (ANOVA), or analysis of covariance (ANCOVA) methods. In stage 1 of the two–stage approach, we can specify the subject-specific summary measure, which represents the difference score between baseline and endpoint measures, ordinary least squares (OLS), or generalized least squares (GLS) regression analysis. Each provides an estimate of the difference score, or the rate of change (slope of a regression line) relating the repeated measurements to corresponding assessment times.

Once we reduce the repeated measurements from each individual to a single summary statistic, we can apply the sample size formula for cross–sectional studies to estimate the sample size for repeated measurement studies. That is, an equation to estimate the sample size for testing the significance of the difference between treatment groups can be adapted to the repeated measurements design via linear trend analysis or difference score analysis, with and without the baseline outcome score entered as a covariate.

The sample size formulas presented here are simplified versions of Overall and Doyle [10], which provided equations for calculating $z_{1-\beta}$ to test the "equal change hypothesis" using simple difference scores or ordinary least squares (OLS) regression slopes or generalized least squares (GLS) regression slopes as dependent variables. Let Δ^* be the effect size, which is the clinically or scientifically meaningful difference between treatment means expressed in standard deviation units. The general power equation for testing "equal change hypothesis" between two treatment groups is of the form

$$z_{1-\beta} = \Delta^* \sqrt{\frac{n}{2}} - z_{1-\alpha/2},$$

where $z_{1-\beta}$ and $z_{1-\alpha/2}$ are the $100(1-\beta)$th and $100(1-\alpha/2)$th percentiles of the standard normal distribution, n is the sample size per groups, and Δ^* is the effect size which depends on the particular definition of change:

$$\Delta^* = \frac{|a'd|}{\sigma\sqrt{a'Ca}},$$

where a' is the contrast vector for difference score, OLS, or GLS analysis, d is the vector of differences between group means, C is the within–subject correlation matrix for the repeated measurements, and σ is the within–subject standard deviation which is assumed to be constant across the repeated measurements. Sample size (n) or power $(1-\beta)$ estimates for difference score, OLS, and GLS analyses can be obtained by considering different a' linear contrasts from the following equation:

$$z_{1-\beta} = \frac{|a'd|}{\sigma\sqrt{a'Ca}}\sqrt{\frac{n}{2}} - z_{1-\alpha/2}. \tag{3.2}$$

3.3.1   Difference Score Analysis

Suppose that the outcome of primary interest is the change score from baseline to endpoint between two treatment groups. Let $\bar{Y}_{1m}$ and $\bar{Y}_{2m}$ be the means at the last measurement in treatment groups 1 and 2, where m is the number of measurements per subject. The sample size for testing the difference in change scores between two treatment groups can be obtained using preliminary estimates of the clinically or scientifically meaningful difference between endpoint means, the within–subject error variance at cross–section in time, and the baseline–to–endpoint correlation (ρ). In a randomized controlled clinical trial, the expected value of the difference in means at baseline between two groups is zero due to randomization in a randomization trial. Thus, a preliminary estimate of the final endpoint mean difference alone is adequate for testing difference scores. Since the within–subject variances at baseline can be assumed equal or comparable to that at subsequent assessment periods, a single cross–sectional variance estimate can be used. Therefore, the sample size for the baseline–to–endpoint change scores can be estimated using only the clinically meaningful difference between treatment group means at the last measurement ($\bar{Y}_{1m}$ and $\bar{Y}_{2m}$), the within–subject baseline–to–endpoint correlation (ρ), and a common within–subject variance at a single assessment period (σ^2). The variance of the difference score between the first and the last measurements can be written as

$$\sigma_d^2 = 2(1 - \rho)\sigma^2.$$

From Equation (3.1), the required sample size is

$$n = \frac{2\sigma_d^2(z_{1-\alpha/2} + z_{1-\beta})^2}{(\bar{Y}_{1m} - \bar{Y}_{2m})^2}.$$

The sample size formula for the difference score can be written as

$$n = \frac{4(1 - \rho)\sigma^2(z_{1-\alpha/2} + z_{1-\beta})^2}{(\bar{Y}_{1m} - \bar{Y}_{2m})^2}. \tag{3.3}$$

The sample size for testing difference scores between two treatment groups is equal to that for testing the significance of the difference between endpoint means alone multiplied by $2(1 - \rho)$. The variance (σ^2) can be obtained from the variance in previous comparable studies or pilot studies.

The sample size can be also easily estimated using Equation (3.2). The contrast vector a has the value -1 for the first measurement, 1 for the last measurement, and 0 for all the measurements between the first and last measurements of the vector. For example, the contrast vector can be written as $a' = [-1, 0, 0, 0, 1]$ for a total of 5 measurements. Then, Equation (3.2) can be expressed as

$$z_{1-\beta} = \frac{|\bar{Y}_{1m} - \bar{Y}_{2m}|}{\sigma}\sqrt{\frac{n}{4(1 - \rho)}} - z_{1-\alpha/2}, \tag{3.4}$$

which yields the same sample size estimate as that in Equation (3.3).

3.3.2   Difference Score Analysis with Baseline Score Adjustment

The baseline score is frequently entered as a covariate to adjust for baseline differences in an observational study or non–randomized clinical trials. The inclusion of baseline score as a covariate can be applied to difference score analysis. Let ρ be the correlation between baseline and endpoint measurements. Then, the proportion of variance accounted for by linear regression on the baseline covariate is equal to ρ^2. The error variance remaining after eliminating the variance associated with the baseline score is $\sigma_c^2 = (1 - \rho^2)\sigma^2$, where σ^2 is the common within–subject variance. The sample size estimate per group for testing difference scores with the baseline score entered as a covariate becomes

$$n = \frac{2(1 - \rho^2)\sigma^2(z_{1-\alpha/2} + z_{1-\beta})^2}{(\bar{Y}_{1m} - \bar{Y}_{2m})^2}. \tag{3.5}$$

The sample size for testing difference scores between two treatment groups without baseline score adjustment is equal to that for testing the significance of the difference score between treatment groups with baseline score adjustment multiplied by $2/(1+\rho)$. Overall and Starbuck [17] provided detailed discussion on difference scores with and without the baseline score as a covariate.

The sample size estimate can be also obtained using Equation (3.2). Here, $|a'd| = |\bar{Y}_{1m} - \bar{Y}_{2m}|$ and $\sigma\sqrt{a'Ca} = \sigma\sqrt{1 - \rho^2}$. Thus, for difference scores with baseline adjustment, the sample size can be estimated by solving the following equation:

$$z_{1-\beta} = \frac{|\bar{Y}_{1m} - \bar{Y}_{2m}|}{\sigma\sqrt{1 - \rho^2}}\sqrt{\frac{n}{2}} - z_{1-\alpha/2}.$$

3.3.2.1   Example

The Brief Psychiatric Rating Scale (BPRS) is a widely used instrument for assessing the positive, negative, and affective symptoms of individuals who have psychotic disorders, especially schizophrenia. It has proven particularly valuable for documenting the efficacy of treatment in patients who have moderate to severe disease. Scores of BPRS have a potential range from 0 to 108, with higher scores indicating more severe pathology. There will be seven assessments including assessments at baseline (week zero) and at weeks 1, 2, 3, 4, 5, and 6. The primary aim of the study is to compare the changes in BPRS total scores from baseline to week 6 between a placebo and a new therapy for schizophrenia. The investigator wants to estimate the sample size needed to detect the difference of 9 in BPRS score with a common standard deviation of 18 with a two–sided significance level of 5% and 80% power. Cohen [18] has characterized a difference of one–half standard deviation as a treatment effect of "medium magnitude." The within–subject correlation structure of the repeated measurements is assumed to conform to AR(1) correlation with the baseline–to–endpoint correlation equal to 0.5. Thus, $z_{1-\beta} = 0.842, z_{1-\alpha/2} = 1.96, \bar{Y}_{1m} - \bar{Y}_{2m} = 9, \sigma = 18$, and $\rho = 0.5$.

For difference score analysis without using the baseline score as a covariate, the required sample size is the smallest integer greater than or equal to

$$n = \frac{4(z_{1-\alpha/2} + z_{1-\beta})^2\sigma^2(1 - \rho)}{(\bar{Y}_{1m} - \bar{Y}_{2m})^2} = \frac{4(1.96 + 0.842)^2 18^2(1 - 0.5)}{9^2} = 62.8.$$

$$(3.6)$$

Therefore, the required minimum sample size is 63 subjects per treatment group.

For difference score analysis with the baseline score entered as a covariate, the required sample size is the smallest integer greater than or equal to

$$n = \frac{2(z_{1-\alpha/2} + z_{1-\beta})^2\sigma^2(1 - \rho^2)}{(\bar{Y}_{1m} - \bar{Y}_{2m})^2} = \frac{2(1.96 + 0.842)^2 18^2(1 - 0.5^2)}{9^2} = 47.1.$$

$$(3.7)$$

The required sample size is 48 subjects per treatment group.

3.3.3 Difference in Rates of Change Across Time

The slope of regression lines appropriately summarizes the rate of changes of outcomes over time in the subjects if the pattern of each individual profile is linear. For each subject, rates of change or trends can be constructed by applying appropriately chosen weighting coefficients to the repeated measurements. The primary hypothesis of interest is to test whether the rates of changes across time are the same among treatment groups or whether there is no significant interaction effect between time and treatment group. For this purpose, once we get slopes for each subject, we can apply the k–sample test such as Student's t–test or Wilcoxon rank–sum test (for $k = 2$) or one–way ANOVA or Kruskal–Wallis test (for $k \geq 3$) to test if there are significant differences in slopes among treatment groups.

Sample size estimates for testing a difference in the rates of changes can be obtained by using linear trend scores that are constructed by applying linear orthogonal polynomial coefficients to the repeated measurements. The linear orthogonal polynomial coefficients are also called linear contrast coefficients. The linear orthogonal polynomial coefficients are equally spaced, and sum to zero. That is, the requirement on orthogonal polynomial coefficient, a_j, is $\sum_{j=1}^{m} a_j = 0$. For example, when there are five repeated measurements including baseline, linear trend can be constructed by applying the vector of linear orthogonal polynomial coefficients, $a = (-2, -1, 0, 1, 2)'$, for baseline and four subsequent repeated measurements. Once a summary statistic is constructed for linear trend scores using linear orthogonal polynomial coefficients, the sample size formula for cross–sectional study can be applied to repeated measurement data. In the ANOVA model that does not use the baseline score as a covariate, tests of significance on the separately calculated linear trend scores are precisely equivalent to tests for the linear component of the interaction between treatment group and time that can be obtained using orthogonal polynomial transformation of the repeated measurements.

The difference between linear trend means for two treatment groups can be calculated by applying the linearly weighted coefficients to the differences between treatment means at each measurement time point,

$$\sum_{j=1}^{m} a_j d_j = \sum_{j=1}^{m} a_j (\bar{Y}_{1j} - \bar{Y}_{2j}),$$

where d_j is the difference between treatment means at the jth ($j = 1, \ldots, m$) measurement times, a_j is the linear orthogonal contrast at the jth measurement time, and $\bar{Y}_{1j}$ and $\bar{Y}_{2j}$ are the observed means for treatment groups 1 and 2 at the jth measurement time. The variance of a linear combination of repeated measurements, $\sum_{j=1}^{m} a_j d_j$, depends on the variances of the individual measurements and the correlation among measurements. The variance of the linear combination of repeated measurements is

$$V \left(\sum_{j=1}^{m} a_j d_j \right) = \sigma^2 \sum_{j=1}^{m} \sum_{j'=1}^{m} a_j a_{j'} \rho_{jj'},$$

where σ^2 is a common within–subject variance at a single measurement period, a_j and $a_{j'}$ are the linearly weighted coefficients, and $\rho_{jj'}$ is the correlation between the jth and j'th measurements. Suppose that we want to test if the linear trend means between two treatment groups are equal at a two–sided significance level of α. The test statistic

$$Z = \frac{\sum_{j=1}^{m} a_j d_j}{\sqrt{\sigma^2 \sum_{j=1}^{m} \sum_{j'=1}^{m} a_j a_{j'} \rho_{jj'}}}$$

is asymptotically normal with mean 0 and variance 1 under the null hypothesis. Based on the asymptotic result, we reject the null hypothesis of no difference in linear trend means if the absolute value of the test statistic Z is larger than $z_{1-\alpha/2}$, the $100(1-\alpha/2)$th percentile of the standard normal distribution. Therefore, the sample size for testing the difference in linear trends between two groups with a two–sided significance level of α and a power of $1 - \beta$ is

$$n = \frac{2\sigma^2 (z_{1-\alpha/2} + z_{1-\beta})^2 \sum_{j=1}^{m} \sum_{j'=1}^{m} a_j a_{j'} \rho_{jj'}}{(\sum_{j=1}^{m} a_j d_j)^2}. \tag{3.8}$$

The sample size estimate can be also obtained using Equation (3.2). For OLS slope difference without the baseline score entered as a covariate, the sample size estimate can be obtained by solving the following equation

$$z_{1-\beta} = \frac{|a'd|}{\sigma} \sqrt{\frac{n}{2a'Ca}} - z_{1-\alpha/2}, \tag{3.9}$$

where a is the mean-corrected vector of linearly increasing time coefficients, for example, $a' = [-2, -1, 0, 1, 2]$ for a total of five measurements, d is the mean-corrected vector of postulated linearly increasing differences between group means, and C is the within–subject correlation matrix among the repeated measurements.

The power for tests on generalized least squares (GLS) regression slopes can be calculated by substituting $a' = z'C^{-1}$ into calculation of effect size Δ^*. Thus the sample size for GLS slope differences without baseline score entered as a covariate can be estimated by

$$z_{1-\beta} = \frac{|z'C^{-1}d|}{\sigma} \sqrt{\frac{n}{2z'C^{-1}z}} - z_{1-\alpha/2}, \tag{3.10}$$

where z is the vector of linearly increasing time coefficients as used for OLS calculations, $z' = [-2, -1, 0, 1, 2]$ for a total of five measurements, and C^{-1} is the inverse of the within–subject correlation matrix of the repeated measurements.

Chapter 4 includes additional discussion of sample size approaches for the comparisons of rates of change under various correlation structures, and missing data patterns. Chapter 5 also includes additional discussion of sample size approaches for the comparisons of rates of change in two–level longitudinal designs under the assumption of a fixed slope and a random slope.

3.3.4 Difference in Rate of Change across Time with Baseline Covariate

The linear trend scores are not generally independent of baseline values like simple pre–post difference scores. Therefore, the ANCOVA can be used to correct for dependence of linear trend scores on baseline differences as in the case of simple pre–post difference scores.

The variance of the composite trend scores is

$$
V\left(\sum_{j=1}^{m} a_j d_j\right) = \sigma^2 \sum_{j=1}^{m} \sum_{j'=1}^{m} a_j a_{j'} \rho_{jj'},
$$

where σ^2 is a common within–subject variance at a single measurement period, a_j and $a_{j'}$ are the linearly weighted coefficients, and $\rho_{jj'}$ is the correlation between the jth and j'th measurements. The correlation between the baseline measure and a weighted combination of repeated measures is a weighted function of the correlations between the baseline measure and each of the repeated measures divided by the square root of the product of the variances of the baseline scores and composite trend scores. Thus, the correlation between the baseline measure and a weighted combination of repeated measures can be written as

$$
\rho_b = \frac{\sum_{j=1}^{m} a_j \rho_{j1}}{\sqrt{\sum_{j=1}^{m} \sum_{j'=1}^{m} a_j a_{j'} \rho_{jj'}}}, \tag{3.11}
$$

where ρ_{j1} is the jth element in the first column of the matrix of correlations among the repeated measurements with the diagonal element $\rho_{11} = 1$. The error variance remaining after the variance associated with the baseline covariate has been removed is $\sigma_c^2 = (1 - \rho_b^2)\sigma^2$, where σ^2 is the common within–subject variance of the outcome measure at cross–section in time. Therefore, the sample size per group for testing the difference in linear trends with baseline covariate between two groups with a two–sided significance level of α and a power of $1 - \beta$ is

$$
n = \frac{2\sigma^2 (z_{1-\alpha/2} + z_{1-\beta})^2 (1 - \rho_b^2) \sum_{j=1}^{m} \sum_{j'=1}^{m} a_j a_{j'} \rho_{jj'}}{(\sum_{j=1}^{m} a_j d_j)^2}.
$$

The sample size estimate can be also obtained using Equation (3.2). For OLS with the baseline score entered as a covariate, the sample size can be estimated by

$$z_{1-\beta} = \frac{a'd}{\sigma} \sqrt{\frac{n}{2(1-\rho_b^2)a'Ca}} - z_{1-\alpha/2}, \qquad (3.12)$$

where $\rho_b = a'Ca_{(1)}/(a'Ca)^{1/2}$ is the correlation between baseline scores and the time-weighted combination of repeated measurements. For example, with a total of five measurements, $a' = [-2, -1, 0, 1, 2]$ with $a'_{(1)} = [1, 0, 0, 0, 0]$.

For GLS slope differences with the baseline score entered as a covariate, the sample size estimate can be obtained from

$$z_{1-\beta} = \frac{z'C^{-1}d}{\sigma} \sqrt{\frac{n}{2(1-\rho_c^2)z'C^{-1}z}} - z_{1-\alpha/2}, \qquad (3.13)$$

where $\rho_c = z'z_{(1)}/(z'C^{-1}z)^{-1/2}$ is the correlation between baseline scores and the transformed GLS definition of change. For example, z' and $z'_{(1)}$ can be written as $z' = [-2, -1, 0, 1, 2]$ and $z'_{(1)} = [1, 0, 0, 0, 0]$ for a total of five measurements.

Power and sample size estimates for the GLS regression approach those for the simple difference-score analysis as the correlation structure of the repeated measurements approaches a true AR(1) pattern. Power and sample size estimates for the GLS regression approach those for the OLS definition of change when the correlation structure approaches compound symmetry (CS). The difference score and GLS regression analyses have superior power and thus require smaller sample sizes than OLS regression in the presence of an autoregressive correlation structure. The OLS and GLS regression provide higher power in the presence of CS correlation than difference score analysis [15].

Dropouts tend to attenuate the power of tests for evaluating differences in patterns of change across time in a repeated measurements design. Overall *et al.* [19] conducted simulation studies to examine the attenuation of power due to dropouts, and concluded that the common practice of increasing the "dropout free" sample size by the anticipated number of dropouts is a useful rule of thumb.

3.3.4.1   Example

We use the Brief Psychiatric Rating Scale (BPRS) example in Section 3.3.2 to illustrate the sample size calculation needed to detect the difference in the rate of change across time. For the sample size estimate using the rate of change with OLS and GLS analysis, the linear trend can be constructed by applying the vector of linear orthogonal polynomial coefficient, $a' = [-3, -2, -1, 0, 1, 2, 3]$. We chose the difference between treatment means, d_j, by $d' = [0, 1.5, 3.0, 4.5, 6.0, 7.5, 9.0]$. The correlation matrix is 7×7 matrix with AR(1) correlation with $\rho_{ij} = 0.5^{(i-j)/6}$, where ρ_{ij} is the (i,j)th element of the matrix C. The sample size estimates for the rate of change with OLS can be obtained from Equations (3.9) and (3.12). The required sample sizes are 69 subjects and 53 subjects without and with baseline covariate adjustment, and the correlation between baseline score and a weighted combination of repeated measures is $\rho_b = -0.477$. From Equations (3.10) and (3.13), the required sample sizes for the rate of change using GLS are 63 subjects and 47 subjects without and with baseline covariate adjustment, respectively, and the correlation between baseline measure and the transformed GLS definition of change is $\rho_c = -0.500$.

Suppose that the correlation structure follows the compound symmetry correlation with $\rho_{jj'} = 0.5$ for $j \neq j'$, and $\rho_{jj} = 1$. The required sample sizes using OLS rate of change are 41 subjects and 34 subjects without and with baseline covariate adjustment, and the correlation between baseline measure and a weighted combination of repeated measures is $\rho_b = -0.401$. The required sample sizes for the rate of change using GLS are 41 subjects and 34 subjects without and with baseline covariate adjustment, respectively, and the correlation between baseline measure and the transformed GLS definition of change is $\rho_c = -0.401$.

3.3.5   Time–Averaged Difference

Investigators frequently make inferences based on the difference in time–averaged responses between treatment groups across the study period. This type of analysis is used when the outcome to be measured changes over time. For example, blood pressure levels vary depending on the amount of food taken, sleep, and exercise, etc. If only one blood pressure measurement is taken from each subject and the mean blood pressure levels are compared between two treatment groups, the experiment may have a poor performance due to substantial within–subject variation in blood pressure levels over time. By measuring the response (e.g., postprandial reduction in blood pressure) at multiple time points, researchers hope that the time–averaged response (e.g., average postprandial reduction in blood pressure over time points) within each group can provide a more precise assessment of the treatment effect. The time–averaged difference (TAD) is particularly meaningful in cases where the treatment effect has rapid onset and repeated measurements continue across an extended period after a maximum effect is achieved. In these situations, the TAD across time between two treatment groups may provide a more powerful evaluation of treatment efficacy than the within–subject trends or change scores [10].

The TAD across all of the measurements between two treatment groups is often considered to be a meaningful indicator of treatment effect even though the baseline difference between treatment groups is expected to be zero in a randomized clinical trial. Sample size needs to be estimated to provide desired power for testing the significance of the TAD in a repeated measurement trial. The test of significance for the TAD across time in a repeated measurement study is equivalent to a test of significance on composite scores that are obtained by summing or averaging the repeated measurements for each individual. Composite scores for testing the TAD between treatment groups can be obtained by applying equal weights to the repeated measurements (for example, $a_j = 1$ for all repeated measurements). That is, the sample size estimate can be obtained by applying equal weight, $a_j = 1$ to Equation (3.8). Therefore, the required sample size per treatment group is

$$n = 2(z_{1-\alpha/2} + z_{1-\beta})^2 \sigma^2 \sum_{j=1}^{m} \sum_{j'=1}^{m} \rho_{jj'} / \left(\sum_{j=1}^{m} d_j \right)^2, \qquad (3.14)$$

where σ^2 is the within–subject variance at each time, $\rho_{jj'}$ is the (j, j')th element in the $m \times m$ correlation matrix among repeated measurements that define the between-groups effect, and d_j is the anticipated differences between treatment group means at the jth time point. Equation (3.14) shows that the sample size depends on the expected pattern of treatment effects across the repeated measurements. It also shows that the sample size increases as the correlation increases. As a special case, suppose that $d_j = d$ for all $j, (j = 1, \ldots m)$, and the measurements within the subject follows the compound symmetry correlation structure with $\rho_{jj'} = \rho$ for $j \neq j'$. Then, Equation (3.14) simplifies to

$$n = 2(z_{1-\alpha/2} + z_{1-\beta})^2 \sigma^2 \frac{\{1 + (m - 1)\rho\}}{md^2}. \qquad (3.15)$$

Equation (3.15) is an extension to the sample size formula for a single measurement study, with additional terms including the number of repeated measurements (m) per subject and the variance inflation factor $[1 + (m - 1)\rho]$, to account for the degree of clustering among observations from the same subject. As ρ increases, the required sample size n increases. The sample size formula in Equation (3.15) has also been used in sample size calculation for cluster randomization trials [20]. Liu and Wu [21] provided an extension of Equation (3.15) to accommodate unbalanced clinical trials.

Chapter 4 includes additional discussion of sample size approaches for the comparisons of time–averaged responses in two-level longitudinal designs under various correlation structures and missing data patterns.

3.3.6   Time–Averaged Difference with Baseline Covariate

The correlation between baseline scores and the sum of repeated measurements needs to be estimated for the inclusion of a baseline covariate. The correlation (ρ_c) between baseline measurement and the repeated measurements is

$$\rho_c = \frac{\sum_{j=1}^{m} \rho_{j1}}{\sqrt{\sum_{j=1}^{m} \sum_{j'=1}^{m} \rho_{jj'}}},$$

which is obtained by applying equal weights, $a_j = 1$ to Equation (3.11). The error variance remaining after the variance associated with the baseline covariate has been removed is $(1 - \rho_c^2)\sigma^2$, where σ^2 is the common within–subject variance of the outcome measures at cross–section in time. Therefore, the sample size with adjustment using the baseline score as a covariate can be computed by multiplying $(1 - \rho_c^2)$ to the sample size obtained by Equation (3.14). That is, the sample size estimate is given by

$$n = 2(z_{1-\alpha/2} + z_{1-\beta})^2 (1 - \rho_c^2)\sigma^2 \sum_{j=1}^{m} \sum_{j'=1}^{m} \rho_{jj'} \bigg/ \left(\sum_{j=1}^{m} d_j\right)^2, \qquad (3.16)$$

3.3.6.1 Example

Investigators would like to estimate the sample size needed to evaluate the time–averaged difference of heart rate between subjects taking the new drug and subjects taking the standard drug. The heart rate will be measured at baseline followed by three additional measurements 30 minutes apart. Previous studies showed an average heart rate of 100 beats per minute (bpm) with a standard deviation of 12 bpm and an AR(1) correlation with a baseline–to–endpoint correlation of 0.7 with $\rho_{jj'} = (0.7)^{|j-j'|/3}$. Then, the correlation between adjacent measurements is $0.7^{1/3} = 0.8879$. How many subjects are needed to detect a 8% reduction in heart rate at a two–sided 5% significance level and 80% power assuming the treatment difference of 8% ($d_j = d = 100bpm \cdot 0.08 = 8bpm$)? Here, $\sum_{j=1}^{4}\sum_{j'=1}^{4} \rho_{jj'} = 13.88$, and $\sum_{j=1}^{4} d_j = 32$.

The required sample size per treatment group without baseline covariate adjustment is

$$n_1 = 2(z_{1-\alpha/2} + z_{1-\beta})^2 \sigma^2 \sum_{j=1}^{4} \sum_{j'=1}^{4} \rho_{jj'} \bigg/ \left(\sum_{j=1}^{4} d_j\right)^2 = 31. \qquad (3.17)$$

The required sample size per treatment group with baseline covariate adjustment can be obtained by multiplying $(1-\rho_c^2)$ to the above sample size estimate. Thus, the required sample size is $n_2 = n_1(1-\rho_c^2) = 31 \cdot (1-0.243^2) = 29$ subjects per treatment group.

Suppose that the correlation structure among the heart rate measurements follows the CS correlation with $\rho_{ij} = 0.8879$ instead of AR(1) correlation. The required sample size per treatment group without baseline covariate adjustment is

$$n_3 = 2(z_{1-\alpha/2} + z_{1-\beta})^2 \sigma^2 \frac{\{1 + (m-1)\rho\}}{md^2} = 33. \qquad (3.18)$$

The sample size without baseline covariate adjustment can be computed using the software PASS 13 [22]. The required sample size per treatment group with baseline covariate adjustment is $n_4 = n_3(1 - \rho_c^2) = 31$ subjects per treatment group.

3.3.7   Sample Size Estimation under Financial Constraints

In clinical trials, sample size is generally estimated based on the number of subjects necessary to detect clinically meaningful differences between treatment groups. However, budget constraint is a challenge faced by investigators in planning almost every clinical trial. Sample size needs to be estimated by incorporating costs for recruiting and measuring subjects in repeated measurement studies [23, 24, 25, 26, 27].

Lai *et al.* [23] investigated how to determine the optimal number of subjects and number of measurements per subject under financial constraints assuming CS correlation structure and no missing data. They estimated the number of subjects and the number of measurements that minimize the length of confidence intervals of the mean difference between treatment group under given financial constraints. Let Y_{ijk} denote the jth measurement of the ith subject at treatment group k. Let n_1 and n_2 denote the number of subjects in treatment groups 1 and 2, and m denote the number of measurement for each subject. Let C be the total financial resources available for the study, and C_1 and C_2 be the cost of recruiting one study subject and the cost making one measurement on a subject for both treatment groups 1 and 2. Let $\bar{Y}_k = \sum\sum Y_{ijk}/(mn_k)$, $k = 1, 2$. The length of the confidence interval is minimized when the variance of the mean difference is minimized. Thus, we need to find m, n_1, and n_2 that minimize the variance of $(\bar{Y}_1 - \bar{Y}_2)$ under financial constraint $nC_1 + nmC_2 \leq C$, where $n = n_1 + n_2$. Let ρ be the common correlation coefficient between any two measurements within a subject under the CS correlation structure. We assume that Y_{ijk} has a normal distribution with an unknown mean μ_k and an unknown variance σ^2. The variance of $\bar{Y}_{ijk}$ is $\sigma^2\{1 + (m - 1)\rho/(mn_k)\}$. The number of subjects (n_1 and n_2) and the number of measurements (m) can be obtained by using $nC_1 + nmC_2 = C$ and applying the Lagrange multiplier method. The optimal number of subjects and the number of measurements are given by $n_k = C/[(1+d)(C_1+mC_2)]$ and $m = \sqrt{C_1(1 - \rho)/(\rho C_2)}$, where $d = n_1/n_2$ [23]. Lai *et al.* [23] provided the sample size estimates without incorporating statistical power for the trial. Zhang and Ahn [27] provided the sample size estimates for the number of subjects and the number of measurements that maximizes the statistical power of the study for the rate of change under financial constraints incorporating various correlation structures and missing patterns with generalized estimating equation (GEE).

3.3.8   Multiple Treatment Groups

Sample size calculations for two–sample repeated measurements have been reported by many investigators. In contrast, the literature has paid relatively little attention to the design and analysis of K–sample trials in repeated measurements studies where K is 3 or greater. In multi–arm clinical trials, sharing a control arm reduces the total sample sizes [28]. If we test experimental agents A and B separately in randomized clinical trials, then we need four treatment arms of A, B, and two control agents, while a single multi–arm trial will need three treatment arms of A, B, and a single control arm. Assuming no multiplicity adjustment, a single multi–arm trial will reduce the total sample size by 25%. In general, with K experimental agents, there is $100 \cdot (K-1)/(2K)\%$ reduction in total sample size [28]. A multi–arm trial may be more attractive to patient participation since a multi–arm trial leads to a higher probability for the patient to be randomized to an experimental arm than separate two–armed trials.

It seems unlikely that a multi–arm clinical trial will focus on the unordered difference among several unordered treatments. Thus, it may be more practical to work with the simple two–group equation and to consider adjusting critical values for multiple comparisons as required. If several treatments or different dose levels are compared to a single placebo or control arm in a multi–arm trial, the substitution of Dunnett's critical values [29] should be considered instead of $z_{1-\alpha/2}$ in the two–group sample size equation. To evaluate the significance of the difference between each possible pair of K ($K \geq 3$) treatment means, Bonferroni critical value or critical values for some other preferred multiple–test correction should be considered instead of $z_{1-\alpha/2}$ in the two–group sample size equation. Cohen [18] provided recommendation for the adjustment of sample size when the primary hypothesis concerns an unspecified overall difference among $K (\geq 3)$ treatments. An empirical rule of thumb is to increase the sample size per group (n) required to test the largest difference between any two groups by 20% for $K = 3$ groups or 25% for $K = 4$ groups [10].

3.3.9 Time–Averaged Difference of Binary Outcomes

Many papers have been published for the estimation of the sample size required to detect treatment differences in repeated measurement studies when the response variable is continuous. However, there is a limited number of papers that estimate sample size for binary outcomes using summary statistics.

Diggle *et al.* [30] and Leon [31] provided the sample size formula for repeated measurements studies with a balanced design that compares the time–averaged responses of a binary outcome between two treatment groups. A closed–form sample size formula was provided under the assumption of no missing data and a compound symmetry (CS) correlation structure among measurements within a subject. Let p_1 and p_2 be the response rates in treatment groups 1 and 2, respectively. From Diggle *et al.* [30], the sample size needed to test $H_0 : p_1 = p_2$ versus $H_1 : p_1 \neq p_2$ with a two–sided significance level of α and a power of $1 - \beta$ is

$$n = \frac{\{z_{1-\alpha/2}\sqrt{2\bar{p}\bar{q}} + z_{1-\beta}\sqrt{p_1 q_1 + p_2 q_2}\}^2 [1 + (m-1)\rho]}{md^2}, \quad (3.19)$$

where $q_1 = 1 - p_1$, $q_2 = 1 - p_2$, $\bar{p} = (p_1 + p_2)/2$, $\bar{q} = 1 - \bar{p}$, ρ is the intraclass correlation coefficient among measurements within a subject, and d is the clinically meaningful difference to be detected (i.e., $d = p_1 - p_2$).

The sample size estimate is a function of the response rates (p_1 and p_2) for each group, the number of repeated observations per subject (m), and the strength of the association among observations within subjects (ρ). The sample size increases as the correlation (ρ) increases.

3.4   Further Readings

The CONSORT (CONsolidated Standards of Reporting Trials) issued CON-SORT 2010 Statement (http://www.consort-statement.org) to improve the reporting of randomized clinical trial, which has been widely accepted as a guideline for designing and reporting clinical trials, endorsed by prominent general and specialty medical journals, and leading editorial organizations. The CONSORT statement recommends the specification of how and when primary and secondary outcome measures were assessed along with the pre-specification of time point of primary interest.

Sample size and/or power estimation has received extensive attention over the years [10, 32, 33, 34, 35, 36]. In this chapter, we used a summary statistic approach for the estimation of sample size and power in repeated measurement studies since quite often the data for each subject may be effectively summarized by summary measures such as endpoint, difference score, or rate of change.

In many studies, investigators have perceived the number of repeated measurements as a fixed design characteristic. However, the number of repeated measurements is a design choice that can be informed by statistical considerations. Vickers [37] investigated the optimal number of repeated measurements in comparing time–averaged responses between groups by examining the relative benefit of additional repeated measurements on statistical power. Overall [38] investigated how many repeated measurements are recommended using a standard repeated measurements ANOVA to test the significance of the between-groups main effect, the Geisser-Greenhouse corrected groups × times interaction, and the difference in linear trends across time. Monte Carlo simulation results illustrated that increasing the number of repeated measurements generally had negative or neutral effect on power of the tests of significance in the presence of serial dependencies that produced heterogeneous correlations such as AR(1) among the repeated measurements [38, 39].

Zhang and Ahn [40] investigated how many measurements are needed for time–averaged differences in repeated measurement studies using the generalized estimating equations (GEE) approach. They showed that the required sample size always decreases as the number of measurements per subject increases under the compound symmetry (CS) correlation. However, the magnitude of sample size reduction quickly decreases to less than 5% when the number of measurements per subject increases beyond four under the compound symmetry (CS) correlation structure. Under the AR(1) correlation structure, making additional measurements from each subject does not always reduce the required sample size. However, in many longitudinal drug studies, investigators follow patients at intervals not only to estimate the rate of changes in treatment effect, but also to monitor the patients over time with respect to safety and other endpoints. That is, the schedule of assessments is determined not only by the cost, but also by regulatory implications. For example, patients might require one or more follow-up visits after the primary assessments are completed. Blindly adopting the statistically optimal design might prevent investigators from addressing these important questions. One solution is to specify the minimal number of repeated measurements that is required by clinical goals, and then conduct statistical approach to identify the optimal design that maximizes power [27]. Zhang and Ahn [40] showed that by introducing measurement error into the AR(1) model, the counterintuitive behavior disappears. That is, additional measurements per subject result in reduced sample sizes.

When there are only two measurements at baseline and the end of the trial in repeated measurement studies, difference score analysis (difference between baseline and endpoint scores) is less efficient than the endpoint analysis if the correlation between the baseline score and endpoint score is less than 0.5 [8]. For the rate-of-change analysis, Overall [38, 39] and Zhang and Ahn [40] showed that making additional measurements from each subject may increase required sample size under AR(1) correlation structure. Relative efficiency of these summary measures approaches of endpoint analysis, difference score analysis, and rate of change analysis needs to be investigated for various correlation structures and missing data patterns.

Dawson [12] presented the sample size formulas for analyses based on summary statistics with incorporation of missing data caused by staggered entry and random dropouts under the assumption of missing completely at random (MCAR), in which data are missing independently of both observed and unobserved data. Frison and Pocock [8] provided the method for determining sample size and the number of pre- and post-treatment measurements using ANCOVA. Further research is needed for the impact of misspecification using a variety of correlation structures such as AR(1), CS, and damped exponential family of correlation structures.

Bibliography

[1] D.R. Jensen. Efficiency and robustness in the use of repeated measurements. *Biometrics*, 38:813–825, 1982.

[2] A. Munoz, V. Carey, J. P. Schouten, M. Segal, and B. Rosner. A parametric family of correlation structures for the analysis of longitudinal data. *Biometrics*, 48(3):733–742, 1992.

[3] S.L. Simpson, L.J. Edwards, K.E. Muller, P.K. Sen, and M.A. Styner. A linear exponent AR(1) family of correlation structures. *Statistics in Medicine*, 29:1825–1838, 2010.

[4] J.J. Grady and R.W. Helms. Model selection techniques for the covariance matrix for incomplete longitudinal data. *Statistics in Medicine*, 14:1397–1416, 1995.

[5] H.I. Patel and E. Rowe. Sample size for comparing linear growth curves. *Journal of Biopharmaceutical Statistics*, 9(2):339–50, 1999.

[6] C. Ahn and S. Jung. Effect of dropout on sample size estimates for test on trends across repeated measurements. *Journal of Biopharmaceutical Statistics*, 15:33–41, 2005.

[7] Y. Guo, H.L. Logan, D.H. Glueck, and K.E. Muller. Selecting a sample size for studies with repeated measures. *BMC Medical Research Methodology*, 13:100, 2013.

[8] L. Frison and S.J. Pocock. Repeated measures in clinical trials: Analysis using mean summary statistics and its implications for design. *Statistics in Medicine*, 11:1685–1704, 1992.

[9] J.N.S. Matthews, D.G. Altman, M.J. Campbell, and P. Royston. Analysis of serial measurements in medical research. *Journal of American Statistical Association*, 300:230–235, 1990.

[10] J.E. Overall and S.R. Doyle. Estimating sample sizes for repeated measurement designs. *Controlled Clinical Trials*, 15:100–123, 1994.

[11] B.S. Everitt. The analysis of repeated measures: A practical view with examples. *The Statistician*, 44:113–135, 1995.

[12] J.D. Dawson. Sample size calculations based on slopes and other summary statistics. *Biometrics*, 54:323–330, 1998.

[13] H.J. Shouten. Planning group sizes in clinical trials with a continuous outcome and repeated measures. *Statistics in Medicine*, 18:2555–264, 1999.

[14] S. Senn, L. Stevens, and N. Chaturvedi. Repeated measures in clinical trials: Simple strategies for analysis using summary measures. *Statistics in Medicine*, 19:861–877, 1999.

[15] C. Ahn, J.E. Overall, and S. Tonidandel. Sample and power calculations in repeated measurement analysis. *Computational Methods and Programs in Biomedicine*, 64:121–124, 2001.

[16] M. Vossoughi, S. Ayatollahi, M. Towhidi, and F. Ketabchi. On summary measure analysis of linear trend and repeated measurement data: performance comparison with two competing methods. *BMC Medical Research Methodology*, 12:33, 2012.

[17] J. Overall and R. Starbuck. Sample size estimation for randomizaed pre-post designs. *Journal of Psychiatric Research*, 15(1):51–55, 1979.

[18] J. Cohen. *Statistical Power Analysis for the Behavioral Sciences*. L. Erlbaum Associates, 1988.

[19] J.E. Overall, G. Shobaki, C. Shivakumar, and J. Steele. Adjusting sample size for anticipated dropouts in clinical trials. *Psychopharmacology Bulletin*, 34:25–33, 1998.

[20] A. Donner and N. Klar. Pitfalls of and controversies in cluster randomization trials. *American Journal of Public Health*, 94(3):416–422, 2004.

[21] H. Liu and T. Wu. Sample size calculation and power analysis of time-averaged difference. *Journal of Modern Applied Statistical Methods*, 4(2):434–445, 2005.

[22] J. Hintze. *PASS 13*. NCSS, LLC. Kaysville, Utah, USA., 2014.

[23] D. Lai, T.M. King, L.A. Moy, and Q. Wei. Sample size for biomarker studies: More subjects or more measurements per subject? *Annals of Epidemiology*, 13(3):204–208, 2003.

[24] D. Bloch. Sample size requirements and the cost of a randomized trial with repeated measurements. *Statistics in Medicine*, 5:663–667, 1986.

[25] K.J. Lui and W. Cumberland. Sample size requirements for repeated measurements in continuous data. *Statistics in Medicine*, 11:633–641, 1992.

[26] B. Winkens, H. Schouten, G. van Breukelen, and M. Berger. Optimal number of repeated measures and group sizes in clinical trials with highly divergent treatment effects. *Contemporary Clinical Trials*, 27:57–69, 2005.

[27] S. Zhang and C. Ahn. Adding subjects or adding measurements in repeated measurement studies under financial constraints. *Statistics in Biopharmaceutical Research*, 3(1):54–64, 2011.

[28] B. Freidlin, E.L. Korn, R. Gray, and A. Martin. Multi–arm clinical trials of new agents: Some design considerations. *Clinical Cancer Research*, 14:4368–4371, 2008.

[29] C.W. Dunnett. New tables for multiple comparisons with a control. *Biometrics*, 3:1–21, 1964.

[30] P.J. Diggle, P. Heagerty, K.Y. Liang, and S.L. Zeger. *Analysis of Longitudinal Data (2nd ed.)*. Oxford University Press, 2002.

[31] A.C. Leon. Sample size requirements for comparisons of two groups on repeated observations of a binary outcome. *Evaluation and the Health Professions*, 27(1):34–44, 2004.

[32] K.E. Muller, L.M. Lavange, S.L. Ramey, and C.T. Ramey. Power calculations for general linear multivariate models including repeated measures applications. *Journal of American Statistical Association*, 87(420):1209–1226, 1992.

[33] D. Hedeker, R.D. Gibbons, and C. Waternaux. Sample size estimation for longitudinal designs with attrition: Comparing time–related contrasts between two groups. *Journal of Educational and Behavioral Statistics*, 24:70–93, 1999.

[34] J. Rochon. Application of GEE procedures for sample size calculations in repeated measures experiments. *Statistics in Medicine*, 17(14):1643–1658, 1998.

[35] S. Jung and C. Ahn. Sample size estimation for GEE method for comparing slopes in repeated measurements data. *Statistics in Medicine*, 22(8):1305–1315, 2003.

[36] S. Zhang and C. Ahn. Sample size calculation for time–averaged differences in the presence of missing data. *Contemporary Clinical Trials*, 33:550–556, 2012.

[37] A.J. Vickers. How many repeated measures in repeated measures design? statistical issues for comparative trials. *BMC Medical Research Methodology*, 1:1–9, 2003.

[38] J.E. Overall. How many repeated measurements are useful? *Journal of Clinical Psychology*, 52(3):243–252, 1996.

[39] J.E. Overall and S.R. Doyle. Comparative evaluation of two models for estimating sample sizes for tests on trends across repeated measurements. *Controlled Clinical Trials*, 19:188–197, 1998.

[40] S. Zhang and C. Ahn. How many measurements for time–averaged differences in repeated measurement studies. *Contemporary Clinical Trials*, 32:412–417, 2011.

4

Sample Size Determination for Correlated Outcome Measurements Using GEE

4.1 Motivation

Experimental designs that generate correlated measurements of an outcome are widely used in biomedical, social, and behavioral studies. We usually classify such experimental designs into two types: clustered or longitudinal. In a clustered trial, randomization is performed at the level of some aggregate, such as schools, clinics, or communities. It is often adopted due to necessity where the intervention under test is delivered on a group basis, for example, a radio campaign against smoking in a socially and economically disadvantaged community. It can also be employed for administrative convenience. For example, to examine the effectiveness of a new hypertension control strategy taking advantage of electronic medical records (EMR) versus the traditional strategy, randomization is performed on caring physicians, by whom the patients are clustered. The reason is that operationally it is difficult for a physician to simultaneously practice different treatment strategies. As for a longitudinal trial, randomization is performed at the individual level. However, over the study period the outcome variable is measured multiple times from the same individuals, hence giving rise to the issue of within-subject correlation. A key difference between these two types of design is that under the clustered design the measurements within a randomization unit (cluster) are usually considered exchangeable. Thus the compound symmetric correlation structure has been frequently employed. Under the longitudinal design, however, the measurements within a unit (individual) are distinguished by their time stamps due to the potential temporal trend. Thus an autoregressive-type of correlation is more frequently used which assumes the correlation to decay by the temporal distance between measurements. In practice researchers are also likely to encounter correlated outcomes with a hierarchical structure, which is even more complicated. For example, patients can be clustered by physicians, while physicians can be clustered by clinics. In this case the correlation structure contains multiple levels of nested clustering. This topic is further explored in Chapter 6. Another example is that patients are clustered by physicians, but each patient contributes a longitudinal series of outcome measurements. In such cases we encounter a hybrid type of correlation structure.

The two most popular approaches for regression analysis with correlated outcomes are the generalized estimation equation (GEE) and the mixed-effect models (MM). Both can be employed to analyze a wide variety of outcome variables, including continuous, binary, and count variables. The MM approach models the outcome by a likelihood function which belongs to the exponential family. It employs random effects to account for correlation. Statistical theory for MM is well established. See for example Laird and Ware [1] and Jennrich and Schlucter [2]. The model parameters are usually estimated by the maximum likelihood method or by a variant known as the restricted maximum likelihood [3, 4]. The GEE approach was first introduced by Liang and Zeger [5]. It is different from the MM approach in that it does not require a full specification of the joint distribution for the clustered or longitudinal measurements. Instead, it introduces estimating equations that can consistently estimate the regression parameters and their variances as long as the marginal mean model is correctly specified. Whether the structure of "working" correlation is correctly specified does not affect the consistency. The correlation of error terms is taken into account by robust estimation of the variances of the regression coefficients. The MM and GEE approaches are critically different in the specification of correlation. Under MM, random effects are included into the linear predictor model to characterize correlation. The linear predictor is then associated with the mean through a link function. In other words, the correlation structure assumed for the random effects applies to the linear predictor, not to the outcome itself. For example, the linear predictor for a binary outcome is specified on the logit scale. Thus if we assume a compound symmetric correlation structure for the random effects, this structure is actually specified on the logit scale, not on the original binary outcome. The correlation structure of the binary outcome does not have a closed form. Under the GEE, however, the correlation structure is directly specified for the outcome variable. This difference explains why the sample sizes calculated based on the GEE and the MM methods might be different, even under the seemingly "equivalent" assumptions.

Compared with ordinary clinical trials (where the outcome measurements are independent), statisticians are faced with some additional challenges in the design and analysis of clinical trials with correlated outcomes. The intraclass correlation (for clustered trials) or the within-subject correlation (for longitudinal trials) needs to be taken into account, which might follow different correlation structures, such as compound symmetric (CS) and first-order autoregressive (AR(1)). Missing data (caused by dropout of study, missed appointments, etc.) also leads to the unique challenge of "partially" observed data. For example, suppose in a trial subjects are scheduled to measure their responses 6 times. Those who contribute 1 to 5 measurements are considered to be partially observed. Traditionally, when estimating the required sample size, researchers have tried to adjust for missing data by first computing the sample size under complete data (denoted as n_0), and then calculating the actual sample size by inflating n_0 by $1/(1 - q)$, where q is the expected proportion of subjects who will contribute incomplete observations. For clinical trials with correlated outcomes, such an adjustment might be too crude because the actual information loss depends on many factors, including the structure and strength of correlation, as well as the overall proportion and temporal distribution (trend over time) of missing values. As a result, the strategy of inflating n_0 by $1/(1 - q)$ tends to result in a sample size that is greater than what is actually needed. Appropriately accounting for the impact of correlation and missing data is challenging but critical for the design of trials with correlated outcomes, which not only saves time and medical resources, but also reduces the ethical risk of exposing excessive patients to ineffective treatments.

On the other hand, researchers also enjoy certain operational flexibility in designing clinical trials with correlated outcomes. For example, in a longitudinal study, if it is difficult to recruit research subjects due to the rareness of the disease under investigation, to some extent researchers can compensate the lack in the number of unique subjects by increasing the number of longitudinal measurements from each subject. The trade-off between the number of unique subjects and the number of measurements per subject can have profound implication in practice because the cost to recruit an additional subject and the cost to obtain an additional measurement from an existing subject usually differ dramatically. Achieving the optimal trade-off between these two designing parameters helps researchers improve the financial efficiency of clinical trials.

In this chapter, we discuss sample size calculation for clinical trials with correlated outcomes based on the GEE approach. Sample size determination using the mixed-effect model approach will be discussed in Chapter 5 and 6. A fundamental rule in sample size calculation is to align the model of sample size calculation with that of planned data analysis. Sticking to this rule helps researchers avoid making what Kimball [6] referred to as the type III error, that is, getting the right answer to a wrong problem. Thus the sample size approaches described in this chapter are appropriate for clinical trials where researchers have decided to analyze the data using the GEE method at the design stage.

4.2   Review of GEE

In this section we briefly describe the GEE method in the context of a longitudinal randomization trial, where m outcome measurements will be obtained from each subject during follow-up. Let y_{ij} be the jth ($j = 1, \ldots, m$) response observed from the ith subject. The vector of covariates associated with y_{ij} is denoted by $x_{ij} = (x_{ij1}, \ldots, x_{ijh})'$, which is of length h and contains information such as the treatment received and the time of measurement. The collection of responses from the ith subject is denoted by vector $\boldsymbol{y}_i = (y_{i1}, \ldots, y_{im})'$. The estimating equation requires specification of the first two moments of the outcome variable. The expectation (first moment) of y_{ij} is modeled by

$$\mu_{ij} = E(y_{ij} \mid \boldsymbol{x}_{ij}) = g(\boldsymbol{x}_{ij}'\boldsymbol{\beta}), \qquad (4.1)$$

where $\boldsymbol{\beta} = (\beta_1, \ldots, \beta_h)'$ is the vector of regression coefficients and $g(\cdot)$ is a known link function. It is also assumed that y_{ij} are correlated within, but independent across, subjects. The covariance matrix of $\boldsymbol{y}_i$, denoted as $\boldsymbol{W}_i = \mathrm{Cov}(\boldsymbol{y}_i)$, is modeled by

$$\boldsymbol{W}_i = \psi \boldsymbol{S}_i^{1/2} \boldsymbol{R}(\xi) \boldsymbol{S}_i^{1/2}. \qquad (4.2)$$

Here $\boldsymbol{S}_i$ is a diagonal matrix defined as $\boldsymbol{S}_i = \mathrm{diag}(s(\mu_{i1}), \ldots, s(\mu_{im}))$ with some known function $s(\cdot)$ and ψ is the dispersion parameter, such that $\mathrm{Var}(y_{ij}) = \psi s(\mu_{ij})$ is the variance of y_{ij}. The matrix $\boldsymbol{R}(\xi)$ is called the working correlation matrix, characterized by parameter ξ. Note that $\boldsymbol{R}(\xi)$ does not necessarily equal the true correlation matrix, which is denoted as $\boldsymbol{R}_0$. Liang and Zeger [5] pointed out that the large-sample distribution of estimator $\hat{\boldsymbol{\beta}}$ does not depend on whether one knows the true value of dispersion parameter ψ. For convenience we follow their practice of assuming ψ to be known.

The preceding model description is rather generic. It can be readily adapted to make statistical inference for a wide variety of outcome measurements: For a classic linear regression model with a continuous outcome, $g(\cdot)$ is the identity function, $s(\cdot) = 1$, and ψ is the residual variance. For a logistic regression model with a binary outcome, $g(\cdot)$ is the logistic function, $s(\mu) = \mu(1 - \mu)$, and $\psi = 1$. For a Poisson regression model with a count outcome, we have $g(\cdot) = \log(\cdot)$, $s(\mu) = \mu$, and $\psi = 1$. As for $\boldsymbol{R}(\xi)$, there are many different models that can be employed to account for various correlation structures. For example, the CS structure and autoregression of different orders have been widely used. In contrast to the mixed-effect model approach, the GEE method does not require the specification of the full likelihood. Only the models of the first two moments, mean (4.1) and covariance (4.2), are needed for statistical inference.

Defining $\boldsymbol{\mu}_i(\boldsymbol{\beta}) = (\mu_{i1}, \ldots, \mu_{im})'$, the estimator $\hat{\boldsymbol{\beta}}$ is obtained as the solution to the general estimating equation

$$\sum_{i=1}^{n} \boldsymbol{D}_i' \boldsymbol{W}_i^{-1} [\boldsymbol{y}_i - \boldsymbol{\mu}_i(\boldsymbol{\beta})] = \boldsymbol{0}. \qquad (4.3)$$

Here $\boldsymbol{D}_i$ is a $m \times h$ gradient matrix defined as $\boldsymbol{D}_i = \partial \boldsymbol{\mu}_i(\boldsymbol{\beta})/\partial \boldsymbol{\beta}$. The calculation of $\hat{\boldsymbol{\beta}}$ usually requires some numeric algorithms such as the Newton-Raphson algorithm. Liang and Zeger [5] presented an attractive feature of GEE that the consistency of estimators only requires the correct specification of the mean function $\mu_{ij} = g(\boldsymbol{x}_{ij}'\boldsymbol{\beta})$. The working correlation matrix $\boldsymbol{R}(\xi)$ may be misspecified.

Specifying an independent working correlation structure, $R(\xi) = I$, has been a convenient and probably the most common choice for analyzing correlated data due to the ease of computation. See, for example, Crowder [7]. It has also been demonstrated that in many situations the resulting $\hat{\beta}$ is highly efficient [5, 8]. In general, if there is no strong prior knowledge about the true correlation, setting $R(\xi) = I$ can be a reasonable option. Computationally, it is the simplest and most stable (in terms of the convergence of the iterative algorithm to solve the estimating equation). While correct specification of the correlation structure will improve efficiency, the true correlation structure is most likely unknown in practice. This property provides a great peace of mind for researchers to know that the estimates of the mean component will remain consistent even if the specified working correlation deviates from the true one.

Under mild regularity conditions, $\sqrt{n}(\hat{\beta} - \beta)$ approximately follows a normal distribution with mean $\mathbf{0}$ and the variance is estimated by a robust formula (commonly called the sandwich formula), $\Sigma_n = A_n^{-1}(\hat{\beta})V_n(\hat{\beta})A_n^{-1}(\hat{\beta})$, where

$$A_n(\hat{\beta}) = \sum_{i=1}^{n} D_i'(\hat{\beta})W_i^{-1}(\hat{\beta})D_i(\hat{\beta})$$

and

$$V_n(\hat{\beta}) = \sum_{i=1}^{n} D_i'(\hat{\beta})W_i^{-1}(\hat{\beta})W_i^*(\hat{\beta})W_i^{-1}(\hat{\beta})D_i(\hat{\beta}). \qquad (4.4)$$

Here $W_i^*(\hat{\beta}) = \hat{\epsilon}_i^{\otimes 2}$ with $\hat{\epsilon}_i = y_i - \mu_i(\hat{\beta})$ and $u^{\otimes 2} = uu'$ for a vector u. $A_n^{-1}(\hat{\beta})$ is the model-based covariance estimator, whose (asymptotic) validity depends on the correct specification of $R(\xi)$. If $R(\xi)$ is correctly specified, it can be shown that the robust covariance estimator $\Sigma_n(\hat{\beta})$ would be identical to the model-based estimator.

For sample size calculation, the hypothesis to be tested usually takes the form $H_0 : H\beta = h_0$ versus $H_1 : H\beta \neq h_0$ with a $q \times h$ matrix H of rank $q \leq h$ and a conformable vector h_0 of constant terms. Most existing sample size approaches are based on two types of test statistics: the quasi-score statistics and the Wald test statistics. We briefly review these two approaches in the following.

4.2.1 The Wald Test Approach

The Wald statistics has been widely used for sample size calculation [9, 10, 11, 12, 13, 14]. It is computed by

$$W = n(H\hat{\beta} - h_0)' \left[H \cdot \Sigma_n(\hat{\beta}) \cdot H' \right]^{-1} (H\hat{\beta} - h_0)$$

and it asymptotically has a non-central Chi-square distribution with q degrees of freedom and a non-centrality parameter

$$\lambda_n \approx n(H\beta - h_0)' \left[H \cdot \Sigma_0 \cdot H' \right]^{-1} (H\beta - h_0),$$

denoted as $\chi_q^2(\lambda_n)$. Here Σ_0 is the limit of $\Sigma_n(\hat{\beta})$ as $n \to \infty$. Under H_0, we have $\lambda_n = 0$, i.e., the test statistics W has a central chi-square distribution, denoted as $\chi_q^2(0)$. Given the nominal type I error α, the critical value $c_{1-\alpha}$ is the $100(1-\alpha)$th percentile of the $\chi_q^2(0)$ distribution. Together with the nominal power, denoted by $1 - \gamma$, the sample size can be obtained as the solution to

$$\int_{c_{1-\alpha}}^{\infty} dF(q, \lambda_n) = 1 - \gamma,$$

where $F(q, \lambda_n)$ is the cumulative distribution function of $\chi_q^2(\lambda_n)$. In essence, the Wald test statistics is constructed based on the approximate normal distribution of vector $\hat{\beta}$. For hypotheses with one degree of freedom, i.e., $q = 1$ hence H being a vector, the Wald statistics is equivalent to a z statistics,

$$z = \frac{\sqrt{n}(H\hat{\beta} - h_0)}{\sqrt{H \cdot \Sigma_n(\hat{\beta}) \cdot H'}},$$

which follows the standard normal distribution under the null hypothesis.

4.2.2   The Quasi-Score Test Approach

Here we also briefly introduce the quasi-score test approach. To facilitate discussion, the models are expressed differently with additional notations. The following derivations are mostly borrowed from Liu and Liang [15] for illustration. The expectation model is expressed as

$$g(\mu_{ij}) = x'_{ij}\beta + z'_{ij}\lambda,$$

where the vector of regression coefficients are divided into β and λ, β representing the parameters of primary interest with dimension h_1, and λ representing nuisance parameters with dimension h_2, with $h = h_1 + h_2$. The hypotheses to be tested are $H_0 : \beta = \beta_0$ versus $H_1 : \beta \neq \beta_0$. Corresponding to β and λ, we decompose the gradient matrix D_i into $D_{i\beta} = \partial\mu_i/\partial\beta$ and $D_{i\lambda} = \partial\mu_i/\partial\lambda$. The quasi-score statistic based on the generalized estimating equation is calculated by

$$T = S'_\beta(\beta_0, \hat{\lambda}_0)Q_0^{-1}S_\beta(\beta_0, \hat{\lambda}_0),$$

where

$$S_\beta(\beta_0, \hat{\lambda}_0) = \sum_{i=1}^{n} D'_{i\beta}(\beta_0, \hat{\lambda}_0)W_i^{-1}(\beta_0, \hat{\lambda}_0)[y_i - \mu_i(\beta_0, \hat{\lambda}_0)],$$

$$Q_0 = \mathrm{Cov}_{H_0}\left[S_\beta(\beta_0, \hat{\lambda}_0)\right],$$

and $\hat{\boldsymbol{\lambda}}_0$ is the estimator of $\boldsymbol{\lambda}$ under H_0 obtained by solving

$$S_\lambda(\boldsymbol{\beta}_0, \hat{\boldsymbol{\lambda}}_0) = \sum_{i=1}^n D'_{i\lambda}(\boldsymbol{\beta}_0, \hat{\boldsymbol{\lambda}}_0) W_i^{-1}(\boldsymbol{\beta}_0, \hat{\boldsymbol{\lambda}}_0)[\boldsymbol{y}_i - \boldsymbol{\mu}_i(\boldsymbol{\beta}_0, \hat{\boldsymbol{\lambda}}_0)] = 0.$$

Under some regularity conditions, T converges to a chi-square distribution with h_1 degree of freedom as $n \to \infty$. Under the alternative hypothesis H_a, where $\boldsymbol{\beta} = \boldsymbol{\beta}_a$ and $\boldsymbol{\lambda} = \boldsymbol{\lambda}_a$, T approximately follows a non-central chi-square distribution. The non-central parameter is $\boldsymbol{\eta}'\boldsymbol{\Sigma}_1^{-1}\boldsymbol{\eta}$, which can be calculated by the procedure developed by Self and Mauritsen [16] and Liu and Liang [15]. Briefly, $\boldsymbol{\eta}$ is the expectation of $S_\beta(\boldsymbol{\beta}, \hat{\boldsymbol{\lambda}}_0)$ under H_1, which is approximated by

$$\boldsymbol{\eta} = E_{H_1}[S_\beta(\boldsymbol{\beta}, \hat{\boldsymbol{\lambda}}_0)] \approx \sum_{i=1}^n P_i^* W_i^{-1}(\boldsymbol{\mu}_i^{(1)} - \boldsymbol{\mu}_i^*),$$

where $\mu_{ij}^{(1)} = g^{-1}(\boldsymbol{x}'_{ij}\boldsymbol{\beta}_a + \boldsymbol{z}'_{ij}\boldsymbol{\lambda}_a)$ and $\mu_{ij}^* = g^{-1}(\boldsymbol{x}'_{ij}\boldsymbol{\beta}_0 + \boldsymbol{z}'_{ij}\boldsymbol{\lambda}_0^*)$. Here

$$P_i^* = D'_{i\beta}(\boldsymbol{\beta}_0, \boldsymbol{\lambda}_0^*) - I_{\beta\lambda}^* I_{\lambda\lambda}^{*-1} D'_{i\lambda}(\boldsymbol{\beta}_0, \boldsymbol{\lambda}_0^*),$$

$$I_{\beta\lambda}^* = \sum_{i=1}^n D'_{i\beta}(\boldsymbol{\beta}_0, \boldsymbol{\lambda}_0^*) W_i^{-1}(\boldsymbol{\beta}_0, \boldsymbol{\lambda}_0^*) D_{i\lambda}(\boldsymbol{\beta}_0, \boldsymbol{\lambda}_0^*),$$

$$I_{\lambda\lambda}^* = \sum_{i=1}^n D'_{i\lambda}(\boldsymbol{\beta}_0, \boldsymbol{\lambda}_0^*) W_i^{-1}(\boldsymbol{\beta}_0, \boldsymbol{\lambda}_0^*) D_{i\lambda}(\boldsymbol{\beta}_0, \boldsymbol{\lambda}_0^*).$$

Note that $\boldsymbol{\lambda}_0^*$ is the limiting value of $\hat{\boldsymbol{\lambda}}_0$ under given $\boldsymbol{\beta}_a$ and $\boldsymbol{\lambda}_a$ as $n \to \infty$, which is the solution to

$$\lim_{n\to\infty} \frac{1}{n} E_{H_1}[S_\lambda(\boldsymbol{\beta}_0, \boldsymbol{\lambda}_0^*); \boldsymbol{\beta}_a, \boldsymbol{\lambda}_a] = 0.$$

$\boldsymbol{\Sigma}_1$ is the covariance of $S_\beta(\boldsymbol{\beta}, \hat{\boldsymbol{\lambda}}_0)$ under H_1, which is approximated by

$$\boldsymbol{\Sigma}_1 = \text{Cov}_{H_1}\left[S_\beta(\boldsymbol{\beta}_0, \hat{\boldsymbol{\lambda}}_0)\right] \approx \sum_{i=1}^n P_i^* W_i^{-1}(\boldsymbol{\beta}_0, \boldsymbol{\lambda}_0^*) \text{Cov}_{H_1}(\boldsymbol{y}_i) W_i^{-1}(\boldsymbol{\beta}_0, \boldsymbol{\lambda}_0^*) P_i^{*'}.$$

The testing power can be approximated from the non-central chi-square distribution. Conversely, given specified type I error and power, the corresponding non-central parameter $\boldsymbol{\eta}'\boldsymbol{\Sigma}_1^{-1}\boldsymbol{\eta}$ can be computed, from which the sample size solution can be obtained. It is noteworthy that although the hypotheses to be tested only involve parameter $\boldsymbol{\beta}$, in sample size or power calculation, the true values of both $\boldsymbol{\beta}$ and $\boldsymbol{\lambda}$ under the alternative hypothesis need to be specified. More details of the quasi-score test approach for sample size calculation can be found in Liu and Liang [15].

In practice, most clinical trials are designed to test a one-degree-of-freedom hypothesis, for example, to compare the mean effect between two treatments in a clustered trial or to compare the rates of change (slope) in a longitudinal trial. In the rest of this chapter, we will present several applications of the GEE sample size approach to clinical trials with various types of outcomes, including continuous, binary, and count data, with the parameter of interest being slope or time averaged difference (TAD). Our discussion is based on the z statistics to test hypotheses with one degree of freedom, which is equivalent to the Wald test approach. We demonstrate that this sample size approach is flexible to accommodate missing data with various missing patterns, arbitrary correlations structures, as well as financial constraints.

4.3   Compare the Slope for a Continuous Outcome

In this section we discuss the GEE sample size method for clinical trials with a continuous outcome which is longitudinally measured over the follow-up period of a pre-specified length. This type of longitudinal trials have also been called clinical trials with repeated measurements [17]. The primary goal is to compare the slopes (rates of change) between two arms, which we generically refer to as the treatment and the control. The sample size calculation for such trials is complicated by the need to appropriately account for within-subject correlation and missing data. Overall et al. [18] presented empirically defined drop-out correction coefficients to adjust sample sizes that have been initially calculated assuming complete data. Here we present a closed-form sample size formula for clinical trials with repeated measurements which is theoretically derived, and is flexible to account for missing data and within-subject correlation.

Suppose n subjects are enrolled in a randomized trial. We use y_{ij} to denote the outcome measured at time t_j $(j = 1, \ldots, m)$ from subject i. Without loss of generality, we normalize the total duration of follow-up to 1 by setting $t_j = (j-1)/(m-1)$. Thus $t_1 = 0$ and $t_m = 1$. The expectation of y_{ij} is modeled by

$$\mu_{ij} = E(y_{ij}) = \beta_1 + \beta_2 r_i + \beta_3 t_{ij} + \beta_4 r_i t_{ij}, \tag{4.5}$$

where $\boldsymbol{\beta} = (\beta_1, \ldots, \beta_4)'$ is the vector of regression coefficients and r_i is the treatment indicator (0 for control and 1 for treatment). Parameters β_1 and β_2 are the intercept and slope of the control group, respectively, while parameters β_3 and β_4 model the differences in intercept and slope between the two groups. In longitudinal clinical trials, usually β_4 is of the primary interest, which represents the treatment effect on the rate of change. The variance of y_{ij} is modeled by $\text{Var}(y_{ij}) = \sigma^2$. It is also assumed that the outcome measurements are independent across subjects but correlated within subjects, $\text{Cov}(y_{ij}, y_{ij'}) = \rho_{jj'}$ $(j \neq j')$ with $\rho_{jj} = 1$. Note that in the above description we only specify models for the first two moments, the mean and covariance matrix of $y_i = (y_{i1}, \ldots, y_{im})'$. Researchers do not have to make any assumption about the distribution of y_i.

To facilitate derivation, Jung and Ahn [13] adopted an equivalent parameterization of model (4.5):

$$\mu_{ij} = b_1 + b_2(r_i - \bar{r}) + b_3 t_{ij} + b_4(r_i - \bar{r})t_{ij}.$$

It can be shown that $b_1 = \beta_1 + \bar{r}\beta_2$, $b_2 = \beta_2$, $b_3 = \beta_3 + \bar{r}\beta_4$, and $b_4 = \beta_4$. We use $\boldsymbol{x}_{ij} = (1, r_i - \bar{r}, t_{ij}, (r_i - \bar{r})t_{ij})'$ to denote the vector of covariates.

The GEE estimator $\hat{\boldsymbol{b}} = (\hat{b}_1, \hat{b}_2, \hat{b}_3, \hat{b}_4)'$ is the solution to

$$n^{-1/2} \sum_{i=1}^{n} \sum_{j=1}^{m} (y_{ij} - \boldsymbol{x}_{ij}'\boldsymbol{b})\boldsymbol{x}_{ij} = \boldsymbol{0}. \tag{4.6}$$

Note that (4.6) is a special case of (4.3) for continuous outcomes with an independent working correlation matrix, with $\boldsymbol{W}_i = \boldsymbol{I}$ and $\boldsymbol{D}_i = \partial\mu_i/\partial\boldsymbol{b} = \boldsymbol{x}_{ij}$. Here $\boldsymbol{\mu}_i = (\mu_{i1}, \ldots, \mu_{im})'$. The expression of $\hat{\boldsymbol{b}}$ is

$$\hat{\boldsymbol{b}} = \left(\sum_i \sum_j \boldsymbol{x}_{ij}^{\otimes 2} \right)^{-1} \sum_i \sum_j \boldsymbol{x}_{ij} y_{ij}. \tag{4.7}$$

By Liang and Zeger [5], the vector $\sqrt{n}(\hat{\boldsymbol{b}} - \boldsymbol{b})$ approximately follows a normal distribution with mean $\boldsymbol{0}$ and variance $\boldsymbol{\Sigma}_n = \boldsymbol{A}_n^{-1}\boldsymbol{V}_n\boldsymbol{A}_n^{-1}$, where

$$\boldsymbol{A}_n = n^{-1} \sum_{i=1}^{n} \sum_{j=1}^{m} \boldsymbol{x}_{ij}^{\otimes 2},$$

$$\boldsymbol{V}_n = n^{-1} \sum_{i=1}^{n} \left(\sum_{j=1}^{m} \boldsymbol{x}_{ij}\hat{\epsilon}_{ij} \right)^{\otimes 2},$$

and $\hat{\epsilon}_{ij} = y_{ij} - \boldsymbol{x}_{ij}'\hat{\boldsymbol{b}}$. The expression of $\boldsymbol{\Sigma}_n$ is a special case of (4.4) for continuous outcomes with an independent working correlation matrix. To test the null hypothesis $H_0 : b_4 = 0$, the test statistic is defined as

$$z = \sqrt{n}|\hat{b}_4|/\hat{\sigma}_4.$$

Here $\hat{\sigma}_4$ is the (4,4)th component of $\boldsymbol{\Sigma}_n$ and z has a standard normal distribution under H_0. We reject the null hypothesis if $z > z_{1-\alpha/2}$. Here α is the pre-specified two-sided type I error and $z_{1-\alpha/2}$ is the $100(1-\alpha/2)$th percentile of the standard normal distribution.

4.3.1 Sample Size Calculation

We first discuss sample size calculation to test hypotheses $H_0 : b_4 = 0$ versus $H_1 : b_4 = \beta_{40}$ with a two-sided type I error α and power $1 - \gamma$, under the ideal scenario that every subject contributes complete data.

Let A and V denote the limits of A_n and V_n, respectively, as $n \to \infty$. Then the limit of Σ_n is $\Sigma = A^{-1}VA^{-1}$. Let σ_4^2 be the (4,4)th component of Σ. We can solve the sample size n from equation

$$P\left(\frac{|\hat{b}_4|}{\hat{\sigma}_4/\sqrt{n}} > z_{1-\alpha/2} \mid H_a\right) = 1 - \gamma.$$

Utilizing the approximate normality of $\hat{b}_4$, the above equation can be simplified to $-\sqrt{n}\beta_{40}/\sigma_4 + z_{1-\alpha/2} = -z_{1-\gamma}$. From which the sample size solution is

$$n = \frac{\sigma_4^2(z_{1-\alpha/2} + z_{1-\gamma})^2}{\beta_{40}^2}.$$

An attractive feature of this sample size approach is that a closed form can be derived for σ_4^2, even in the presence of missing data and arbitrary correlations structures. This extension, however, requires the assumption of missing completely at random (MCAR). That is, the events that lead to any particular measurement being missing are independent both of observable and unobservable parameters of interest, and occur entirely at random [19].

To accommodate the potential occurrence of missing data, we introduce indicator Δ_{ij}, which takes value 1 if subject i has an outcome measurement at t_j and takes value 0 otherwise. Under MCAR, $(\Delta_{i1}, \ldots, \Delta_{im})$ are independent of $(y_{i1}, \ldots, y_{im})$. Then we have the more general formulas of A_n and V_n that account for missing data:

$$A_n = n^{-1}\sum_{i=1}^{n}\sum_{j=1}^{m}\Delta_{ij}\begin{pmatrix} 1 & r_i - \bar{r} & t_j & (r_i - \bar{r})t_j \\ r_i - \bar{r} & (r_i - \bar{r})^2 & (r_i - \bar{r})t_j & (r_i - \bar{r})^2 t_j \\ t_j & (r_i - \bar{r})t_j & t_j^2 & (r_i - \bar{r})t_j^2 \\ (r_i - \bar{r})t_j & (r_i - \bar{r})^2 t_j & (r_i - \bar{r})t_j^2 & (r_i - \bar{r})^2 t_j^2 \end{pmatrix},$$

and

$$V_n = n^{-1}\sum_{i=1}^{n}\sum_{j=1}^{m}\sum_{j'=1}^{m}\Delta_{ij}\Delta_{ij'}\hat{\epsilon}_{ij}\hat{\epsilon}_{ij'}$$

$$\times \begin{pmatrix} 1 & r_i - \bar{r} & t_{j'} & (r_i - \bar{r})t_{j'} \\ r_i - \bar{r} & (r_i - \bar{r})^2 & (r_i - \bar{r})t_{j'} & (r_i - \bar{r})^2 t_{j'} \\ t_j & (r_i - \bar{r})t_j & t_j t_{j'} & (r_i - \bar{r})t_j t_{j'} \\ (r_i - \bar{r})t_j & (r_i - \bar{r})^2 t_j & (r_i - \bar{r})t_j t_{j'} & (r_i - \bar{r})^2 t_j t_{j'} \end{pmatrix}.$$

Let $\delta_j = P(\Delta_{ij} = 1)$ be the marginal probability of subjects with observations at t_j, and $\delta_{jj'} = E(\Delta_{ij}\Delta_{ij'})$ be the joint probability of subjects with observations at both t_j and t'_j (note that $\delta_{jj} = \delta_j$). Then it can be shown that

$$A = \lim_{n \to \infty} A_n = \mu_0 \begin{pmatrix} 1 & 0 & \mu_1 & 0 \\ 0 & \sigma_r^2 & 0 & \mu_1\sigma_r^2 \\ \mu_1 & 0 & \mu_2 & 0 \\ 0 & \mu_1\sigma_r^2 & 0 & \mu_2\sigma_r^2 \end{pmatrix}$$

and

$$V = \lim_{n\to\infty} V_n = \sigma^2 \begin{pmatrix} \eta_0 & 0 & \eta_1 & 0 \\ 0 & \eta_0\sigma_r^2 & 0 & \eta_1\sigma_r^2 \\ \eta_1 & 0 & \eta_2 & 0 \\ 0 & \eta_1\sigma_r^2 & 0 & \eta_2\sigma_r^2 \end{pmatrix},$$

where $\sigma_r^2 = \bar{r}(1 - \bar{r})$, $\mu_0 = \sum_{j=1}^m \delta_j$, $\mu_k = \mu_0^{-1}\sum_{j=1}^m \delta_j t_j^k$ for $k = 1, 2$, $\eta_0 = \sum_{j=1}^m \sum_{j'=1}^m \delta_{jj'}\rho_{jj'}$, $\eta_1 = \sum_{j=1}^m \sum_{j'=1}^m \delta_{jj'}\rho_{jj'}t_j$, $\eta_2 = \sum_{j=1}^m \sum_{j'=1}^m \delta_{jj'}\rho_{jj'}t_j t_{j'}$. Defining $\sigma_t^2 = \mu_2 - \mu_1^2$, it can be shown after a few steps of algebra that

$$A^{-1} = \frac{1}{\mu_0\sigma_r^2\sigma_t^2} \begin{pmatrix} \mu_2\sigma_r^2 & 0 & -\mu_1\sigma_r^2 & 0 \\ 0 & \mu_2 & 0 & -\mu_1 \\ -\mu_1\sigma_r^2 & 0 & \sigma_r^2 & 0 \\ 0 & -\mu_1 & 0 & 1 \end{pmatrix}.$$

Thus we can obtain a closed-form expression of the (4,4)th component of $\Sigma = A^{-1}VA^{-1}$,

$$\sigma_4^2 = \frac{\sigma^2 s_t^2}{\mu_0^2\sigma_r^2\sigma_t^4}, \tag{4.8}$$

where $s_t^2 = \eta_2 - 2\mu_1\eta_1 + \mu_1^2\eta_0 = \sum_{j=1}^m \sum_{j'=1}^m \delta_{jj'}\rho_{jj'}(t_j - \mu_1)(t_{j'} - \mu_1)$. The closed-form sample size formula is

$$n - \frac{\sigma^2 s_t^2(z_{1-\alpha/2} + z_{1-\gamma})^2}{\beta_{40}^2\mu_0^2\sigma_r^2\sigma_t^4}, \tag{4.9}$$

which is easy to calculate even with a scientific calculator. The advantage of (4.9) is that the within-subject correlation is incorporated as $\rho_{jj'}$, which is flexible enough to accommodate arbitrary correlation structures. Similarly, missing data are incorporated as marginal and joint observation probabilities, δ_j and $\delta_{jj'}$, which can accommodate arbitrary missing patterns.

4.3.2 Impact of Correlation

For sample size calculation, the true correlation coefficients $(\rho_{jj'})$ need to be specified. An unstructured $m \times m$ correlation matrix contains $m(m - 1)/2$ unique parameters, which in practice will be difficult to fully specify. As a compromise, researchers have tried to use various models, which are charac-terized by a few parameters, to approximate the true correlation matrix. Two of the most widely used correlation models are the compound symmetry (CS) and the first-order auto-regressive (AR(1)). Under the CS model, we have $\rho_{jj'} = \rho$ $(j \neq j')$. It assumes a constant correlation between measurements from the same subject, regardless of their temporal distance between mea-surement times. Effectively it assumes the measurements to be exchangeable within a subject. Hence besides trials with longitudinal measurements, the

CS model has been used extensively in clustered randomization trials to account for correlation among measurements obtained from the same cluster. Under the AR(1) model we have $\rho_{jj'} = \rho^{|t_j - t_{j'}|}$, which assumes the within-subject correlation to decay as the temporal distance between two measurements increases. Both the CS and AR(1) models are fully specified with a single parameter ρ.

In reality, the number of possible correlation structures can be infinite. For example, the correlation between two measurements may decay as they become further apart. However, the decay may not occur strictly at the rate dictated by the power of $|t_j - t_{j'}|$, like that under the AR(1) model. Munoz et al. [20] introduced a damped exponential family of correlation structures. It utilizes two parameters (ρ, θ) to model $\rho_{jj'}$:

$$\rho_{jj'} = \rho^{|t_j - t_{j'}|^\theta}.$$

This method provides a rich family of correlation structures where CS and AR(1) are special cases with $\theta = 0$ and $\theta = 1$, respectively.

It is also worthwhile to investigate the impact of parameter ρ. For example, can we have some straightforward conclusion such as a greater value of ρ always leads to a greater sample size requirement? First we observe that in the sample size formula (4.9), the correlation coefficients $\rho_{jj'}$ only impact sample size through s_t^2. Specifically, $\rho_{jj'}$ only impacts the second term of

$$s_t^2 = \sum_{j=1}^{m} \delta_j (t_j - \mu_1)^2 + 2 \sum_{j=1}^{m-1} \sum_{j'=j+1}^{m} \delta_{jj'} \rho_{jj'} (t_j - \mu_1)(t_{j'} - \mu_1).$$

We first investigate the simple case of no missing data ($\delta_1 = \cdots = \delta_m = 1$). Under the CS correlation structure ($\theta = 0$ such that $\rho_{jj'} = \rho$ for $j \neq j'$), the second term of s_t^2 becomes

$$\rho \left\{ \sum_{j=1}^{m} (t_j - \mu_1) \right\}^2 - \rho \sum_{j=1}^{m} (t_j - \mu_1)^2 = -\rho \sum_{j=1}^{m} (t_j - \mu_1)^2, \qquad (4.10)$$

which is a decreasing function in ρ. It means that a larger within-subject correlation is associated with a smaller sample size in the test of slopes. More specifically, (4.10) represents a negative linear relationship between sample size and correlation. Under the AR(1) correlation structure ($\theta = 1$), the relationship between s_t^2 and ρ is not so obvious, which is numerically explored as follows: We set $m = 6$ and $t_k = (k - 1)/(m - 1)$ for $k = 1, \ldots, m$. Because it is reasonable to assume a positive within-subject correlation in most clinical trials, we let the value of correlation parameter ρ range from 0 to 1. Figure 4.1 shows the curves of s_t^2 versus ρ under different values of θ from the damped exponential family. It indicates that when there is no missing data, a larger value of ρ is always associated with a smaller s_t^2, and in turn, a smaller

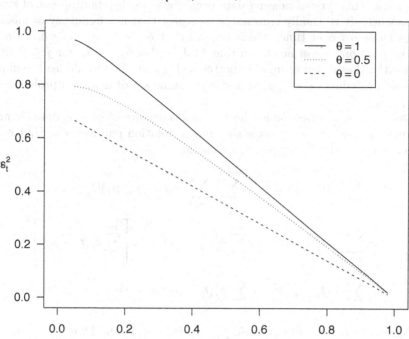

FIGURE 4.1
Numerical study to explore the relationship between s_t^2 and ρ, under the scenario of complete data and various values of θ from the damped exponential family. $\theta = 1$ corresponds to AR(1) and $\theta = 0$ corresponds to CS. The measurement times are normalized such that $t_m - t_1 = 1$. Hence $\rho_{1m} = \rho$ under all values of θ.

sample size requirement. It also shows a trend that with all the other parameters being the same, a larger power parameter θ leads to a larger sample size.

4.3.3 Impact of Missing Data

Missing data can occur for different reasons, including loss to follow-up, missed appointments, equipment failure, etc. They lead to various missing patterns, which are characterized by marginal (δ_j) and joint ($\delta_{jj'}$) observant probabilities. One frequently encountered missing pattern is called the independent missing (IM) pattern. Under this pattern missing values occur independently over $(t_1, \ldots, t_m)$. Thus we have $\delta_{jj'} = \delta_j \delta_{j'}$ for $j \neq j'$. Note that $\delta_{jj} = \delta_j$.

Another is called the monotone missing (MM) pattern. Under this pattern a subject who misses a measurement at t_j would miss all subsequent measurements. This type of missing data occurs when subjects drop out of study permanently. It is usually reasonable to expect that the frequency of missing values increases over time, which implies that $\delta_1 \geq \delta_2 \geq \ldots \geq \delta_m$. Under this assumption, it can be shown that MM implies $\delta_{jj'} = \delta_{j'}$ for $j \leq j'$. It is noteworthy that different missing patterns (IM and MM) would lead to different joint probabilities $\delta_{jj'}$, even under the same set of marginal probabilities $(\delta_1, \ldots, \delta_m)$.

One interesting question is whether the occurrence of missing data changes the relationship between sample size and correlation parameter ρ. Under the IM pattern and the CS correlation structure,

$$s_t^2 = \sum_{j=1}^{m} \delta_j (t_j - \mu_1)^2 + 2\rho \sum_{j=1}^{m-1} \sum_{j'=j+1}^{m} \delta_j \delta_{j'} (t_j - \mu_1)(t_{j'} - \mu_1).$$

$$= \sum_{j=1}^{m} \delta_j (t_j - \mu_1)^2 - \rho \sum_{j=1}^{m} \delta_j^2 (t_j - \mu_1)^2 + \rho \left[\sum_{j=1}^{m} \delta_j (t_j - \mu_1) \right]^2$$

$$= \sum_{j=1}^{m} \delta_j (t_j - \mu_1)^2 - \rho \sum_{j=1}^{m} \delta_j^2 (t_j - \mu_1)^2.$$

We have the last equality because $\sum_{j=1}^{m} \delta_j (t_j - \mu_1) = 0$. Thus s_t^2 (or equivalently, the sample size) remains a decreasing linear function of ρ.

In Figure 4.1 we have numerically explored the impact of correlation ρ under complete data. It is interesting to see whether the impact of ρ is modified by the presence of missing data. We present the result of another numerical study in Figure 4.2, where the design parameters are the same as those in Figure 4.1, except that missing data occur following a linear trend of observant probabilities, $\delta_j = 1 - 0.06(j - 1)$ for $j = 1, \ldots, m$. With $m = 6$, it implies a dropout rate of 0.3 at the end of study. We also consider two missing patterns, IM and MM. To facilitate comparison, Figures 4.1 and 4.2 are constructed on the same scale. It shows that in general a larger correlation ρ leads to a smaller sample size requirement in the test of slopes, with or without missing data. Expression (4.9) suggests that with all the other parameters being the same, the only term that is affected by different missing patterns is s_t^2. Within each missing pattern, a larger value of θ is associated with a larger value of s_t^2 (thus a larger sample size). We also observe that under each θ, the red curve is always above the corresponding black one, i.e., the IM missing pattern leads to a larger sample size than MM. This observation agrees with the intuition that MM might be a more serious type of missing problem, under which the occurrence of missing values tends to concentrate in a subgroup of subjects who drop out of study permanently, resulting in a greater information loss.

It is also likely that in a clinical trial missing values occur following a mixed pattern, which we denote by MIX. For example, some subjects might

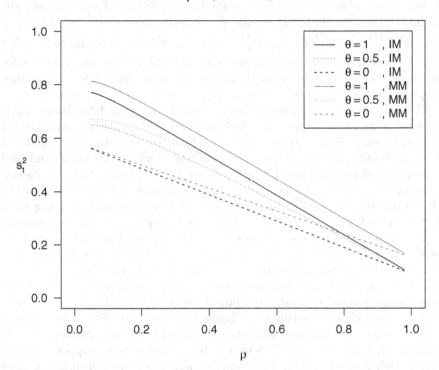

FIGURE 4.2
Numerical study to explore the relationship between s_t^2 and ρ, under the scenario of incomplete data and various values of θ from the damped exponential family. $\theta = 1$ corresponds to AR(1) and $\theta = 0$ corresponds to CS. IM and MM represent the independent and monotone missing pattern, respectively. The measurement times are normalized such that $t_m - t_1 = 1$. Hence $\rho_{1m} = \rho$ under all values of θ.

drop out of study permanently while others might miss a few appointments randomly over the study period. Hence it might be too restrictive to assume that missing values occur following a single pattern. Here we demonstrate that the sample size formula (4.9) is flexible enough to accommodate this mixture type of missing patterns. Because different missing patterns might be associated with different marginal probabilities, we use $(\delta_1^{(IM)}, \ldots, \delta_m^{(IM)})$ and $(\delta_1^{(MM)}, \ldots, \delta_m^{(MM)})$ to denote the marginal observant probabilities under the IM and MM patterns, respectively. We also use $\delta_{jj'}^{(IM)}$ and $\delta_{jj'}^{(MM)}$ to denote the corresponding joint probabilities under each pattern, as described above.

Suppose in a clinical trial the proportions of subjects who would potentially follow the IM and MM patterns are w and $(1-w)$, respectively. Then a more general sample size formula to accommodate a mixture of missing patterns can be obtained by replacing δ_j and $\delta_{jj'}$ in (4.9) with $\delta_j^{(MIX)} = w\delta_j^{(IM)} + (1-w)\delta_j^{(MM)}$ and $\delta_{jj'}^{(MIX)} = w\delta_{jj'}^{(IM)} + (1-w)\delta_{jj'}^{(MM)}$. Here the superscript (MIX) indicates that the marginal and joint probabilities are calculated under the mixture of missing patterns. It is easy to show that with all the other designing parameters being the same, we have $n^{(IM)} \leq n^{(MIX)} \leq n^{(MM)}$.

There are many other methods to account for missing data. Dawson [21] presented a sample size approach which first divides the data into strata based on their missing data patterns. Then standard test statistics are calculated within each stratum. The stratum-specific test statistics are then combined into an overall test statistic based on a weighting scheme. The sample size requirement is determined based on the overall test statistic. It should be pointed out that Dawson's "missing pattern" has a different meaning from what we have been using in this section. For example, according to Dawson, subjects who contribute the first measurement only and those who contribute the first two measurements only are considered to follow two "missing patterns," and they form two strata. In expression (4.9), however, we consider them to follow the same MM pattern in the calculation of $\delta_{jj'}$. Thus Dawson's approach is most suitable to clinical trials with an MM pattern, where the number of possible patterns are limited (at most m). For a clinical trial with an IM pattern, however, the total number of possible patterns is $2^m - 1$, which might lead to an insufficient number of subjects to support reliable inference within each stratum. Rochon [10] also tried to incorporate missing data into the sample size formula of Liu and Liang [15], but failed to provide a closed form that covers general situations.

4.3.4   Trade-Off between n and m

For clinical trials with longitudinal measurements, the sample size (n) is usually calculated treating the number of measurements per subject (m) as a given configuration of the experimental design [22]. In practice, however, the number of measurements to be obtained from each subject is another designing parameter that is within the control of researchers, but so far it has received limited attention in the experimental design community. Exploring the trade-off between the number of unique subjects (n) and the number of measurements per subject (m) in a clinical trial is meaningful because the cost to enroll an additional subject and the cost to make an additional measurement from an existing subject can be drastically different. For example, to conduct a clinical trial in a rare disease population, it might be difficult to enroll a large number of subjects. To a certain extent, researchers have the option of increasing the number of measurements from each subject to compensate for the lack of unique subjects. The trade-off between n and m, however, is complicated.

It depends on the strength and structure of within-subject correlation. It is further complicated by the potential presence of missing data.

We first discuss a simpler scenario where the correlation structure is CS and there is no missing data ($\delta_j = \delta_{jj'} = 1$). Under this scenario, three terms in sample size (4.9), (μ_0, σ_t^2, s_t^2), depend on m, which affect the sample size requirement to detect treatment effect β_{40} given type I error α and power $1 - \gamma$. It can be shown that $\mu_0 = m$,

$$\sigma_t^2 = \sum_{j=1}^{m} t_j^2/m - \left(\sum_{j=1}^{m} t_j/m\right)^2 = \frac{m+1}{12(m-1)},$$

and

$$\begin{aligned}
s_t^2 &= \sum_{j=1}^{m}(t_j - \mu_1)^2 + 2\rho \sum_{j=1}^{m} \sum_{j'=j+1}^{m}(t_j - \mu_1)(t_{j'} - \mu_1) \\
&= (1-\rho)\sum_{j=1}^{m}(t_j - \mu_1)^2 + \rho \sum_{j=1}^{m}\sum_{j'=1}^{m}(t_j - \mu_1)(t_{j'} - \mu_1) \\
&= (1-\rho)\sum_{j=1}^{m}(t_j - \mu_1)^2 + \rho\left(\sum_{j=1}^{m}(t_j - \mu_1)\right)^2 \\
&= (1-\rho)\cdot m\sigma_t^2 + 0 \\
&= \frac{(1-\rho)m(m+1)}{12(m-1)}.
\end{aligned}$$

To analytically explore the trade-off between n and m, in the following we use $n\{m\}$ to denote the sample size requirement under a given number of repeated measurements (m) from each subject. Then by plugging the preceding expressions of μ_0, σ_t^2, and s_t^2 into (4.9), we have

$$n\{m\} = \frac{48\sigma^2(1-\rho)(m-1)}{m(m+1)} \cdot \frac{(z_{1-\alpha/2} + z_{1-\gamma})^2}{\beta_{40}^2} \propto \frac{m-1}{m(m+1)}.$$

That is, under no missing data and the CS correlation structure, the trade-off between n and m does not depend on the correlation coefficient ρ. We can explore the trade-off by

$$\frac{n\{m+1\}}{n\{m\}} = \frac{m^2}{(m-1)(m+2)}.$$

We have some interesting observations about $n\{m+1\}/n\{m\}$. First, when $m = 2$, $n\{m+1\}/n\{m\}=1$. That is, when comparing slopes in a clinical trial where there is no missing data and the observations follow the CS correlation structure, making 2 or 3 measurements within the same follow-up period will achieve exactly the same power. Second, it can be shown that

$n\{m+1\}/n\{m\} < 1$ for any $m > 2$. Thus increasing the number of measurements from each subject always leads to a smaller sample size requirement. We have $n\{m\} \to 0$ as $m \to \infty$. Thus theoretically we can continue increasing m to reduce n. In practice, however, it can be shown that $n\{m+1\}/n\{m\}$ quickly converges to 1 as m increases.

The trade-off between n and m is much more complicated in the presence of missing data and under other correlation structures. In these scenarios, numerical studies will have to be performed to explore $n\{m+1\}/n\{m\}$. In Figure 4.3 we explore the relationship between $n\{m+1\}/n\{m\}$ and m over different combinations of missing data and correlation structures. We set $\rho = 0.3$, a dropout rate of 0.3 at the end of study, and a linear trend in the observant probabilities, i.e., $\delta_j = 1 - 0.3 * (j-1)/(m-1)$ for $j = 1, \ldots, m$. The IM and MM missing patterns and the CS and AR(1) correlation structures are considered. For reference, the curves corresponding to the complete data scenarios are also included. We have several observations. First, under the CS correlation structure, $n\{m+1\}/n\{m\}$ is always below 1 for $m > 2$, suggesting that increasing m (the number of repeated measurements from each subject) always reduces sample size requirement. Under the AR(1) correlation structure, however, increasing m does not always lead to a smaller sample size. For the scenarios that we have considered, under complete data or the MM missing pattern, increasing m actually leads to a larger sample size requirement under the AR(1) structure. It is only under IM that an increasing m is associated with a smaller sample size.

Figure 4.3 represents the particular scenarios that $\rho = 0.3$ and the observant probabilities follow a linear trend with a dropout rate of 0.3. We have explored other scenarios and it is generally true that under the CS correlation structure, increasing the number of measurements from each subject can reduce sample size requirement. Under the AR(1) correlation structure, however, numerical studies are required to determine whether it is advantageous to increase m. Also, as m increases, $n\{m+1\}/n\{m\}$ quickly converges to 1. Thus it is only to a limited extent that researchers can feasibly compensate the lack of unique subjects by increasing the number of repeated measurements. It should be pointed out that the preceding discussion only applies to the scenario where the length of follow-up is fixed, where we set $t_m - t_1 = 1$ and ρ_{1m}. If the increase in m is accompanied by a prolonged follow-up period, a greater saving in sample size can be achieved. However, evaluating the trade-off under this scenario is further complicated by an additional factor of study duration.

4.3.5   Compare the slope among $K \geq 3$ Groups

Many randomized trials compare the efficacy of $K \geq 3$ treatments simultaneously. For example, Liu and Dahlberg [23] reported that 24 phase III trials had more than two treatment groups out of 112 phase III trials in progress at that time conducted by the Southwest Oncology Group. Makuch and Simon

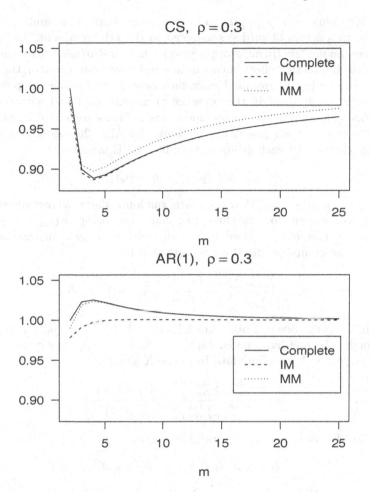

FIGURE 4.3
A numerical study to explore $\frac{n\{m+1\}}{n\{m\}}$ under missing data and different correlation structures. The vertical axis is $\frac{n\{m+1\}}{n\{m\}}$. "Complete" indicates the scenario of complete data. "IM" and "'MM" indicate the independent and monotone missing patterns, respectively, with marginal observant probabilities computed by $\delta_j = 1 - 0.3 * (j - 1)/(m - 1)$.

[24] also cited that out of 60 ongoing randomized breast cancer studies, a third of them involved three or more treatment groups. Despite the wide practice of clinical trials with $K \geq 3$ treatments, relatively little effort has been directed to theoretically investigating the design and analysis of such trials, especially in the context of repeated measurements.

Let y_{kij} be the observation of a continuous outcome observed at time t_j $(j = 1, \ldots, m)$ from the ith $(i = 1, \ldots, n_k)$ subject of the kth $(k = 1, \ldots, K)$

group. We define $n = \sum_{k=1}^{K} n_k$ to be the total sample size, and $r_k = n_k/n$ to be the proportion of subjects assigned to the kth treatment. To compare the slopes among treatment groups, traditionally researchers have conducted tests on the interaction effect between time and treatment, fitted on the overall data set containing K groups. Under such models, evaluating the power and sample size requirement in the presence of missing data and within-subject correlation can be difficult. To provide a closed-form sample size formula that is easy to calculate for practitioners, Jung and Ahn [25] adopted a slightly different scheme. For each group k, they fit a GEE model with

$$\mu_{kij} = E(y_{kij}) = \theta_k + \beta_k t_j,$$

and $y_{kij} = \mu_{kij} + \epsilon_{kij}$. Here ϵ_{kij} are random effects whose distribution is unknown, except for the first two moments being $E(\epsilon_{kij}) = 0$ and $\text{Cov}(\epsilon_{kij}, \epsilon_{kij'}) = \sigma^2 \rho_{jj'}$. Here $\rho_{jj'}$ is the within-subject correlation (with $\rho_{jj} = 1$). The group-specific slope is estimated by

$$\hat{\beta}_k = \frac{\sum_{i=1}^{n_k} \sum_{j=1}^{m} (t_j - \bar{t}) y_{kij}}{\sum_{i=1}^{n_k} \sum_{j=1}^{m} (t_j - \bar{t})^2}, k = 1, \ldots, K, \qquad (4.11)$$

which is a special case of Equation (4.7). Here $\bar{t} = \sum_{j=1}^{m} t_j/m$. On the other hand, under the null hypothesis, $H_0 : \beta_1 = \beta_2 = \cdots = \beta_k$, the common slope can be estimated by pooling data from the K groups,

$$\hat{\beta} = \frac{\sum_{k=1}^{K} \sum_{i=1}^{n_k} \sum_{j=1}^{m} (t_j - \bar{t}) y_{kij}}{\sum_{k=1}^{K} \sum_{i=1}^{n_k} \sum_{j=1}^{m} (t_j - \bar{t})^2}. \qquad (4.12)$$

We make inference based on the random vector

$$\boldsymbol{B} = \sqrt{n}(\hat{\beta}_1 - \hat{\beta}, \cdots, \hat{\beta}_{K-1} - \hat{\beta})'.$$

Here we define $\boldsymbol{B}$ as a vector of length $(K-1)$ instead of $\sqrt{n}(\hat{\beta}_1 - \hat{\beta}, \ldots, \hat{\beta}_K - \hat{\beta})'$ because the K elements in the latter vector is subject to a linear constraint, $\sum_{k=1}^{K} r_k(\hat{\beta}_k - \hat{\beta}) = 0$, which would lead to a covariance matrix that is not of full rank. Based on (4.11) and (4.12), it can be shown that

$$\sqrt{n}(\hat{\beta}_k - \hat{\beta}) = \sqrt{n}(\beta_k - \bar{\beta}) + \frac{1}{\sqrt{n}} \sum_{l=1}^{K} \sum_{i=1}^{n_l} \sum_{j=1}^{m} (\xi_{kl} w_{lij} - \tilde{w}_{lij}) \epsilon_{lij}, \qquad (4.13)$$

where $\bar{\beta} = \sum_{k=1}^{K} r_k \beta_k$, ξ_{kl} is an indicator variable taking value 1 if $k = l$ and 0 otherwise, and

$$w_{kij} = \frac{t_j - \bar{t}}{n^{-1} \sum_{i=1}^{n_k} \sum_{j=1}^{m} (t_j - \bar{t})^2},$$

$$\tilde{w}_{kij} = \frac{t_j - \bar{t}}{n^{-1} \sum_{k=1}^{K} \sum_{i=1}^{n_k} \sum_{j=1}^{m} (t_j - \bar{t})^2}.$$

Under H_0, the first term on the right-hand side of (4.13) equals 0. By applying the central limit theorem to (4.13), under H_0, $\boldsymbol{B}$ is approximately normal with mean $\boldsymbol{0}$ and variance $\boldsymbol{V}$ that can be consistently estimated by $\hat{\boldsymbol{V}} = (\hat{v}_{kk'})$ with

$$\hat{v}_{kk'} = \frac{1}{n-2K} \sum_{l=1}^{K} \sum_{i=1}^{n_l} \left\{ \sum_{j=1}^{m} (\xi_{kl} w_{lij} - \tilde{w}_{lij}) \hat{\epsilon}_{lij} \right\} \left\{ \sum_{j=1}^{m} (\xi_{k'l} w_{lij} - \tilde{w}_{lij}) \hat{\epsilon}_{lij} \right\},$$

where $\hat{\epsilon}_{kij} = y_{kij} - \hat{\theta}_k - \hat{\beta}_k t_j$. In the denominator of $\hat{v}_{kk'}$, $2K$ corresponds to the number of parameters $\{(\theta_k, \beta_k), 1 \le k \le K\}$ estimated to obtain the residuals $\hat{\epsilon}_{kij}$. We reject H_0 with type I error α if $\boldsymbol{B}' \hat{\boldsymbol{V}}^{-1} \boldsymbol{B} > \chi_{K-1,1-\alpha}^2$ where $\chi_{v,1-\alpha}^2$ is the $100(1-\alpha)$th percentile of a chi-square distribution with v degree of freedom.

For sample size calculation, given the true slopes β_k ($k = 1, \ldots, K$), $\boldsymbol{B}' \hat{\boldsymbol{V}}^{-1} \boldsymbol{B}$ approximately has a non-central chi-square distribution with $K-1$ degrees of freedom and noncentrality parameter $n \boldsymbol{\eta}' \boldsymbol{V}^{-1} \boldsymbol{\eta}$, where $\boldsymbol{V} = \lim_{n \to \infty} \hat{\boldsymbol{V}}$ and $\boldsymbol{\eta} = (\beta_1 - \bar{\beta}, \ldots, \beta_{K-1} - \bar{\beta})'$. Let $\chi_{K-1}^2(U)$ denote a non-central chi-square distribution with $K-1$ degrees of freedom and non-centrality parameter U. Under type I error α and power $1-\gamma$, the sample size n is obtained by first solving U from the following equation,

$$1 - \gamma = P\left(\chi_{K-1}^2(U) > \chi_{K-1,1-\alpha}^2\right).$$

Denoting the solution by $U = U(K-1, \alpha, \gamma)$, the required sample size is

$$n = \frac{U(K-1, \alpha, \gamma)}{\boldsymbol{\eta}' \boldsymbol{V}^{-1} \boldsymbol{\eta}}. \tag{4.14}$$

The value of $U(K-1, \alpha, \gamma)$ can be obtained through numerical search. One option is to employ the SAS function CNONCT to obtain the solution. In the following we derive the expression of $\boldsymbol{V}$ or $\boldsymbol{V}^{-1}$ in the presence of missing data and arbitrary correlation structure, which eventually leads to a closed-form sample size formula. The derivation is based on the assumption of missing completely at random. Let Δ_{kij} be the indicator which takes value 0/1 if a subject's outcome measurement at t_j is missing/observed. We also define $\delta_j = E(\Delta_{kij})$ to be the probability of a subject having an observation at time t_j, and $\delta_{jj'} = E(\Delta_{kij} \Delta_{kij'})$ to be the probability of a subject having observations at both t_j and $t_{j'}$. Then the general formula of $\hat{v}_{kk'}$ that accommodates missing data is

$$\hat{v}_{kk'} = \frac{1}{n-2K} \sum_{l=1}^{K} \sum_{i=1}^{n_l} \left\{ \sum_{j=1}^{m} \Delta_{lij} (\xi_{kl} w_{lij} - \tilde{w}_{lij}) \hat{\epsilon}_{lij} \right\}$$

$$\times \left\{ \sum_{j=1}^{m} \Delta_{lij} (\xi_{k'l} w_{lij} - \tilde{w}_{lij}) \hat{\epsilon}_{lij} \right\}.$$

As $n \to \infty$, it converges to

$$v_{kk} = \frac{(1 - r_k)\sigma^2}{r_k \bar{m}^2 \sigma_t^4} \sum_{j=1}^{m} \sum_{j'=1}^{m} \delta_{jj'} \rho_{jj'} (t_j - \bar{t})(t_{j'} - \bar{t})$$

and

$$v_{kk'} = -\frac{\sigma^2}{\bar{m}^2 \sigma_t^4} \sum_{j=1}^{m} \sum_{j'=1}^{m} \delta_{jj'} \rho_{jj'} (t_j - \bar{t})(t_{j'} - \bar{t}) \text{ for } k \neq k'.$$

Let $s_t^2 = \sum_{j=1}^{m} \sum_{j'=1}^{m} \delta_{jj'} \rho_{jj'} (t_j - \bar{t})(t_{j'} - \bar{t})$. Then we have

$$V = \frac{\sigma^2 s_t^2}{\bar{m}^2 \sigma_t^4} \left\{ \text{diag}(1/r_1, \cdots, 1/r_{K-1}) - \mathbf{1}\mathbf{1}' \right\},$$

where $\mathbf{1} = (1, \ldots, 1)'$ is a unity vector of length $(K-1)$. The inverse of V can be readily calculated,

$$V^{-1} = \frac{\bar{m}^2 \sigma_t^4}{\sigma^2 s_t^2} \left\{ \text{diag}(r_1, \ldots, r_{K-1}) + r_k^{-1} \mathbf{r}\mathbf{r}' \right\},$$

where $\mathbf{r} = (r_1, \ldots, r_{K-1})'$. Thus we have a closed-form expression of the non-centrality parameter:

$$n\boldsymbol{\eta}'V^{-1}\boldsymbol{\eta} = \frac{n\bar{m}^2 \sigma_t^4}{\sigma^2 s_t^2} \left\{ \sum_{k=1}^{K-1} r_k(\beta_k - \bar{\beta})^2 + r_K^{-1} \left(\sum_{k=1}^{K-1} r_k(\beta_k - \bar{\beta}) \right)^2 \right\}. \quad (4.15)$$

A closed-form sample size formula based on (4.14) can then be obtained.

Expression (4.15) can be simplified in some special cases, which have been frequently adopted by practitioners to design clinical trials comparing $K \geq 3$ treatments:

- Special Case 1
  For two-sample testing ($K = 2$), $\beta_1 - \bar{\beta} = r_2(\beta_1 - \beta_2)$ and hence

  $$\boldsymbol{\eta}'V^{-1}\boldsymbol{\eta} = \frac{\bar{m}^2 \sigma_t^4 r_1 r_2(\beta_1 - \beta_2)^2}{\sigma^2 s_t^2}.$$

  It leads to a sample size formula identical to (4.9).

- Special Case 2
  Under a balanced allocation, $r_1 = \cdots = r_k = 1/K$, we have

  $$\boldsymbol{\eta}'V^{-1}\boldsymbol{\eta} = \frac{\bar{m}^2 \sigma_t^4}{K \sigma^2 s_t^2} \left\{ \sum_{k=1}^{K-1} (\beta_k - \bar{\beta})^2 + \left(\sum_{k=1}^{K-1} (\beta_k - \bar{\beta}) \right)^2 \right\}.$$

- Special Case 3
  Where patients in groups 1 to $K - 1$ receive experimental treatments with similar efficacy while patients in group K serve as the control, then we have $\beta_1 = \cdots = \beta_{K-1} = \beta_K + d$. Here d represents the common treatment effect compared with the control. In this case we have $\bar{\beta} = \beta_K + (1 - r_K)d$ and $\beta_k - \bar{\beta} = r_K d$. Under a balanced allocation,

$$\eta' V^{-1} \eta = \frac{r_K(1 - r_K)\bar{m}^2 d^2 \sigma_t^4}{\sigma^2 s_t^2}.$$

- Special Case 4
  Where the treatments assigned to groups 1 to $K - 1$ are ordered by their expected treatment effects, and group K serves as the control. For example, this scenario is applicable to trials where the dosage of an experimental drug increases over the groups, with a constant improvement in efficacy between consecutive dosages. In this case we have $\beta_k = \beta_K + kd$ for $k = 1, \ldots, K - 1$. Assuming a balanced allocation, $r_1 = \cdots = r_K = 1/K$, we have $\bar{\beta} = \beta_K + d(K - 1)/2$ and $\beta_k - \bar{\beta} = d[k - (K - 1)/2]$. Then it can be shown that

$$\eta' V^{-1} \eta = \frac{(K^2 - 1)\bar{m}^2 d^2 \sigma_t^4}{12\sigma^2 s_t^2}.$$

The closed-form sample size formula that we have presented here is derived based on a set of models fitted on individual treatment groups. Although still unbiased and consistent, parameters obtained under this modeling strategy are less efficient than those under the traditional strategy where a single model with time-treatment interactions is fitted on all groups. Here a small sacrifice in efficiency is made to achieve improved usability with a closed-from sample size formula.

4.3.6 Financial Constraint

Budget constraint is a challenge faced by investigators in planning almost every clinical trial. When designing a study with repeated measurements, investigators need to decide how to make full use of the limited financial resources: increasing the number of participating subjects (n) or increasing the number of repeated measurements per subject (m). The ultimate goal is to maximize the testing power. The trade-off between n and m has been a classic design problem in repeated measurement studies [26, 27]. Ahn and Jung [28, 29] presented a method to assess the relative benefit of adding a subject versus adding a measurement in terms of the GEE estimator of slope coefficients. However, their approach did not take cost constraint into consideration, which is equivalent to assuming an infinite financial resource. In this section we present an approach that combines the GEE estimator of the slope coefficient with the financial constraint in clinical trials where the primary interest is to test the difference in slope between two treatments.

The cost incurred by an additional subject is usually different from that incurred by an additional measurement from an existing subject. Several papers have provided the sample size estimator for repeated measurement studies incorporating the costs of recruiting and measuring subjects [30, 31, 32, 33, 34].

The determination of n and m is complicated by the presence of missing data and different types of correlation structures among repeated measurements. Here our goal is to derive the optimal combination of n and m such that the GEE estimator of the treatment effect achieves the smallest variance within a particular budget. In other words, the optimal combination of n and m maximizes the power in comparing the treatment effect (represented by slope) when researchers are faced with a budget constraint.

Our discussion is based on the linear regression model (4.5). We use the same notations as in Sections 4.3.1 without repeating their definitions. In practice, the number of subjects and the number of repeated measures per subject are often restricted by budget constraints. We use C_1 and C_2 to denote the costs of recruiting one study subject and the cost of making one measurement on an existing subject, respectively. We assume the costs to be identical between the control and treatment groups. For simplicity, we further assume that there is no overhead cost. We denote the upper limit of total budget by C. Thus the mathematical expression of the budget constraint is $nC_1 + n(\sum_{j=1}^{m} \delta_j)C_2 \leq C$, which takes into account the possibility that some measurements might be missing. Recall that δ_j is the proportion of subjects who have an observation at time t_j. The goal is to find (n^0, m^0), the optimal combination of (n, m), that minimizes σ_4^2/n within a budget of C. The computation of σ_4^2 in (4.8) does not involve sample size n. It only involves m. Hence for a given m, a larger n always leads to a smaller σ_4^2/n and a greater power. We define an integer function of m such that

$$n(m) = \max_{n \in I^+} \left\{ n : n \leq C / \left(C_1 + \sum_{j=1}^{m} \delta_j C_2 \right) \right\}. \qquad (4.16)$$

Here I^+ denotes the set of positive integers. Thus for investigators, the action space determined by the budget constraint is $\{(n(m), m) : m \geq 2\}$. Furthermore, in medical studies investigators might also face logistic constraints. For example, within the study period, the maximum feasible number of measurements is limited at m_{max}. Then we have

$$(n^0, m^0) \in \mathcal{R}, \text{ where } \mathcal{R} \equiv \{(n(m), m) : 2 \leq m \leq m_{max}\}. \qquad (4.17)$$

In Equation (4.8) the terms depending on m are s_t^2, μ_0, and σ_t^2. Thus to find (n^0, m^0), our goal of minimizing σ_4^2/n is equivalent to minimizing

$$Q = \frac{s_t^2}{\mu_0^2 \sigma_t^4 n}, \qquad (4.18)$$

under constraint (4.17). For the time being we ignore the integer constraint for n, and plug $n = C/(C_1 + \sum_{j=1}^{m} \delta_j C_2)$ into (4.18). We have

$$Q = \frac{C_2}{C} \cdot \frac{s_t^2(\sum_{j=1}^{m} \delta_j + U)}{\mu_0^2 \sigma_t^4}, \qquad (4.19)$$

where $U = C_1/C_2$. Thus the optimal combination (n^0, m^0) can be obtained by first identifying m^0 which minimizes $s_t^2(\sum_{j=1}^{m} \delta_j + U)/(\mu_0^2 \sigma_t^4)$, and then obtaining n^0 by $n^0 = n(m^0)$. Zhang and Ahn [35] provided the following fact:

Fact 1   *For a financial constraint represented by (C, C_1, C_2) and a given set of design parameters, the optimal number of repeated measurements (m^0) is only affected by the cost ratio $U = C_1/C_2$. Changing the total budget C only affects the sample size n^0.*

This fact is generally applicable to scenarios with missing data and various correlation structures. In the special case of no missing data ($\delta_j = 1, j = 1, \ldots, m$) and the CS correlation structure ($\rho_{jj'} = \rho$ for $j \neq j'$), (n^0, m^0) can be obtained analytically.

Fact 2   *The optimal solution under no missing data and the CS correlation structure is*

$$m^0 = \begin{cases} 2, & \text{if } C_1/C_2 \leq 2, \\ 2 \text{ or } m_{max}, & \text{if } C_1/C_2 > 2, \end{cases}$$

and $n^0 = n(m^0)$.

Proof.   See Appendix A.1.   □

When $C_1/C_2 \leq 2$, the testing power is always maximized when investigators take $m^0 = 2$ measurements from each subject. When $C_1/C_2 > 2$, the maximum power is achieved at the two extremes, either $m^0 = 2$ or $m^0 = m_{max}$.

Under missing data or other correlation structures such as AR(1), the expression of Q in (4.18) is too complicated to explore (n^0, m^0) analytically. However, one reasonable intuition is that as $U = C_1/C_2$ increases, the value of m^0 is non-decreasing given a fixed combination of the other design factors (including missing pattern, correlations structure, dropout rate, and ρ), the reason being that it becomes more and more expensive to enroll a new subject than to obtain an additional measurement from an existing subject. The numerical studies conducted by Zhang and Ahn [35] are consistent with this intuition.

4.3.7   Example

To employ the sample size formula (4.9), the designing parameters σ^2 and ρ need to be specified, which in practice usually can be estimated based on data collected from previous studies. Let $\hat{\epsilon}_{ij}$ be the residuals obtained from the analysis of a previous study. Then the variance can be estimated by

$$\hat{\sigma}^2 = \frac{\sum_{i=1}^{n} \sum_{j=1}^{m} \hat{\epsilon}_{ij}^2}{nm}.$$

For the CS correlation structure, a consistent estimator of ρ can be obtained by

$$\hat{\rho} = \frac{2 \sum_{i=1}^{n} \sum_{j=1}^{m-1} \sum_{j'=j+1}^{m} \hat{\epsilon}_{ij} \hat{\epsilon}_{ij'}}{\hat{\sigma}^2 nm(m-1)}.$$

For the AR(1) correlation structure, the consistent estimator of ρ can be obtained from the equation

$$\sum_{i=1}^{n} \sum_{j=1}^{m-1} \sum_{j'=j+1}^{m} \left(\hat{\epsilon}_{ij} \hat{\epsilon}_{ij'} - \hat{\sigma}^2 \rho^{|t_{ij}-t_{ij'}|} \right) = 0.$$

Suppose we apply the GEE sample size method to a labor pain study [36]. Eighty-three women in labor were randomly assigned to either receive a pain medication (43 women) or a placebo (40 women). Self-reported pain scores were collected at 30-minute intervals, with 0 indicating no pain and 100 indicating extreme pain. A total of $m = 6$ scores were planned to be collected from each woman. However, numerous missing values occurred and they followed the monotone missing pattern. From a previous study it is estimated that $\sigma^2 = 815.84$. Researchers would like to detect a difference of $\sigma = 28.6$ in the mean pain score between the two groups at the end of follow-up. With the total duration of follow-up normalized at $t_6 - t_1 = 1$, we set $\beta_{40} = \sigma = 28.6$. A balanced design will be adopted, hence $\bar{r} = 0.5$. Also, from the previous study the marginal observation probabilities are $\delta_1 = (\delta_1, \delta_2, \delta_3, \delta_4, \delta_5, \delta_6)' = (1, 0.9, 0.78, 0.67, 0.54, 0.41)'$. From the above specifications, we have $\sigma_r^2 = 0.25$, $\mu_0 = 4.31$, $\mu_1 = 0.40$, $\mu_2 = 0.27$, and $\sigma_t^2 = 0.11$.

Under CS, we estimate that $\rho = 0.4$ and $s_t^2 = 0.38$, so we need a sample size of $n = 77$ to detect a difference in slope of size $\beta_{40} = 28.6$ with a power of 0.9 at a two-sided type I error of 0.05. Under the AR(1) correlation structure, the estimated correlation parameter is $\rho = 0.25$. It is calculated that $s_t^2 = 0.41$ and the corresponding sample size is $n = 117$.

The marginal observant probabilities δ_1 that we have assumed approximately follow a linear trend over the study period, i.e., the observant probabilities decrease steadily over time (see Figure 4.4). In reality, the observant probabilities can follow various trends, and their impact on the sample size requirement can be different. In Figure 4.4 we present another two sets of observant probabilities:

$$\delta_2 = (1, 0.95, 0.9, 0.85, 0.63, 0.41)',$$
$$\delta_3 = (1, 0.8, 0.6, 0.54, 0.48, 0.41)'.$$

Both δ_2 and δ_3 has the same dropout rate of $1 - 0.41 = 0.59$ as δ_1 at the end of study but they show different trends. Under δ_2, the observant probabilities are relatively stable initially, but drop quickly afterward. It is opposite under δ_3, where the probabilities decrease quickly at the beginning but plateau later. Traditionally researchers have tried to adjust sample size for missing data

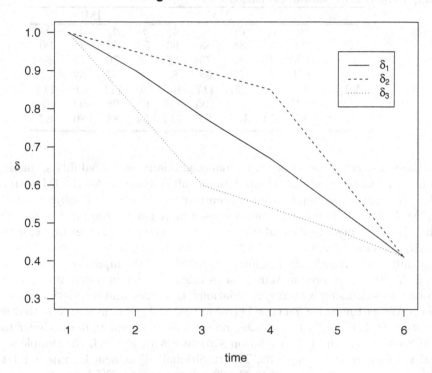

FIGURE 4.4
Different trends in the marginal observant probabilities. δ_1 approximately follows a linear trend. δ_2 is relatively steady initially but drops quickly afterward. δ_3 drops quickly from the beginning but plateaus.

by first calculating the sample size under complete data, denoted as n_0, then computing the final sample size by $n_0/(1-q)$, where q is the expected dropout rate at the end of study (in our example, $q = 0.59$). Such an adjustment for missing data might be too crude. The reasons are: (1) Subjects who dropped out at t_j might have partial observations measured at $t_1, \ldots, t_{j-1}$. It has been a routine practice (such as in SAS PROCs MIXED and GENMOD) to include partial observations into statistical analysis. Under the traditional adjustment of $n_0/(1-q)$, however, the contribution from partial observations are effectively excluded from sample size consideration. (2) The impact of missing patterns also need to be considered. Due to the same dropout rate at the end of study, the traditional adjustment leads to a final sample size that is the same for δ_1, δ_2, and δ_3. However, from Figure 4.4 it is easy to see that the overall proportion of missing values is the greatest under δ_3.

TABLE 4.1

Sample sizes under various scenarios

	ρ	MM				IM			
		δ_0	δ_1	δ_2	δ_3	δ_0	δ_1	δ_2	δ_3
CS	0.1	54	88	83	93	54	86	81	90
	0.25	45	82	75	88	45	76	72	80
	0.4	36	77	68	83	36	67	62	71
AR(1)	0.1	80	127	117	135	80	111	108	114
	0.25	68	117	105	126	68	98	94	101
	0.4	54	105	92	114	54	84	80	87

Furthermore, given the same set of marginal observant probabilities, missing values tend to be evenly distributed among all subjects under the IM pattern while missing values tend to be concentrated in a subset of subjects under the MM pattern. The latter usually causes more severe information loss. (3) The impact of missing data also depends on correlation. For example, if the longitudinal measurements are highly correlated, missing a few measurements may only lead to small information loss, and thus little impact on sample size.

In Table 4.1 we present sample sizes calculated under different observant probabilities, missing patterns, correlations structures, and correlation parameter ρ. We set the two-sided type I error at 0.05 and power at 0.9. Note that we use $\delta_0 = (1, 1, 1, 1, 1, 1)$ to indicate the scenario of complete data. Under the MM pattern and the CS correlation structure with $\rho = 0.4$, the sample size under complete data is $n_0 = 36$. The traditional adjustment for missing data will require a sample size of 36/0.41=88, which is 6% to 29% larger than what is actually needed under $\delta_1 - \delta_3$. Similarly, under the MM pattern and the AR(1) correlation structure with $\rho = 0.25$, we have $n_0 = 68$ and a traditionally adjusted sample size of 68/0.41=166. The actually needed samples under δ_1, δ_2, and δ_3 are 117, 105, and 126, respectively. That is, the GEE approach achieves a saving of 31% to 58% in sample size. This approach is advantageous because all relevant factors, including the specific trend of observant probabilities, the missing patterns, and the strength and structure of correlation, are appropriately accounted for in the adjustment for missing data. From Table 4.1 we also observe that, with all other design parameters being the same, sample sizes under the IM pattern are generally smaller than those under the MM pattern. Also, a stronger correlation is generally associated with a smaller sample size for the test of slopes.

4.4   Test the TAD for a Continuous Outcome

In controlled clinical trials, sometimes researchers measure the response multiple times, hoping that the time-averaged response can be more precise than a single measurement. Comparing two treatments based on the time-averaged difference (TAD), defined as the difference in the time-averaged responses between two treatments, might offer a greater testing power. Care must be taken in the analysis of such studies because of the correlation introduced when several measurements are taken from the same individual. The analysis might be further complicated by the occurrence of missing data.

Testing the slope (the rate of change) is appropriate when the treatment is slow to take effect so the responses tend to change at a certain rate during the study period, for example, the change in weekly blood pressure measurements during an two-month physical training intervention. Testing the TAD, on the other hand, is more appropriate when the treatment takes effect quickly and tends to maintain the responses around a desired level, for example, the change in blood pressures under a combined regimen of renin-angiotensin system inhibitors and diuretics. Diggle et al. [37] provided closed-form sample size formulae for comparing the rates of change and time-averaged responses between two groups based on continuous outcomes. Their derivation assumed no missing data, an equal number of subjects between two groups, and the CS within-subject correlation structure. Liu and Wu [38] extended the sample size formula for TAD to unbalanced clinical trials. Here we present the derivation of a closed-form sample size formula that is flexible enough to accommodate arbitrary missing patterns and correlation structures for the test of TAD.

Let y_{ij} denote the continuous response measurement obtained at time t_j $(j = 1, \ldots, m)$ from subject i $(i = 1, \ldots, n)$, where m is the planned number of repeated measurements per subject, and n is the total number of subjects enrolled. The length of the follow-up period is normalized such that $T = t_m - t_1 = 1$. The subjects are randomly assigned to the control or treatment group, indicated by $r_i = 0$ or 1, respectively. We use $\bar{r} = \sum_{i=1}^{n} r_i/n$ to denote the proportion of subjects assigned to the treatment group. To make inference based on TAD between two groups, we assume the following GEE model,

$$\mu_{ij} = E(y_{ij}) = \beta_1 + \beta_2 r_i,$$

with variance $\mathrm{Var}(y_{ij}) = \sigma^2$ and within-subject correlation $\mathrm{Cov}(y_{ij}, y_{ij'}) = \rho_{jj'}$ (with $\rho_{jj} = 1$). It is also assumed that the responses are independent across subjects. This is a special case of the generic GEE model presented in Section 4.2. The parameter β_1 models the intercept effect while β_2 is the TAD between two groups. The primary interest is to test hypothesis $H_0 : \beta_2 = 0$.

To simplify derivation, we rewrite the model as

$$\mu_{ij} = b_1 + b_2(r_i - \bar{r}),$$

with $b_1 = \beta_1 + \beta_2 \bar{r}$ and $b_2 = \beta_2$. Define the vector of covariates $\boldsymbol{x}_{ij} = (1, r_i - \bar{r})'$ and assume an independent working correlation. The GEE estimator $\hat{\boldsymbol{b}} = (\hat{b}_1, \hat{b}_2)'$ is solved from

$$n^{-1/2} \sum_{i=1}^{n} \sum_{j=1}^{m} (y_{ij} - \boldsymbol{x}'_{ij} \boldsymbol{b}) \boldsymbol{x}_{ij} = 0.$$

The solution is

$$\hat{\boldsymbol{b}} = \left(\sum_{i=1}^{n} \sum_{j=1}^{m} \boldsymbol{x}_{ij}^{\otimes 2} \right)^{-1} \sum_{i=1}^{n} \sum_{j=1}^{m} \boldsymbol{x}_{ij} y_{ij}.$$

Thus $\sqrt{n}(\hat{\boldsymbol{b}} - \boldsymbol{b})$ approximately has a normal distribution with mean 0 and variance $\boldsymbol{\Sigma}_n = \boldsymbol{A}_n^{-1} \boldsymbol{V}_n \boldsymbol{A}_n^{-1}$, where

$$\boldsymbol{A}_n = n^{-1} \sum_{i=1}^{n} \sum_{j=1}^{m} \boldsymbol{x}_{ij} \boldsymbol{x}_{ij}' = n^{-1} \sum_{i=1}^{n} \sum_{j=1}^{m} \left(\begin{array}{cc} 1 & r_i - \bar{r} \\ r_i - \bar{r} & (r_i - \bar{r})^2 \end{array} \right),$$

$$\boldsymbol{V}_n = n^{-1} \sum_{i=1}^{n} \left(\sum_{j=1}^{m} \boldsymbol{x}_{ij} \hat{\epsilon}_{ij} \right)^{\otimes 2}$$

$$= n^{-1} \sum_{i=1}^{n} \sum_{j=1}^{m} \sum_{j'=1}^{m} \hat{\epsilon}_{ij} \hat{\epsilon}_{ij'} \left(\begin{array}{cc} 1 & r_i - \bar{r} \\ r_i - \bar{r} & (r_i - \bar{r})^2 \end{array} \right),$$

and $\hat{\epsilon}_{ij} = y_{ij} - \boldsymbol{x}_{ij}' \hat{\boldsymbol{b}}$. We reject $H_0 : b_2 = 0$, which is equivalent to $H_0 : \beta_2 = 0$, if the absolute value of $\sqrt{n} \hat{b}_2 / \hat{\sigma}_2$ is greater than $z_{1-\alpha/2}$. Here $\hat{\sigma}_2^2$ is the (2,2)th component of $\boldsymbol{\Sigma}_n$ and α is the significance level. Let σ_2^2 be the true variance of GEE estimator $\hat{b}_2$. Given type I error α and power $1 - \gamma$, and true TAD β_{20}, the required sample size is

$$n = \frac{\sigma_2^2 (z_{1-\alpha/2} + z_{1-\gamma})^2}{\beta_{20}^2}.$$

4.4.1   A General Sample Size Formula

Here we present a closed-form expression of σ_2^2 that accommodates missing data and various correlation structures. It leads to a closed-form sample size formula for the test of TAD. First we introduce indicator Δ_{ij} which takes value 1/0 if the jth response from the ith subject (y_{ij}) is observed/missing. Similar to Section 4.3, we define marginal probability $\delta_j = E(\Delta_{ij})$ and joint probability $\delta_{jj'} = E(\Delta_{ij} \Delta_{ij'})$. Then the general expressions of $\boldsymbol{A}_n$ and $\boldsymbol{V}_n$ that accommodate the presence of missing data are

$$\boldsymbol{A}_n = n^{-1} \sum_{i=1}^{n} \sum_{j=1}^{m} \Delta_{ij} \left(\begin{array}{cc} 1 & r_i - \bar{r} \\ r_i - \bar{r} & (r_i - \bar{r})^2 \end{array} \right)$$

and

$$\boldsymbol{V}_n = n^{-1} \sum_{i=1}^{n} \sum_{j=1}^{m} \sum_{j'=1}^{m} \Delta_{ij} \Delta_{ij'} \hat{\epsilon}_{ij} \hat{\epsilon}_{ij'} \left(\begin{array}{cc} 1 & r_i - \bar{r} \\ r_i - \bar{r} & (r_i - \bar{r})^2 \end{array} \right).$$

As $n \to \infty$, we have $\boldsymbol{\Sigma}_n \to \boldsymbol{\Sigma} = \boldsymbol{A}^{-1} \boldsymbol{V} \boldsymbol{A}^{-1}$, where

$$\boldsymbol{A}_n \to \boldsymbol{A} = \bar{m} \left(\begin{array}{cc} 1 & 0 \\ 0 & \sigma_r^2 \end{array} \right) \text{ and } \boldsymbol{V}_n \to \boldsymbol{V} = \sigma^2 \left(\begin{array}{cc} \eta & 0 \\ 0 & \eta \sigma_r^2 \end{array} \right),$$

with $\sigma_r^2 = \bar{r}(1-\bar{r})$, $\bar{m} = \sum_{j=1}^m \delta_j$, and $\eta = \sum_{j=1}^m \sum_{j'=1}^m \delta_{jj'} \rho_{jj'}$. It is easy to show that the (2,2)th component of Σ is

$$\sigma_2^2 = \frac{\sigma^2 \eta}{\bar{m}^2 \sigma_r^2}.$$

Thus the required sample size is

$$n = \frac{\sigma^2 \eta (z_{1-\alpha/2} + z_{1-\gamma})^2}{\beta_2^2 \bar{m}^2 \sigma_r^2}. \tag{4.20}$$

This closed-form sample size formula provides a flexible approach to sample size estimation because it can accommodate a broad spectrum of experimental designs, missing patterns, and correlation structures through the specification of $(\bar{r}, \delta_j, \delta_{jj'}, \rho_{jj'})$. For example, $\bar{r} - 0.5$ indicates a balanced design while $\bar{r} = 0.33$ implies a 1:2 randomization ratio between the treatment and control group. We can use $\delta_j = 1 - \theta t_j$ with $\theta \geq 0$ to model the trend that the proportion of missing data increases steadily over time. Other types of temporal trends (e.g., missing probability accelerates or plateaus over time) can also be specified. The specific values of $\delta_{jj'}$ are determined based on the missing patterns. Finally, arbitrary types of within-subject correlation can be represented by parameters $\rho_{jj'}$.

This sample size approach is also applicable to the test of treatment effect using clustered randomization trials, where m is the cluster size and ρ is the intracluster correlation. Subjects within each cluster are usually considered exchangeable hence the CS correlation structure is most frequently used.

4.4.2   Missing Data

Compared with the traditional adjustment approach for missing data, $n_0/(1-q)$, where n_0 is the sample size calculated assuming complete data and q is the expected dropout rate at the end of study, the sample size formula (4.20) makes a better use of information to account for missing data. In $\bar{m} = \sum_{j=1}^m \delta_j$, it considers the whole trajectory of observation probabilities over the study period instead of the dropout rate at the end of study only. In $\eta = \sum_{j=1}^m \sum_{j'=1}^m \delta_{jj'} \rho_{jj'}$, the correlation ($\rho_{jj'}$) between pairs of measurements are taken into account as well as the the probabilities ($\delta_{jj'}$) that these pairs are actually observed in the trial. The following theorem indicates that, under realistic scenarios, the traditional adjustment for missing data always leads to an overestimated sample size to maintain the nominal power and type I error.

Theorem 1   *In designing a clinical trial to detect TAD between the control and treatment groups, the sample size estimated by (4.20) is smaller than that obtained by traditional adjustment for missing data,*

$$n \leq n_0/\delta_m, \tag{4.21}$$

as long as the following two conditions hold:

1. *The probabilities of observation are non-increasing over time,* $\delta_1 \geq \delta_2 \geq \cdots \geq \delta_m$.

2. *The within-subject correlation is non-negative, $\rho_{jj'} \geq 0$. The equality sign in (4.21) only holds for complete data ($\delta_1 = \cdots = \delta_m = 1$) with independent measurements within subjects ($\rho_{jj'} = 0$ for $j \neq j'$).*

Proof.　　See Appendix A.2.　　　　　　　　　　　　　　　　　$\square$

It is noteworthy that Theorem 1 holds for any missing pattern or correlation structure.

It is also noteworthy that given the same set of marginal observant probabilities δ, the missing pattern (IM or MM) only affects one term, $\eta = \sum_{j=1}^{m} \sum_{j'=1}^{m} \delta_{jj'} \rho_{jj'}$, in (4.20) through $\delta_{jj'}$. Assuming that $\delta_1 \geq \delta_2 \geq \cdots \geq \delta_m$, we have

$$\left(\delta_{jj'}^{(IM)} = \delta_j \delta_{j'} \right) \leq \left(\delta_{jj'}^{(MM)} = \delta_{j'} \right), \quad (j \leq j').$$

Here we use superscripts $^{(IM)}$ and $^{(MM)}$ to denote parameters under the IM and MM patterns, respectively. Hence we have $\eta^{2(IM)} \leq \eta^{2(MM)}$ if $\rho_{jj'} \geq 0$. As a result, we have the conclusion that with all the other design parameters being the same, the sample size required under MM is always larger than that under IM when testing the TAD with respect to a continuous outcome. This conclusion applies to all correlation structures within the damped exponential family as described in Section 4.3.2.

It is also straightforward that correlation affects sample size (4.20) only through η. Following a similar argument, we can show that conditional on $\rho \geq 0$, a larger ρ is always associated with a larger sample size for the test of TAD, regardless of correlation structures or missing patterns.

4.4.3　Trade-off between n and m

In many comparative trials, investigators use repeated-measurement design to decrease the variance of averaged responses due to within-subject variability, and hope to decrease the total number of subjects required. For example, Heiberg et al. [39] showed that there was considerable fluctuation in self-reported health status measured on patients with rheumatoid arthritis. Mowinckel et al. [40] demonstrated that the between-subject variation was reduced in patients with rheumatoid arthritis by using repeated measurements, and they empirically determined the optimal number of measurements that effectively reduces the number of subjects required for the clinical trial. There has been little guidance in the methodological literature about determining the optimal number of repeated measurements. Note that in Section 4.3.6 we investigate the trade-off between n and m under a financial constraint. The discussion in this section, however, focuses on the mathematical impact on sample size induced by additional measurements per subject.

4.4.3.1   Under the CS correlation structure

Under the CS within-subject correlation structure, no missing data, and a balanced design ($\bar{r} = 0.5$), sample size (4.20) can be simplified as

$$n = 4(z_{1-\alpha/2} + z_{1-\gamma})^2 \sigma^2 [1 + (m-1)\rho]/(m\beta_2^2). \tag{4.22}$$

Equation (4.22) has also been widely used to estimate the sample size for clustered randomization trials [41], where $[1 + (m-1)\rho]$ is called the variance inflation factor to account for the degree of clustering among observations from the same cluster. It is obvious that sample size (4.22) increases as ρ increases. As for its trade-off with the number of repeated measurements per subject, we use $n(m)$ to indicate that the sample size is a function of m. Letting $C = 4(z_{1-\alpha/2} + z_{1-\gamma})^2 \sigma^2 / \beta_2^2 > 0$, it is easy to show that

$$n(m+1) - n(m) = C \cdot \left\{ \frac{1 + m\rho}{m+1} - \frac{1 + (m-1)\rho}{m} \right\} = C \cdot \frac{-1 + \rho}{m(m+1)} < 0$$

for any $-1 < \rho < 1$. Thus sample size (4.22) decreases as m increases. However, the marginal reduction, $C \cdot \frac{-1+\rho}{m(m+1)}$, quickly diminishes as m increases. Furthermore, it can be shown that

$$\lim_{m \to \infty} n = 4(z_{1-\alpha/2} + z_{1-\gamma})^2 \sigma^2 \rho / \beta_2^2, \tag{4.23}$$

which does not go to 0. That is, it is only to a limited extent that increasing the number of measurements per subject can compensate for the lack of unique subjects. Equation (4.23) provides a lower bound for required sample size for a given variance σ^2, within-subject correlation ρ, and treatment effect β_2. It is also noteworthy that, in contrast to (4.23), the sample size for the comparison of slope does go to 0 as $m \to \infty$. This observation suggests that sample sizes calculated for the test of slopes have very different properties than sample sizes calculated for the test of TAD. Separated investigations should be conducted to explore their performance.

4.4.3.2   Under the AR(1) correlation structure

Here we assume that the correlation between y_{ij} and y_{ik} is $\rho^{|k-j|/(m-1)}$, where ρ is the correlation between the first and last measurements. With no missing data and a balanced design ($\bar{r} = 0.5$), sample size (4.20) can be simplified as

$$n = 4(z_{1-\alpha/2} + z_{1-\gamma})^2 \sigma^2 \left\{ m + 2 \sum_{j=1}^{m-1} (m-j)\rho^{j/(m-1)} \right\} /(m^2 d^2). \tag{4.24}$$

As a special case, the AR(1) correlation is equivalent to the CS correlation when $m = 2$. For $\rho > 0$, it is easy to show that the sample size increases with ρ. However, equation (4.24) is too complicated to derive the analytic relationship between n and m. Zhang and Ahn [42] conducted numerical studies.

Their observation includes that for $\rho \geq 0.4$, the sample size increases as m increases beyond 2. When ρ is smaller, such as $\rho = 0.2$, the sample size increases as m increases beyond 3. The relative sample size reduction is only 2% under $\rho = 0.2$ even when m increases from 2 to 3. Their results show that under AR(1), in most cases, increasing m might actually lead to an increased sample size. One important message is that researchers should be cautious about assuming an AR(1) model in repeated-measurement studies, whether in clinical trial design or in data analysis. For example, if a clinician assumes an AR(1) structure with a moderate level of correlation, then any design scheme that has more than 2 measurements per subject will be considered theoretically suboptimal. Note that when the data suffer from missingness or unstructured measurement error, increasing the number of measurements might be beneficial [43]. Numerical studies are needed to explore the properties of the sample size in this case.

4.4.4   Compare $K \geq 3$ groups

Zhang and Ahn [44] investigated sample size calculation for clinical trials where $K \geq 3$ groups are tested simultaneously with respect to TAD. Let n_k $(k = 1, \ldots, K)$ be the number of subjects in each treatment arm. Then $n = \sum_{k=1}^{K} n_k$ is the total sample size. We define $r_k = n_k/n$ to be the proportion of patients assigned to the kth treatment. To detect TAD among the K treatment arms, we consider the following model,

$$y_{kij} = \beta_k + \epsilon_{kij}, \tag{4.25}$$

where β_k indicates the group-specific treatment effect and ϵ_{kij} $(j = 1, \ldots, m)$ is a zero-mean error term with variance σ^2. Here m is the total number of measurements from each subject. We assume independence of ϵ_{kij} across subjects, and the within-subject correlation is defined as $\rho_{jj'} = \mathrm{Corr}(\epsilon_{kij}, \epsilon_{kij'})$, with $\rho_{jj} = 1$. We are interested in testing the null hypothesis $H_0 : \beta_1 = \cdots = \beta_K$.

We first describe the testing procedure without missing data. Under an independent working correlation structure, the GEE estimator of β_k is

$$\hat{\beta}_k = \frac{\sum_{i=1}^{n_k} \sum_{j=1}^{m} y_{kij}}{n_k J} = \frac{\sum_{i=1}^{n_k} \mathbf{1}' \mathbf{y}_{ki}}{\sum_{i=1}^{n_k} \mathbf{1}' \mathbf{1}}.$$

Here $\mathbf{y}_{ki} = (y_{ki1}, \ldots, y_{kim})'$ is the vector of repeated measurements from the same patient and $\mathbf{1}$ is a vector with all m elements being 1.

Under the null hypothesis, we use β to denote the common value of treatment effects, $\beta_1 = \beta_2 = \cdots = \beta_K = \beta$. The GEE estimator of β is obtained by pooling observations from all K groups,

$$\hat{\beta} = \frac{\sum_{l=1}^{K} \sum_{i=1}^{n_l} \sum_{j=1}^{m} y_{lij}}{nm} = \frac{\sum_{l=1}^{K} \sum_{i=1}^{n_l} \mathbf{1}' \mathbf{y}_{li}}{\sum_{l=1}^{K} \sum_{i=1}^{n_l} \mathbf{1}' \mathbf{1}}.$$

We define a vector $B = \sqrt{n}(\hat{\beta}_1 - \hat{\beta}, \ldots, \hat{\beta}_{K-1} - \hat{\beta})'$. Note that we define B as a vector of length $(K-1)$ to avoid singularity caused by the linear constraint $\hat{\beta} = \sum_{k=1}^{K} r_k \hat{\beta}_k$. Following model (4.25), we have

$$\hat{\beta}_k - \hat{\beta} = \beta_k - \beta + \frac{\sum_{i=1}^{n_k} \mathbf{1}' \epsilon_{ki}}{\sum_{i=1}^{n_k} \mathbf{1}' \mathbf{1}} - \frac{\sum_{l=1}^{K} \sum_{i=1}^{n_l} \mathbf{1}' \epsilon_{li}}{\sum_{l=1}^{K} \sum_{i=1}^{n_l} \mathbf{1}' \mathbf{1}}.$$

Here $\epsilon_{ki} = (\epsilon_{ki1}, \ldots, \epsilon_{kim})'$. Under H_0, the central limit theorem suggests that as $n \to \infty$, vector B is approximately normal with a mean 0 and a $(K-1) \times (K-1)$ variance matrix $W = (w_{kh})$, with

$$w_{kh} = E\left[n \left(\frac{\sum_{i=1}^{n_k} \mathbf{1}' \epsilon_{ki}}{\sum_{i=1}^{n_k} \mathbf{1}' \mathbf{1}} - \frac{\sum_{l=1}^{K} \sum_{i=1}^{n_l} \mathbf{1}' \epsilon_{li}}{\sum_{l=1}^{K} \sum_{i=1}^{n_l} \mathbf{1}' \mathbf{1}} \right) \cdot \left(\frac{\sum_{i=1}^{n_h} \mathbf{1}' \epsilon_{hi}}{\sum_{i=1}^{n_h} \mathbf{1}' \mathbf{1}} - \frac{\sum_{l=1}^{K} \sum_{i=1}^{n_l} \mathbf{1}' \epsilon_{li}}{\sum_{l=1}^{K} \sum_{i=1}^{n_l} \mathbf{1}' \mathbf{1}} \right) \right].$$

In practice, W can be consistently estimated based on empirical error vector $\hat{\epsilon}_{ki} = y_{ki} - \mathbf{1}\hat{\beta}_k$. We reject H_0 with type I error α if $B'\hat{W}^{-1}B > \chi^2_{K-1,1-\alpha}$. Here $\chi^2_{K-1,1-\alpha}$ is the $100(1-\alpha)$th percentile of a chi-square distribution with $(K-1)$ degrees of freedom.

4.4.4.1   Sample size and power calculation

Under the true treatment effects β_k $(k = 1, \ldots, K)$, as $n \to \infty$, $B'\hat{W}^{-1}B$ approximately has a non-central chi-square distribution with $(K-1)$ degrees of freedom and a non-centrality parameter $n\eta'W^{-1}\eta$, where $W = \lim_{n\to\infty} \hat{W}$ and

$$\eta = (\beta_1 - \bar{\beta}, \cdots, \beta_{K-1} - \bar{\beta}) \tag{4.26}$$

with $\bar{\beta} = \sum_{k=1}^{K} r_k \beta_k$. Let $\chi^2_{K-1}(U)$ denote a non-central chi-square random variable with $(K-1)$ degrees of freedom and a non-centrality parameter U. Under type I error α and power $1-\gamma$, the sample size is calculated by first solving for U from

$$1 - \gamma = P(\chi^2_{K-1}(U) > \chi^2_{K-1,1-\alpha}).$$

Denoting the solution as $U = U(K-1, \alpha, \gamma)$, the required sample size is

$$n = \frac{U(K-1, \alpha, \gamma)}{\eta'W^{-1}\eta}. \tag{4.27}$$

The following theorem derives the closed-form expression of W or W^{-1} in the presence of missing data and arbitrary correlation structures, which leads to a flexible sample size formula that can be applied to a variety of realistic scenarios. The derivation requires the MCAR assumption. Let $\Delta_{kij} = 0/1$ indicate that a subject's outcome at t_j is missing/observed. We define $\delta_j = E(\Delta_{kij})$ to be the probability of a subject having an observation at t_j, and $\delta_{jj'} = E(\Delta_{kij}\Delta_{kij'})$ to be the joint probability of a subject having observations at both t_j and $t_{j'}$.

Theorem 2 *As $n \to \infty$, the $(K-1) \times (K-1)$ variance matrix $\hat{W}$ converges to*

$$
W = \frac{s}{\bar{m}} \begin{pmatrix}
(1/r_1 - 1) & -1 & -1 & \cdots & -1 \\
-1 & (1/r_2 - 1) & -1 & \cdots & -1 \\
-1 & -1 & (1/r_3 - 1) & \cdots & -1 \\
\cdots & \cdots & \cdots & \cdots & \cdots \\
-1 & -1 & -1 & \cdots & (1/r_{K-1} - 1)
\end{pmatrix}.
$$

Furthermore, we have $W^{-1} = \frac{\bar{m}^2}{s}[diag(r) + r_K^{-1}rr']$. *Here* $s = \sigma^2 \sum_{j=1}^{m} \sum_{j'=1}^{m} \delta_{jj'}\rho_{jj'}$, $\bar{m} = \sum_{j=1}^{m} \delta_j$, *and* $r = (r_1, \ldots, r_{K-1})'$.

Proof. See the Appendix of Zhang and Ahn [44]. $\qquad\qquad\square$

From Theorem 2 we have

$$
\eta'W^{-1}\eta = \frac{\bar{m}^2}{s}\left[\sum_{k=1}^{K-1} r_k\eta_k^2 + r_k^{-1}\left(\sum_{k=1}^{K-1} r_k\eta_k \right)^2 \right]. \tag{4.28}
$$

Plugging (4.28) into (4.27), we have the final sample size formula accounting for missing data and arbitrary correlation in the test of K-group TAD.

It is obvious that the proposed sample size is superior to traditional methods in the adjustment for missing data, as described by Theorem 1, in the special case of $K-$group comparisons of time-averaged responses.

4.4.4.2 Special cases

The sample size requirement also depends on the alternative hypothesis. One frequently assumed alternative hypothesis is that among the K groups, one receives a control treatment and the others receive different experimental treatments with similar efficacy. Without loss of generality, let β_K denote the control treatment effect, and $\beta_1 = \cdots = \beta_{K-1} = \beta_K + d$ denote the experimental treatment effect. This specification implies that $\bar{\beta} = \beta_K + (1 - r_K)d$ and $\eta_k = r_K d$ for $k = 1, \ldots, K - 1$. Then the sample size formula is

$$
n = \frac{U(K-1, \alpha, \gamma)sr_K}{\bar{m}^2 d^2 (1 - r_K) \sum_{k=1}^{K-1} r_k^2}.
$$

Another widely used alternative hypothesis is that the experimental groups $(k = 1, \ldots, K-1)$ are ordered in treatment effect. For example, the dosage of an experimental drug increases over the groups, with a constant improvement in efficacy between consecutive dosages. In this case we have $\beta_k = \beta_K + kd$ for $k = 1, \ldots, K - 1$. Here we only present the special case of a balanced design, $r_1 = \cdots = r_K = 1/K$. First we have $\bar{\beta} = \beta_K + d(K-1)/2$ and $\eta_k = d[k - (K-1)/2]$ for $k = 1, \ldots, K-1$. The required sample size is

$$n = \frac{12U(K-1, \alpha, \gamma)s}{\bar{m}^2 d^2 (K^2 - 1)}.$$

4.5 Compare the Slope for a Binary Outcome

In this section we consider sample size calculation for clinical trials where researchers are interested in comparing the slopes between two treatments based on a longitudinally measured binary outcome. Lui [45] proposed a sample size formula for correlated binary observations with the Markov dependency given subject-specific probabilities without accounting for missing data. The subject-specific probabilities are assumed to arise from a beta-binomial distribution. Jung and Ahn [48] presented sample size formulae for repeated binary outcomes incorporating missing patterns and correlation structures based on the GEE method. Lipsitz and Fitzmaurice [47] proposed a sample size formula for comparing response probabilities of treatment groups when each subject receives all different treatments. These studies are focused on testing the marginal binomial probabilities. In this section we present the application of the GEE sample size method to clinical trials comparing the slopes between two treatments.

Suppose that n_k subjects are allocated to treatment k ($k = 1, 2$). Thus $n = n_1 + n_2$ is the total sample size. For subject i ($i = 1, \ldots, n_k$) in group k, let y_{kij} denote a binary response variable at measurement time t_j ($j = 1, \ldots, m$) with $p_{kij} = E(y_{kij})$ modeled as

$$\log\left(\frac{p_{kij}}{1 - p_{kij}}\right) = a_k + \beta_k t_j. \tag{4.29}$$

Here m is the number of repeated measurements from each subject. It is assumed that observations are correlated within the same subject, $\rho_{jj'} = \text{Corr}(y_{kij}, y_{kij'})$, and independent across different patients. Parameter β_k is the rate of change (slope) in log-odds, which represents the treatment effect. Model (4.29) can be expressed as

$$p_{kij} = p_{kj} = p_j(a_k, \beta_k) = \frac{e^{a_k + \beta_k t_j}}{1 + e^{a_k + \beta_k t_j}}.$$

Here we assume the slope to be homogeneous within a treatment group, hence patient subscript i is omitted for response rates. The GEE estimator based on the independent working correlation can be solved from $U(a_k, \beta_k) = 0$, where

$$U(a_k, \beta_k) = \frac{1}{\sqrt{n_k}} \sum_{i=1}^{n_k} \sum_{j=1}^{m} [y_{kij} - p_j(a_k, \beta_k)] \begin{pmatrix} 1 \\ t_j \end{pmatrix}.$$

The equation can be solved using the Newton-Raphson algorithm: at the lth iteration,

$$\begin{pmatrix} \hat{a}_k^{(l)} \\ \hat{\beta}_k^{(l)} \end{pmatrix} = \begin{pmatrix} \hat{a}_k^{(l-1)} \\ \hat{\beta}_k^{(l-1)} \end{pmatrix} + n_k^{-1/2} \boldsymbol{A}_k^{-1}(\hat{a}_k^{(l-1)}, \hat{\beta}_k^{(l-1)}) U(\hat{a}_k^{(l-1)}, \hat{\beta}_k^{(l-1)})$$

where

$$\boldsymbol{A}_k(a_k, \beta_k) = -n_k^{-1/2} \frac{\partial U(a_k, \beta_k)}{\partial(a_k, \beta_k)}$$

$$= \frac{1}{n_k} \sum_{i=1}^{n_k} \sum_{j=1}^{m} p_j(a_k, \beta_k)[1 - p_j(a_k, \beta_k)] \begin{pmatrix} 1 & t_j \\ t_j & t_j^2 \end{pmatrix}.$$

By Liang and Zeger [5],

$$\sqrt{n_k} \begin{pmatrix} \hat{a}_k - a_k \\ \hat{\beta}_k - \beta_k \end{pmatrix} \rightarrow N(0, \boldsymbol{\Sigma}_k)$$

in distribution as $n \rightarrow \infty$. Here $\boldsymbol{\Sigma}_k$ is consistently estimated by $\hat{\boldsymbol{\Sigma}}_k = \hat{\boldsymbol{A}}_k^{-1}(\hat{a}_k, \hat{\beta}_k)\hat{\boldsymbol{V}}_k\hat{\boldsymbol{A}}_k^{-1}(\hat{a}_k, \hat{\beta}_k)$ where

$$\hat{\boldsymbol{V}}_k = \frac{1}{n_k} \sum_{i=1}^{n_k} \left\{ \sum_{j=1}^{m} \hat{\epsilon}_{kij} \begin{pmatrix} 1 \\ t_j \end{pmatrix} \right\}^{\otimes 2}.$$

Here $\hat{\epsilon}_{kij} = y_{kij} - p(\hat{a}_k, \hat{\beta}_k)$.

We are often interested in comparing the rates of change in the binomial probabilities of longitudinal measurements. Let $\hat{v}_k^2$ be the (2,2)th element of $\hat{\boldsymbol{\Sigma}}_k$. Then we reject $H_0 : \beta_1 = \beta_2$ if

$$\frac{|\hat{\beta}_1 - \hat{\beta}_2|}{\sqrt{\hat{v}_1^2/n_1 + \hat{v}_2^2/n_2}} > z_{1-\alpha/2}. \tag{4.30}$$

Note that for hypothesis testing, we do not have to specify the true correlation structure. It is reflected in the variance estimator $\hat{v}_k^2$ (actually, in $\hat{\boldsymbol{\Sigma}}_k$). For sample size calculation, however, we need to specify the true correlation.

4.5.1   Sample size and power calculation

Let $\boldsymbol{A}_k$ and $\boldsymbol{V}_k$ denote the limits of $\hat{\boldsymbol{A}}_k$ and $\hat{\boldsymbol{V}}_k$, as $n_k \rightarrow \infty$. Then $\hat{\boldsymbol{\Sigma}}_k$ converges to $\boldsymbol{\Sigma}_k = \boldsymbol{A}_k^{-1}\boldsymbol{V}_k\boldsymbol{A}_k^{-1}$. Let v_k^2 be the (2,2)th element of $\boldsymbol{\Sigma}_k$, and $r_k = n_k/n$ be the proportion of patients allocated to group k ($k = 1, 2$). Under the alternative hypothesis, $H_1 : \beta_2 - \beta_1 = d$, in order to achieve power $1 - \gamma$ at type I error α, the required sample size can be solved from

$$1 - \gamma = P\left(\frac{|\hat{\beta}_1 - \hat{\beta}_2|}{\sqrt{\hat{v}_1^2/n_1 + \hat{v}_2^2/n_2}} > z_{1-\alpha/2} \mid H_1\right)$$

$$\approx P\left(\frac{\sqrt{n}|\beta_1 - \beta_2|}{\sqrt{v_1^2/r_1 + v_2^2/r_2}} > z_{1-\alpha/2} \mid H_1\right).$$

We can obtain

$$n = \frac{(z_{1-\alpha/2} + z_{1-\gamma})^2(v_1^2/r_1 + v_2^2/r_2)}{d^2}. \tag{4.31}$$

The above sample size formula requires the expression of v_k^2 under H_1. Furthermore, we would like to derive a sample size that accommodates missing data, which is frequently encountered by researchers in practice. Similar to Section 4.3, we introduce indicator variable Δ_{kij} which takes value 1 if subject i from group k has an observation at time t_j and 0 otherwise. We also define marginal probability $\delta_j = E(\Delta_{kij})$ and joint probability $\delta_{jj'} = E(\Delta_{kij}\Delta_{kij'})$. Then the general expressions of $\hat{A}_k$ and $\hat{V}_k$ in the presence of missing data is

$$\hat{A}_k(\hat{a}_k, \hat{\beta}_k) = \frac{1}{n_k}\sum_{i=1}^{n_k}\sum_{j=1}^{m}\Delta_{kij}p_j(\hat{a}_k, \hat{\beta}_k)[1 - p_j(\hat{a}_k, \hat{\beta}_k)]\begin{pmatrix} 1 & t_j \\ t_j & t_j^2 \end{pmatrix}$$

and

$$\hat{V}_k = \frac{1}{n_k}\sum_{i=1}^{n_k}\left\{\sum_{j=1}^{m}\Delta_{kij}\hat{e}_{kij}\begin{pmatrix} 1 \\ t_j \end{pmatrix}\right\}^{\otimes 2}.$$

As $n_k \to \infty$, $\hat{A}(\hat{a}_k, \hat{\beta}_k)$ and $\hat{V}_k$ converge to

$$A_k = \sum_{j=1}^{m}\delta_j p_{kj}(1 - p_{kj})\begin{pmatrix} 1 & t_j \\ t_j & t_j^2 \end{pmatrix}$$

$$V_k = \sum_{j=1}^{m}\sum_{j'=1}^{m}\delta_{jj'}\rho_{jj'}\sqrt{p_{kj}(1 - p_{kj})p_{kj'}(1 - p_{kj'})}\begin{pmatrix} 1 & t_j \\ t_{j'} & t_j t_{j'} \end{pmatrix},$$

respectively. Jung and Ahn [48] showed that the (2,2)th element of $\Sigma_k = A_k^{-1}V_kA_k^{-1}$ has a closed form:

$$v_k^2 = \frac{s_k^2 + c_k^2}{s_k^4}, \tag{4.32}$$

where

$$s_k^2 = \sum_{j=1}^{m}\delta_j p_{kj}q_{kj}(t_j - \tau_k)^2$$

$$c_k^2 = \sum\sum_{j \neq j'}\delta_{jj'}\rho_{jj'}\sqrt{p_{kj}q_{kj}p_{kj'}q_{kj'}}(t_j - \tau_k)(t_{j'} - \tau_k)$$

with

$$\tau_k = \frac{\sum_{j=1}^{m} \delta_j p_{kj} q_{kj} t_j}{\sum_{j=1}^{m} \delta_j p_{kj} q_{kj}}.$$

Here we introduce notation $q_{kj} = 1 - p_{kj}$. Note that s_k^2 is proportional to a weighted variance of measurement time t_j, with weights proportional to $\delta_j p_{kj} q_{kj}$.

4.5.2　Impact of Design Parameters

It is noteworthy that for clinical trials with a continuous outcome, the sample size requirement depends on the difference in treatment effect and the variance, but not on the baseline mean in the control group. For a binary outcome Y, however, its variance and mean (response rate) are both determined by the same parameter (p), $\mathrm{Var}(Y) = p(1-p)$. Thus the control response rate needs to be specified as a designing factor for sample size calculation. Importantly, the variance $p(1-p)$ is not a monotone function of p. That is, under the logistic regression model (4.29), the means of y_{kij}, $p_{kij} = E(y_{kij})$, change monotonically over $t_j (j = 1,\dots, m)$, but the variances of y_{kij}, $\mathrm{Var}(y_{kij}) = p_{kij}(1 - p_{kij})$, do not. In fact, the variances follow a $\cap$ shape, achieving maximum at $p_{kij} = 0.5$. This special property of a binary variable makes it difficult to provide a straightforward summary of the association between various factors and sample size requirement.

With all other parameters fixed, the sample size (4.31) is minimized at $r_1 = (1 + \sqrt{v_2^2/v_1^2})^{-1}$, which is equal to 0.5 if $v_1^2 = v_2^2$, larger than 0.5 if $v_1^2 > v_2^2$. Under H_1, v_1^2 and v_2^2 are usually unequal, so the total sample size n is minimized when a larger portion of patients is allocated to the group with a larger variance.

Recall that for the test of slopes of a continuous outcome, we conclude that the correlation parameter ρ is negatively associated with sample size requirement under complete data and the CS correlation structure. For a binary outcome, ρ impacts sample size only through term c_k^2. Under complete data and the CS structure, it is expressed as

$$c_k^2 = \rho \sum \sum_{j \neq j'} w_{kj} w_{kj'} (t_j - \tau_k)(t_{j'} - \tau_k)$$

$$= \rho \left\{ \left[\sum_{j-1}^{m} w_{kj}(t_j - \tau_k) \right]^2 - \sum_{j=1}^{m} w_{kj}^2(t_j - \tau_k)^2 \right\},$$

where $w_j = \sqrt{p_{kj} q_{kj}}$ and $\tau_k = \sum_{j=1}^{m} w_j^2 t_j / \sum_{j=1}^{m} w_j^2$. Because $\sum_{j=1}^{m} w_{kj}(t_j - \tau_k) \neq 0$, we do not have a straightforward relationship between ρ and c_k^2.

It can also be shown that sample size has no straightforward relationship with missing pattern or correlation structure. Thus for clinical trials to compare the slopes of binary outcomes, numerical studies are needed to explore the impact of different design parameters under particular scenarios.

4.5.3 Application Example

Suppose researchers would like to design a clinical trial to assess the efficacy of an experimental drug in preventing the occurrence of pulmonary fibrosis in subjects with scleroderma. Patients will be randomly assigned to either receive the drug or placebo with a randomization ratio of 1:1. For each subject, the presence or absence of pulmonary fibrosis will be assessed $m = 6$ times, at baseline, and at months 6, 12, 18, 24, and 30. The primary interest is to detect the difference in the rate of change between the two groups. The targeted power and two-sided type I error are 0.8 and 0.05, respectively.

From a previous observational study [49], GENISOS (Genetics versus Environment in Scleroderma Outcome Study), the within-subject correlation is assume to have an AR(1) structure with $\rho = 0.8$. In the placebo group it is assumed that the proportion of patients with pulmonary fibrosis is 75% at baseline (t_1) and 50% at the 30th month (t_6). The experimental drug is expected to prevent or delay the occurrence of pulmonary fibrosis, i.e., the proportion of subjects with pulmonary fibrosis is expected to remain at 75% at the 30th month under treatment. Thus the true parameter values are set at $a_1 = a_2 = \text{logit}(0.75) = 1.1$, $\beta_1 = \text{logit}(0.5) - \text{logit}(0.75) = -1.1$, and $\beta_2 = 0$. Finally, the marginal observant probabilities are assumed to be $(\delta_1, \delta_2, \delta_3, \delta_4, \delta_5, \delta_6) = (1, 0.95, 0.9, 0.85, 0.8, 0.75)$. The missing data are expected to occur following the independent missing pattern, hence $\delta_{jj'} = \delta_j \delta_{j'}$ for $j \neq j'$. According to the sample size formula presented in Section 4.5.1, we have $v_1^2 = 0.305$ and $v_2^2 = 0.353$, and the required sample size is 215.

If we assume the monotone missing pattern, it can be shown that $v_1^2 = 0.324$, and $v_2^2 = 0.308$, and the sample size is $n = 229$.

4.6 Test the TAD for a Binary Outcome

In this section we present the sample size calculation for clinical trials that compare the time averaged responses of a binary outcome between two groups. The key difference between the test of TAD and the test of slope is that one focuses on the mean while the other focuses on the temporal trend. As a result, the sample sizes of these two types of tests have rather different properties.

Let y_{ij} be the binary response obtained at time t_j $(j = 1, \ldots, m)$ for subject i $(i = 1, \ldots, n)$. We use $r_i = 0/1$ to indicate that subject i belongs to the treatment/control group, and $\bar{r} = E(r_i)$ is the proportion of subjects randomly assigned to the treatment group. To evaluate the TAD between two groups, we model y_{ij} with the following logistic model: $y_{ij} \sim \text{Bernoulli}(p_{ij})$

and

$$\log\left(\frac{p_{ij}}{1-p_{ij}}\right) = \beta_1 + \beta_2 r_i. \tag{4.33}$$

Here β_1 models the time-averaged response on the log-odds scale for the control group, and β_2 is the log odds ratio between the treatment and control, representing the treatment effect. Our primary interest is to test the null hypothesis $H_0 : \beta_2 = 0$. To facilitate derivation, we reparameterize (4.33) as

$$\log\left(\frac{p_{ij}}{1-p_{ij}}\right) = b_1 + b_2(r_i - \bar{r}), \tag{4.34}$$

where $b_1 = \beta_1 + \beta_2 \bar{r}$ and $b_2 \equiv \beta_2$. Hence testing $b_2 = 0$ is equivalent to testing $\beta_2 = 0$. From (4.34) we have

$$p_{ij}(\boldsymbol{b}) = \frac{e^{\boldsymbol{b}'\boldsymbol{Z}_{ij}}}{1 + e^{\boldsymbol{b}'\boldsymbol{Z}_{ij}}},$$

where $\boldsymbol{b} = (b_1, b_2)'$ and $\boldsymbol{Z}_{ij} = (1, r_i - \bar{r})'$.

Under an independent working correlation, the GEE estimator $\hat{\boldsymbol{b}}$ is obtained as the solution to

$$\boldsymbol{U}_n(\boldsymbol{b}) = \frac{1}{\sqrt{n}}\sum_{i=1}^{n}\sum_{j=1}^{m}[y_{ij} - p_{ij}(\boldsymbol{b})]\boldsymbol{Z}_{ij} = 0.$$

The Newton-Raphson algorithm can be employed: at the lth iteration,

$$\hat{\boldsymbol{b}}^{(l)} = \hat{\boldsymbol{b}}^{(l-1)} + n^{-1/2}\boldsymbol{A}_n^{-1}(\hat{\boldsymbol{b}}^{(l-1)})\boldsymbol{U}_n(\hat{\boldsymbol{b}}^{(l-1)})$$

with

$$\boldsymbol{A}_n(\boldsymbol{b}) = -n^{-1/2}\frac{\partial \boldsymbol{U}_n(\boldsymbol{b})}{\partial \boldsymbol{b}} = \frac{1}{n}\sum_{i=1}^{n}\sum_{j=1}^{m}p_{ij}q_{ij}\begin{pmatrix} 1 & r_i - \bar{r} \\ r_i - \bar{r} & (r_i - \bar{r})^2 \end{pmatrix}.$$

Here $q_{ij} = 1 - p_{ij}$. As $n \to \infty$, $\sqrt{n}(\hat{\boldsymbol{b}} - \boldsymbol{b})$ approximately follows the $N(0, \boldsymbol{\Sigma})$ distribution, and $\boldsymbol{\Sigma}$ can be consistently estimated by $\boldsymbol{\Sigma}_n = \boldsymbol{A}_n^{-1}(\hat{\boldsymbol{b}})\boldsymbol{V}_n(\hat{\boldsymbol{b}})\boldsymbol{A}_n^{-1}(\hat{\boldsymbol{b}})$, with

$$\boldsymbol{V}_n(\hat{\boldsymbol{b}}) = \frac{1}{n}\sum_{i=1}^{n}\left(\sum_{j=1}^{m}\hat{\epsilon}_{ij}\boldsymbol{Z}_{ij}\right)^{\otimes 2}.$$

Here $\hat{\epsilon}_{ij} = y_{ij} - p_{ij}(\hat{\boldsymbol{b}})$. We reject $H_0 : b_2 = 0$ if $|\sqrt{n}\hat{b}_2/\hat{\sigma}_2| > z_{1-\alpha/2}$, where $\hat{\sigma}_2^2$ is the (2,2)th element of $\boldsymbol{\Sigma}_n$.

Under the alternative hypothesis $b_2 = \beta_{20}$, we would like to estimate the sample size required to achieve a power of $1 - \gamma$ with a significance level of α. Let $\boldsymbol{A}$ and $\boldsymbol{V}$ denote the limits of $\boldsymbol{A}_n$ and $\boldsymbol{V}_n$, respectively. Then as $n \to \infty$,

Σ_n converges to $\Sigma = A^{-1}\Sigma A^{-1}$. Let σ_2^2 be the (2,2)th element of Σ. The required sample size is

$$n = \frac{\sigma_2^2(z_{1-\alpha/2} + z_{1-\gamma})^2}{\beta_{20}^2}. \qquad (4.35)$$

In the following, we show that a closed-form expression can be derived for σ_2^2 under the MCAR (missing completely at random) assumption and arbitrary correlation structures. Letting $\Delta_{ij} = 0/1$ denote that the outcome from subject i is missing/observed at time t_j, then the general expressions of $A_n(\hat{b})$ and $V_n(\hat{b})$ in the presence of missing data are

$$A_n(\hat{b}) = \frac{1}{n}\sum_{i=1}^{n}\sum_{j=1}^{m}\Delta_{ij}p_{ij}q_{ij}\begin{pmatrix} 1 & r_i - \bar{r} \\ r_i - \bar{r} & (r_i - \bar{r})^2 \end{pmatrix},$$

$$V_n(\hat{b}) = \frac{1}{n}\sum_{i=1}^{n}\left(\sum_{j=1}^{m}\Delta_{ij}\hat{\epsilon}_{ij}Z_{ij}\right)^{\otimes 2}.$$

The limit of $A_n(\hat{b})$ and $V_n(\hat{b})$ are

$$A = (1-\bar{r})p_1q_1\sum_{j=1}^{m}\delta_j\begin{pmatrix} 1 & -\bar{r} \\ -\bar{r} & \bar{r}^2 \end{pmatrix} + \bar{r}p_2q_2\sum_{j=1}^{m}\delta_j\begin{pmatrix} 1 & 1-\bar{r} \\ 1-\bar{r} & (1-\bar{r})^2 \end{pmatrix}$$

and

$$V = (1-\bar{r})p_1q_1\sum_{j=1}^{m}\sum_{j'=1}^{m}\delta_{jj'}\rho_{jj'}\begin{pmatrix} 1 & -\bar{r} \\ -\bar{r} & \bar{r}^2 \end{pmatrix}$$

$$+ \bar{r}p_2q_2\sum_{j=1}^{m}\sum_{j'=1}^{m}\delta_{jj'}\rho_{jj'}\begin{pmatrix} 1 & 1-\bar{r} \\ 1-\bar{r} & (1-\bar{r})^2 \end{pmatrix}.$$

Here we define $p_1 = e^{\beta_1}/(1 + e^{\beta_1})$ and $p_2 = e^{\beta_1+\beta_2}/(1 + e^{\beta_1+\beta_2})$ to be the true response rate in the control and treatment group, respectively. We also define $q_1 = 1 - p_1$ and $q_2 = 1 - p_2$. We use $\rho_{jj'} = \text{Corr}(y_{ij}, y_{ij'})$ to denote within-subject correlation, with $\rho_{jj'} = 1$. Finally we define $\delta_j = E(\Delta_{ij})$ and $\delta_{jj'} = E(\Delta_{ij}\Delta_{ij'})$ to be the marginal and joint probability of observing the outcomes. Note that $\delta_{jj} = \delta_j$.

After some algebra, it can be shown that the (2,2)th element of $\Sigma = A^{-1}VA^{-1}$ is

$$\sigma_2^2 = \frac{\tau\sum_{j=1}^{m}\sum_{j'=1}^{m}\delta_{jj'}\rho_{jj'}}{(\sum_{j=1}^{m}\delta_j)^2\sigma_r^2p_1q_1p_2q_2}, \qquad (4.36)$$

where $\tau = (1-\bar{r})p_1q_1 + \bar{r}p_2q_2$ is effectively the pooled variance from the control and treatment groups and $\sigma_r^2 = \bar{r}(1-\bar{r})$. Plugging (4.36) into (4.35), we obtain a general sample size formula to detect the TAD of a binary outcome between two groups, which accommodates missing data, various correlation structures, and unbalanced design.

The expression of σ_2^2 is relatively simple, which allow us to assess the impact of various design parameters analytically. First we consider the damped exponential family of correlation structures [20],

$$\rho_{jj'} = \rho^{|t_j - t_{j'}|^\theta}.$$

Equation (4.36) suggests that within-subject correlation affects sample size only through the numerator of σ_2^2. For any given $\theta \geq 0$, $\rho_{jj'}$ is an increasing function of ρ. Thus as ρ increases, the sample size (4.35) to detect the TAD of a binary outcome increases. Note that this property is generally applicable to clinical trials with or without missing data, and it is shared by the sample size to test TAD for a continuous outcome. For the test of slopes, under the relatively simpler case of no missing data and the CS correlation structure, we have the conclusion that a greater value of ρ is associated with a smaller sample size. This property applies to both continuous and binary outcomes.

It is also noteworthy that given parameter ρ and the duration of follow-up fixed at $t_m - t_1 = 1$, $\rho_{jj'}$ (and eventually, the sample size) is an increasing function in θ. Thus with all other parameters fixed, the AR(1) correlation structure would lead to a larger sample size than the CS structure.

It can also be shown that given the non-increasing marginal observation probabilities ($\delta_1 \geq \cdots \geq \delta_m$) and $\rho \geq 0$, the monotone missing pattern is always associated with a larger sample size than the independent missing pattern because $(\delta_{jj'}^{(MM)} = \delta_{j'}) \geq (\delta_{jj'}^{(IM)} = \delta_j \delta_{j'})$ for $j' \geq j$.

Finally, it is important to note that given the same difference in the response rate, $d = p_2 - p_1$, the required sample size might be different depending on the baseline level p_1, which impacts the sample size calculation through term $p_1 q_1$. In other words, detecting the improvement from $p_1 = 0.1$ to $p_2 = 0.3$ is not equivalent to testing the improvement from $p_1 = 0.2$ to $p_2 = 0.4$. They require different sample sizes despite the same magnitude of difference to be detected. For binary outcomes, it has also been popular to specify treatment effect by odds ratio or relative risk. No matter how the treatment effect is specified, the requirement to set the baseline level p_1 as the design parameter remains the same. For clinical trials with a continuous outcome, however, only the difference in treatment effect is required for sample size calculation. The baseline mean is considered a nuisance parameter.

4.7   Compare the Slope for a Count Outcome

Clinical trials with count data as the primary outcome are frequently conducted in various medical fields. For example, in a clinical trial of epileptics, patients were randomly assigned either to an anti-epileptic drug or a placebo in addition to the standard chemotherapy. The number of epileptic seizures were recorded at baseline and at four consecutive two-week intervals for every patient. Researchers are interested in detecting the difference in the rate of change (slope) between the two treatment arms. Patel and Rowe [50] proposed a sample size formula for comparing the slopes of two linear curves of repeated count outcomes. Ogungbenro and Aarons [51] proposed sample size calculation for clinical trials with a repeatedly measured count outcome using the GEE method based on the non-central Wald test. A Poisson mixed-effect model was employed. One limitation of the above two methods is that they are only applicable to the ideal scenario of complete data. In this section we present a closed-form sample size formula for comparing the rates of change in a repeatedly measured count outcome using the GEE method, which is flexible enough to accommodate arbitrary missing patterns and correlations structures.

For treatment group k ($k = 1, 2$), let y_{kij} be the count response obtained at time t_j ($j = 1, \ldots, m$) from subject i ($i = 1, \ldots, n_k$). Hence $n = n_1 + n_2$ is the total sample size, and $r_k = n_k/n$ is the proportion of subjects assigned to treatment k. We model y_{kij} by a Poisson distribution

$$f(y_{kij}) = \frac{e^{-\mu_{kij}} \mu_{kij}^{y_{kij}}}{y_{kij}!}. \tag{4.37}$$

Here $\mu_{kij} = E(y_{kij})$ is expressed as

$$g(\mu_{kij}) = a_k + \beta_k t_j, \tag{4.38}$$

where $g(\mu) = \log(\mu)$ is a log-transformation, a_k is the intercept for each group, and β_k represents the rate of change (on the log scale). Thus $\mu_{kij} = \exp(a_k + \beta_k t_j)$. Repeated measurements from the same subject are assumed to be correlated, with $\rho_{jj'} = \mathrm{Corr}(y_{kij}, y_{kij'})$ being the within-subject correlation. The GEE estimator based on an independent working correlation is the solution to $U(a_k, \beta_k) = 0$ where

$$U(a_k, \beta_k) = \frac{1}{\sqrt{n_k}} \sum_{i=1}^{n_k} \sum_{j=1}^{m} [y_{kij} - \mu_{kij}(a_k, \beta_k)] \begin{pmatrix} 1 \\ t_j \end{pmatrix}.$$

The Newton-Raphson algorithm can be employed: at the lth iteration,

$$\begin{pmatrix} \hat{a}_k^{(l)} \\ \hat{\beta}_k^{(l)} \end{pmatrix} = \begin{pmatrix} \hat{a}_k^{(l-1)} \\ \hat{\beta}_k^{(l-1)} \end{pmatrix} + n_k^{-\frac{1}{2}} \hat{A}_k^{-1} \left(\hat{a}_k^{(l-1)}, \hat{\beta}_k^{(l-1)} \right) U \left(\hat{a}_k^{(l-1)}, \hat{\beta}_k^{(l-1)} \right),$$

where

$$\hat{A}_k(\hat{a}_k, \hat{\beta}_k) = -n_k^{-\frac{1}{2}} \frac{\partial U(a_k, \beta_k)}{\partial(a_k, \beta_k)} = \frac{1}{n_k} \sum_{i=1}^{n_k} \sum_{j=1}^{m} \mu_{kij}(\hat{a}_k, \hat{\beta}_k) \begin{pmatrix} 1 & t_j \\ t_j & t_j^2 \end{pmatrix}.$$

By Liang and Zeger [5], as $n \to \infty$,

$$\sqrt{n_k} \begin{pmatrix} \hat{a}_k - a_k \\ \hat{\beta}_k - \beta_k \end{pmatrix} \to N(0, \boldsymbol{\Sigma}_k).$$

The variance matrix $\boldsymbol{\Sigma}_k$ can be consistently estimated by $\hat{\boldsymbol{\Sigma}}_k = \boldsymbol{\hat{A}}^{-1}(\hat{a}_k, \hat{\beta}_k)\boldsymbol{\hat{V}}_k\boldsymbol{\hat{A}}^{-1}(\hat{a}_k, \hat{\beta}_k)$ where

$$\hat{V}_k = \frac{1}{n_k} \sum_{i=1}^{n_k} \left\{ \sum_{j=1}^{m} \hat{\epsilon}_{kij} \begin{pmatrix} 1 \\ t_j \end{pmatrix} \right\}^{\otimes 2},$$

and $\hat{\epsilon}_{kij} = y_{kij} - \mu_{kij}(\hat{a}_k, \hat{\beta}_k)$.

To detect the difference in the rate of change between the two groups, we reject the null hypothesis $H_0 : \beta_1 = \beta_2$, in favor of $H_1 : \beta_1 \neq \beta_2$, if

$$\left| \frac{\hat{\beta}_1 - \hat{\beta}_2}{\sqrt{\hat{v}_1^2/n_1 + \hat{v}_2^2/n_2}} \right| > z_{1-\alpha/2},$$

where $\hat{v}_k^2$ is the (2,2)th element of $\hat{\boldsymbol{\Sigma}}_k$.

On the other hand, given the true difference $d = \beta_2 - \beta_1$, to achieve a power of $1 - \gamma$ with a type I error α, the required sample size is

$$n = \frac{(z_{1-\alpha/2} + z_{1-\gamma})^2(v_1^2/r_1 + v_2^2/r_2)}{d^2}. \tag{4.39}$$

In the following we illustrate that closed-form expressions can be derived for v_1^2 and v_2^2 in the presence of missing data and arbitrary correlation structures. To accommodate missing data, we introduce indicator Δ_{kij} which takes value 1 if subject i from the kth group has an observation at time t_j, and 0 otherwise. Then the general expressions of $\hat{A}_k$ and $\hat{V}_k$ in the presence of missing data are

$$\hat{A}_k(\hat{a}_k, \hat{\beta}_k) = \frac{1}{n_k} \sum_{i=1}^{n_k} \sum_{j=1}^{m} \Delta_{kij}\mu_{kij}(\hat{a}_k, \hat{\beta}_k) \begin{pmatrix} 1 & t_j \\ t_j & t_j^2 \end{pmatrix},$$

$$\hat{V}_k = \frac{1}{n_k} \sum_{i=1}^{n_k} \left\{ \sum_{j=1}^{m} \Delta_{kij}\hat{\epsilon}_{kij} \begin{pmatrix} 1 \\ t_j \end{pmatrix} \right\}^{\otimes 2}.$$

Because we assume that all subjects follow the same measurement schedule, the notation of $\mu_{kij}(\cdot)$ can be simplified as $\mu_{kj}(\cdot)$, and we use $\mu_{kj} = \mu_{kj}(a_k, \beta_k)$ to denote the value obtained under the true values (a_k, β_k) from H_1. We also let $\delta_j = E(\Delta_{kij})$ be the proportion of subjects with observations at t_j and $\delta_{jj'} = E(\Delta_{kij}\Delta_{kij'})$ be the proportion of subjects with observations at both

t_j and $t_{j'}$ ($\delta_{jj} = \delta_j$). Then it can be shown that $\hat{A}_k(\hat{a}_k, \hat{\beta}_k)$ and $\hat{V}_k$ converge to

$$A_k = \sum_{j=1}^{m} \delta_{kij}\mu_{kj} \begin{pmatrix} 1 & t_j \\ t_j & t_j^2 \end{pmatrix}$$

and

$$\hat{V}_k = \sum_{j=1}^{m}\sum_{j'=1}^{m} \delta_{jj'}\rho_{jj'}\sqrt{\mu_{kj}\mu_{kj'}} \begin{pmatrix} 1 & t_j \\ t_j' & t_jt_{j'} \end{pmatrix},$$

respectively. Here we use the fact that for Poisson random variables, $\mathrm{Var}(y_{kij}) = \mu_{kj}$. After some algebra, it can be shown that the (2,2)th element of Σ_k is expressed as

$$v_k^2 = \frac{s_k^2 + c_k^2}{s_k^4}, \tag{4.40}$$

where

$$s_k^2 = \sum_{j=1}^{m} \delta_j\mu_{kj}(t_j - \tau_k)^2,$$

$$c_k^2 = \sum\sum_{j\neq j'} \delta_{jj'}\rho_{jj'}\sqrt{\mu_{kj}\mu_{kj'}}(t_j - \tau_k)(t_{j'} - \tau_k),$$

with $\tau_k = (\sum_{j=1}^{m} \delta_j\mu_{kj}t_j)/(\sum_{j=1}^{m} \delta_j\mu_{kj})$. Here τ_k is the weighted average of measurement time t_j with weights proportional to $\delta_j\mu_{kj}$. Plugging (4.40) into (4.39), we obtain a closed-form sample size formula to compare the slope of a repeatedly measured count outcome.

4.7.1 Impact of Design Factors

- The Randomization Ratio
  It can be shown that the optimal randomization ratio is achieved when $r_k \propto \sqrt{v_k^2}$, where the variance is minimized.

- The Correlation
  Following the similar argument as in Section 4.5.2, under the CS correlation structure and no missing data, it can be shown that the sample size is a decreasing function of ρ.

- The Control Level
  Similar to a binary outcome, the variance of a count outcome variable also depends on its mean. As a result, the intercept and slope parameters in the control group need to be specified for sample size calculation. However, there is a key difference between the count and binary outcome. For a count outcome Y under the Poisson model, we have $E(Y) = \mathrm{Var}(Y) = \mu$. So model (4.38) implies that both the mean and variance of y_{ij} change monotonically over time $t_1, \ldots, t_j$. For a binary outcome, however, the variance, which is $p(1-p)$, does not have this monotonic property because it follows a $\cap$ shape with respect to its mean level p.

4.8   Test the TAD for a Count Outcome

When testing the TAD of a count response between two groups, the likelihood function is the same as (4.37) for the comparison of slopes. The mean function, however, is modeled differently:

$$g(\mu_{kij}) = \log(\mu_{ij}) = \beta_1 + \beta_2 r_i, \tag{4.41}$$

where parameter β_1 is the time-averaged responses on the log scale for the control group, and β_2 is the TAD between the two groups, and $r_i = 0/1$ is a binary indicator that subject i belongs to the control/treatment group. We define $\bar{r} = E(r_i)$ to be the proportion of subjects randomly assigned to the treatment group. The null hypothesis of interest is $H_0 : \beta_2 = 0$. To simplify derivation, we rewrite (4.41) as

$$g(\mu_{kij}) = b_1 + b_2(r_i - \bar{r}),$$

where $b_1 = \beta_1 + \beta_2\bar{r}$ and $b_2 = \beta_2$. Thus testing $H_0 : \beta_2 = 0$ is equivalent to testing $H_0 : b_2 = 0$. We define

$$\mu_{ij}(\boldsymbol{b}) = e^{\boldsymbol{b}'\boldsymbol{Z}_{ij}},$$

where $\boldsymbol{b} = (b_1, b_2)'$ and $\boldsymbol{Z}_{ij} = (1, r_i - \bar{r})$. The GEE estimator $\hat{\boldsymbol{b}}$ under an independent working correlation structure can be obtained by solving $U_n(\boldsymbol{b}) = 0$ with

$$U_n(\boldsymbol{b}) = \frac{1}{\sqrt{n}} \sum_{i=1}^{n} \sum_{j=1}^{m} (y_{ij} - \mu_{ij}(\boldsymbol{b})) \boldsymbol{Z}_{ij}.$$

The Newton-Raphson algorithm can be employed, at the lth iteration,

$$\hat{\boldsymbol{b}}^{(l)} = \hat{\boldsymbol{b}}^{(l-1)} + n^{-1/2} \boldsymbol{A}_n^{-1}(\hat{\boldsymbol{b}}^{(l-1)}) U_n(\hat{\boldsymbol{b}}^{(l-1)}),$$

where

$$\boldsymbol{A}_n(\boldsymbol{b}) = -n^{-\frac{1}{2}} \frac{\partial U_n(\boldsymbol{b})}{\partial \boldsymbol{b}} = \frac{1}{n} \sum_{i=1}^{n} \sum_{j=1}^{m} \mu_{ij} \begin{pmatrix} 1 & r_i - \bar{r} \\ r_i - \bar{r} & (r_i - \bar{r})^2 \end{pmatrix}.$$

Liang and Zeger [5] proved that as $n \to \infty$, $\sqrt{n}(\hat{\boldsymbol{b}} - \boldsymbol{b}) \to N(0, \boldsymbol{\Sigma})$ in distribution. Here $\boldsymbol{\Sigma}$ is consistently estimated by $\boldsymbol{\Sigma}_n = \boldsymbol{A}_n^{-1}(\hat{\boldsymbol{b}}) \boldsymbol{V}_n \boldsymbol{A}_n^{-1}(\hat{\boldsymbol{b}})$, where

$$\boldsymbol{V}_n = \frac{1}{n} \sum_{i=1}^{n} \left(\sum_{j=1}^{m} \hat{\epsilon}_{ij} \boldsymbol{Z}_{ij} \right)^{\otimes 2}$$

and $\hat{\epsilon}_{ij} = y_{ij} - \mu_{ij}(\hat{\boldsymbol{b}})$. We reject $H_0 : b_2 = 0$ if $|\sqrt{n}\hat{b}_2/\hat{\sigma}_2| > z_{1-\alpha/2}$, where $\hat{\sigma}_2^2$ is the (2,2)th element of $\boldsymbol{\Sigma}$.

4.8.1 Sample Size

Let A and V denote the limits of A_n and V_n, respectively, as $n \to \infty$, then Σ_n converges to $\Sigma = A^{-1}VA^{-1}$. Let σ_2^2 be the (2,2)th element of Σ, to achieve a power of $1 - \gamma$ under the alternative hypothesis $H_1 : \beta_2 = \beta_{20}$, the required sample size is

$$n = \frac{\sigma_2^2 (z_{1-\alpha/2} + z_{1-\gamma})^2}{\beta_{20}^2}. \tag{4.42}$$

In the following we provide a closed-form expression for σ_2^2 in the presence of missing data. First, the general expressions of A_n and V_n are

$$A_n(\hat{b}) = \frac{1}{n} \sum_{i=1}^{n} \sum_{j=1}^{m} \Delta_{ij} \mu_{ij}(\hat{b}) \begin{pmatrix} 1 & r_i - \bar{r} \\ r_i - \bar{r} & (r_i - \bar{r})^2 \end{pmatrix}$$

and

$$V_n = \frac{1}{n} \sum_{i=1}^{n} \left(\sum_{j=1}^{m} \Delta_{ij} \hat{\epsilon}_{ij} Z_{ij} \right)^{\otimes 2}. \tag{4.43}$$

It can be shown that they converge to

$$A = (1 - \bar{r}) e^{\beta_1} \sum_{j=1}^{m} \delta_j \begin{pmatrix} 1 & -\bar{r} \\ -\bar{r} & \bar{r}^2 \end{pmatrix} + \bar{r} e^{\beta_1 + \beta_2} \sum_{j=1}^{m} \delta_j \begin{pmatrix} 1 & 1 - \bar{r} \\ 1 - \bar{r} & (1 - \bar{r})^2 \end{pmatrix}$$

and

$$V = (1 - \bar{r}) e^{\beta_1} \sum_{j=1}^{m} \sum_{j'=1}^{m} \delta_{jj'} \rho_{jj'} \begin{pmatrix} 1 & -\bar{r} \\ -\bar{r} & \bar{r}^2 \end{pmatrix}$$
$$+ \bar{r} e^{\beta_1 + \beta_2} \sum_{j=1}^{m} \sum_{j'=1}^{m} \delta_{jj'} \rho_{jj'} \begin{pmatrix} 1 & 1 - \bar{r} \\ 1 - \bar{r} & (1 - \bar{r})^2 \end{pmatrix},$$

respectively.

It is easy to show that the (2,2)th element of $\Sigma = A^{-1}VA^{-1}$ is

$$\sigma_2^2 = \frac{\tilde{\mu} \sum_{j=1}^{m} \sum_{j'=1}^{m} \delta_{jj'} \rho_{jj'}}{(\sum_{j=1}^{m} \delta_j)^2 \sigma_r^2 e^{\beta_1 + \beta_2}}, \tag{4.44}$$

where $\tilde{\mu} = (1 - \bar{r}) e^{\beta_1} + \bar{r} e^{\beta_1 + \beta_2}$ and $\sigma_r^2 = \bar{r}(1 - \bar{r})$. Thus we obtain a closed-form sample size formula for the test of TAD of a count outcome between two groups.

4.8.2   Impact of Design Factors

- The Control Level
  In (4.44) the term related to the control level (β_1) is

$$\frac{\tilde{\mu}}{e^{\beta_1} e^{\beta_1 + \beta_2}} = \frac{(1 - \bar{r}) + \bar{r} e^{\beta_2}}{e^{\beta_1 + \beta_2}}, \tag{4.45}$$

which is a decreasing function in β_1. Thus the sample size required to detect the same value of TAD (β_2) increases with the mean level in the control group.

- The Correlation
  Under the damped exponential family of correlation structures [20],

$$\rho_{jj'} = \rho^{|t_j - t_{j'}|^\theta}.$$

It is obvious in (4.44) that σ_2^2 is an increasing function of ρ, which implies that the sample size increases as ρ increases, whether there is missing data or not. When we normalize the duration of study at $t_m - t_1 = 1$, it can be shown that the sample size is also an increasing function in θ.

- The Randomization Ratio
  The term in (4.44) related to $\bar{r}$ is

$$\frac{\tilde{\mu}}{\sigma_r^2} = \frac{(1 - \bar{r}) e^{\beta_1} + \bar{r} e^{\beta_1 + \beta_2}}{\bar{r}(1 - \bar{r})}.$$

The optimal $\bar{r}$ to minimize the sample size requirement is not straightforward, which will require exploration based on the design parameters of each trial.

4.8.3   Test TAD among $K \geq 3$ Groups of a Count Outcome

In this section we further extend the sample size formula to the scenario of testing TAD among $K \geq 3$ groups for clinical trials with a count outcome. Let y_{kij} be the count response measured at time t_j ($j = 1, \ldots, m$) for the ith ($i = 1, \ldots, n_k$) subject from the kth ($k = 1, \ldots, K$) group. The total sample size is $n = \sum_{k=1}^{K} n_k$ and $r_k = n_k/n$ is the proportion of subjects assigned to each treatment group. A similar Poisson model will be employed. With $\mu_{kij} = E(y_{kij})$ and a link function $g(\mu) = \log(\mu)$, we have

$$g(\mu_{kij}) = b_k$$

where parameter b_k is the time-averaged response on the log scale under treatment k. Conversely we have $\mu_{kij} = e^{b_k}$. The within-subject correlation is assumed to be $\rho_{jj'} = \mathrm{Corr}(y_{kij}, y_{kij'})$. We define $\boldsymbol{b} = (b_1, \ldots, b_K)'$. Utilizing the independent working correlation structure, the GEE estimator $\hat{b}$ can be obtained by solving $U(b) = 0$, where

$$U(b) = \left\{ \begin{array}{l} \frac{1}{\sqrt{n_1}} \sum_{i=1}^{n_1} \sum_{j=1}^{m} (y_{1ij} - \mu_{1ij}(b)) \\ \cdots \\ \frac{1}{\sqrt{n_1}} \sum_{i=1}^{n_1} \sum_{j=1}^{m} (y_{1ij} - \mu_{1ij}(b)) \end{array} \right\}.$$

We employ the Newton-Raphson algorithm to solve for $\hat{b}$. At the lth iteration,

$$\hat{b}^{(l)} = \hat{b}^{(l-1)} + n^{-\frac{1}{2}} A_n^{-1}(\hat{b}^{(l-1)}) U(\hat{b}^{(l-1)}),$$

where $A_n(b)$ is a $K \times K$ diagonal matrix with the kth diagonal element being $\frac{1}{n_k} \sum_{i=1}^{n_k} \sum_{j=1}^{m} e^{b_k}$. By Liang and Zeger [5], $\hat{b} - b$ approximately has a normal distribution with mean $\mathbf{0}$ and variance matrix $\Sigma_n = \frac{1}{n} W A_n^{-1}(\hat{b}) V_n(\hat{b}) A_n^{-1}(\hat{b}) W$. Here W is diagonal with main diagonal elements $(1/\sqrt{r_1}, \ldots, 1/\sqrt{r_K})$, and $V_n(\hat{b})$ is a diagonal matrix with the kth element being

$$\frac{1}{n_k} \sum_{i=1}^{n_k} \left(\sum_{j=1}^{m} \hat{\epsilon}_{kij} \right)^2.$$

Here $\hat{\epsilon}_{kij} = y_{kij} - \mu_{kij}(\hat{b})$.

In order to test the null hypothesis $H_0 : b_1 = \cdots = b_K$, we construct the following test statistic:

$$T = \frac{C'b}{\sqrt{\text{Var}(C'b)}},$$

where $C = (1, -\frac{1}{K-1}, \ldots, -\frac{1}{K-1})'$. We reject H_0 if $|T| > z_{1-\alpha/2}$, where $z_{1-\alpha/2}$ is the $100(1 - \alpha/2)$th percentile of the standard normal distribution.

4.8.4 Sample Size Calculation

Let A and V be the limit of $A_n(\hat{b})$ and $V_n(\hat{b})$, respectively, as $n \to \infty$. Then Σ_n converges to $\Sigma = \frac{1}{n} W A^{-1} V A^{-1} W$. Given the true values of b, denoted as $b_0 = (b_{10}, \ldots, b_{K0})'$, to achieve a power of $1 - \gamma$ at a type I error of α, the required sample size can be expressed as

$$n = \frac{(z_{1-\alpha/2} + z_{1-\gamma})^2 C' W A^{-1} V A^{-1} W C}{(C'b_0)^2}. \tag{4.46}$$

We employ the similar method to derive a generalize sample size formula to accommodate the existence of missing data. Specifically we introduce $\Delta_{kij} = 1/0$ to indicate whether the ith subject from Group k has an observation at time t_j, as well as the marginal and joint observation probabilities $\delta_j = E(\Delta_{kij})$ and $\delta_{jj'} = E(\Delta_{kij}\Delta_{kij'})$. The derivation details are similar as those in previous sections. It can be shown that the general expression of A, V, and Σ are

$$A = \sum_{j=1}^{m} \delta_j \begin{pmatrix} e^{b_1} & 0 & \cdots & 0 \\ 0 & e^{b_2} & \cdots & 0 \\ \vdots & \vdots & \cdots & \vdots \\ 0 & 0 & \cdots & e^{b_K} \end{pmatrix},$$

$$V = \sum_{j=1}^{m} \sum_{j'=1}^{m} \delta_{jj'} \rho_{jj'} \begin{pmatrix} e^{b_1} & 0 & \cdots & 0 \\ 0 & e^{b_2} & \cdots & 0 \\ \vdots & \vdots & \cdots & \vdots \\ 0 & 0 & \cdots & e^{b_K} \end{pmatrix},$$

and

$$\Sigma = \frac{\sum_{j=1}^{m} \sum_{j'=1}^{m} \delta_{jj'} \rho_{jj'}}{(\sum_{j=1}^{m} \delta_j)^2} \begin{pmatrix} e^{b_1}/r_1 & 0 & \cdots & 0 \\ 0 & e^{b_2}/r_2 & \cdots & 0 \\ \vdots & \vdots & \cdots & \vdots \\ 0 & 0 & \cdots & e^{b_K}/r_K \end{pmatrix}.$$

It can be shown that

$$C'WA^{-1}VA^{-1}WC = \frac{\sum_{j=1}^{m} \sum_{j'=1}^{m} \delta_{jj'} \rho_{jj'}}{(\sum_{j=1}^{m} \delta_j)^2} \left[\frac{e^{-b_1}}{r_1} + \frac{1}{(K-1)^2} \sum_{i=1}^{K} \frac{e^{-b_i}}{r_i} \right].$$
$$(4.47)$$

Plugging (4.47) into (4.46), we obtain the closed-form formula for testing the TAD among $K \geq 3$ groups with respect to a count outcome.

When $K = 2$, sample size (4.46) can be simplified to

$$n = \frac{(z_{1-\alpha/2} + z_{1-\gamma})^2 \sum_{j=1}^{m} \sum_{j'=1}^{m} \delta_{jj'} \rho_{jj'}}{(\sum_{j=1}^{m} \delta_j)^2 (b_1 - b_2)^2} \left(\frac{1}{r_1 e^{b_1}} + \frac{1}{r_2 e^{b_2}} \right),$$

which is identical to the sample size formula for two-group comparison presented in Section 4.8.

4.9   Further Readings

In this chapter we have presented the GEE method to calculate sample size for clinical trials with longitudinal or clustered measurements, where the outcome can be continuous, binary, or count, and the research interest can be the comparison of slopes (rates of change) or time-averaged responses. The challenge for the design and analysis of this type of clinical trials lies in the frequent encounter of missing data and different correlation structures. The sample size formulas we have presented are advantageous in having closed-forms, which are easy to implement in practice. They also provide great flexibility to accommodate arbitrary missing patterns and correlation structures. The "distribution-free" property of GEE is particularly desirable for power and sample size calculation. Without real observations, at the design stage it is usually difficult to verify parametric assumptions and the GEE approach helps ensure a robust design and analysis of the experiment. One potential criticism of the method is the utilization of the independent working correlation. When there is limited information about the actual correlation structure, the independent working correlation can be employed due to its convenience and the additional benefit of computational ease. When there is adequate information about the actual correlation structure, using a correctly specified working correlation can improve the efficiency of the trial. However, it should be noted that using a sophisticated working correlation structure involves estimation of nuisance parameters, which might cause convergence problems in estimating the parameters of primary interest [7].

Most existing sample size methods for clinical trials with longitudinal or clustered measurements are developed under the missing completely at random (MCAR) assumption, which might be unrealistic for some trials. There are several reasons for this limitation. One is that when deviating from the MCAR assumption, an additional missing data model needs to be specified to describe the mechanism of missingness associated with observed (and unobserved) measurements. This additional missing data model will make sample size calculation much more complicated. Closed-form sample size formulas as we have presented are almost impossible to be obtained. Sophisticated numerical studies have to be performed to find the sample size solution, which require significant skills and effort for computational coding and debugging. More importantly, the true missing data mechanism is usually unknown at the design stage of a clinical trial. In fact, the validity of the assumed missing data mechanism is often untestable. Thus even after a sophisticated numerical study, the calculated sample size might still be inaccurate due to the misspecified missing data model. Finally, the missing data mechanism might be different for each clinical study, depending on the disease, the measurement procedure and schedule, as well as the demographic and the social-economical status of the target population, etc. Accounting for the missing data mechanism effectively implies developing a custom-made algorithm for each particular clinical trial, which will be inapplicable to other trials. Consequently, the number of researchers who will be interested in the sample size approach developed for a particular study will be extremely limited. For the above reasons, there has been limited literature on sample size calculation under non-MCAR assumptions. One practical strategy has been to estimate sample size based on a relatively simpler formula derived under the MCAR assumption, and then conduct simulation studies to assess its performance under various possible missing data mechanisms. If deviation from MCAR severely impacts the validity of the calculated sample size, custom-made numerical studies then have to be conducted. In such cases, sample sizes calculated under the MCAR assumption can still be useful as plausible initial values in searching for the final solution.

There is another special case of clinical trials with longitudinal measurements: the pre- and post-intervention study. It has been widely used in medical and social behavioral research [52, 53, 54, 55, 56]. One distinct feature of a pre-post study is that each patient contributes a pair of observations (measured at pre-intervention and post-intervention). The McNemar's test [57] has been the most widely used approach to detecting the intervention effect on a binary outcome in pre-post studies. Sample size calculation for studies involving the McNemar's test has been explored by many researchers. Miettinen [58] and Connor [59] derived sample size formulas through a conditional procedure based on the approximately normal distribution of the McNemar's test statistic given the number of discordant pairs. Shork and Williams [60] presented an exact formula for the unconditional case. Lachin [61] compared different unconditional sample size expressions relative to the exact power function. Lu and Bean [62] investigated sample size requirement for one-sided equivalence of sensitivities based on the McNemar's test. The existing literature, however, have not addressed the issue of incomplete observations frequently encountered by practitioners. Specifically, some subjects might participate in the pre-intervention phase of the study, but then drop out of the study, resulting in missing values for post-intervention measurements. Thus the pre-intervention measurements are observed in all subjects, but the post-intervention measurements are missing in a subgroup of subjects. One disadvantage of the McNemar's test is that it could not use incomplete pairs of observations, which have to be excluded from analysis. The GEE sample size method that we have described in previous sections can be directly applied to pre-post studies, which offers a flexible extension to the McNemar's test in accounting for incomplete observations. Interested readers can find more details about this method in Zhang et al. [63].

In this chapter, the predictors of GEE models only include intercept, treatment, time, and the treatment−time interaction. The reason being that for a randomized trial, factors other than the treatments are probabilistically balanced among groups. Tu et al. [64] provided GEE sample size methods for complicated regression models with covariates in addition to intercept and slope. Under such scenarios, the variance matrix does not have a closed form. The solution provided by the authors was to use Monte Carlo simulation to approximate the variance [65]. Note that the variance matrix depends on the covariates through their distributions, rather than a set of specific values. This sample size approach would be particularly useful for observational studies with longitudinal/clustered measurements. Because the treatments under investigation are not randomly assigned, valid inference on treatment effect needs to control for the imbalance of various factors, such as age, gender, race, etc. Similar sample size calculation for complicated regression models has been investigated by Muller et al. [66] and Rochon [10].

All GEE sample size approaches are developed based on the large sample properties of estimated parameters. In practice, however, they have been frequently employed for the design of relatively smaller clinical trials, which can be due to financial constraint, a rare disease population, or a large expected treatment effect. It is thus important to assess the performance of GEE sample size formulas for moderate or small trials. Zhang and Ahn [67] recognized the relationship between sample size n and treatment effect β_2 expressed in (4.20) for the test of TAD of a continuous outcome. They fixed the sample size at $n = 60$, and solved for the corresponding treatment effect utilizing the transformation,

$$\beta_2 = \sqrt{\frac{\sigma^2 \eta (z_{1-\alpha/2} + z_{1-\gamma})^2}{n \bar{m}^2 \sigma_r^2}},$$

given the other design parameters. Their simulation suggests that at a relatively smaller sample size of $n = 60$, the GEE method can still maintain the power and type I error at their nominal levels. Nonetheless, improving the small-sample performance of sandwich estimator and Wald test remains an active research area [68]. Because the sandwich estimator tends to underestimate the covariance of regression coefficients, Mancl and DeRouen [69] proposed to achieve a better control on the size of the Wald test by correcting the bias of the sandwich estimator. Alternatively, Pan and Wall [70] investigated the variability of the sandwich estimator and constructed an approximate t or F test. For the score test approach, Guo ct al. [71] investigated the small-sample performance of robust score test and its modification in GEE.

In this chapter we have focused on the application of GEE sample size methods for clinical trials with only one level of correlation structure. Teerenstra et al. [72] investigated sample size calculation for randomized trials with a hierarchical structure of correlation based on the GEE method. For example, a trial assesses the efficacy of interventions which are implemented in health care units, aiming at changing the behavior of health care professionals working in the units, and the effects are measured from patients cared by the professionals. They derived a sample size formula to account for two-levels of clustering: one for professionals within each health care unit, and the other for patients within each professional. The formula reveals that the sample size is inflated, relative to the design with completely independent measurements, by a multiplicative term that can be expressed as a product of two variance inflation factors. One quantifies the impact of within-unit correlation and one quantifies the impact of within-professional correlation. The authors employed bias-corrected estimating equations for the correlation parameters in combination with the model-based covariance estimator or the sandwich estimator with a finite sample correction [69, 73]. The impact of missing data or unequal cluster sizes are not investigated. An alternative treatment of sample size calculation for multi-level clustered trial using linear mixed models can be found in Heo and Leon [74].

A potential future development of GEE sample size calculation is the re-estimation of variance matrix and subsequently, the sample size, based on interim data. The estimation of sample size in clinical trials requires knowledge of parameters that involve treatment effect, variability, and correlation (for clustered/longitudinal measurements), which are usually uncertain to researchers at the design stage. This uncertainty is especially pronounced for clinical trials with new classes of study therapies for unfamiliar diseases where natural history data is lacking. Shih and Zhao [75] offered a design with a simple stratification strategy that enables researchers to verify and update the assumption of the response rates given initially in the protocol. The design provides a method to re-estimate the sample size based on interim data while preserving the trial's blinding. Employing a similar strategy to GEE sample size calculations can further improve their flexibility and robustness. More literature on the topic of sample size re-estimation based on interim data can be found in Wittes and Brittain [76], Gould [77], and references therein.

Appendix A.1 Proof of Fact 2

Under no missing data and the CS correlation structure, we have $\mu_0 = m$,

$$\sigma_t^2 = \frac{m+1}{12(m-1)},$$

$$s_t^2 = (1-\rho)\frac{(m+1)m}{12(m-1)}.$$

Plugging μ_0, s_t^2, and σ_t^2 into (4.19), we have

$$Q \propto \frac{(m-1)(m+U)}{m(m+1)}.$$

Setting

$$\frac{\partial \log(Q)}{\partial m} = \frac{1}{m-1} + \frac{1}{m+U} - \frac{1}{m} - \frac{1}{m+1}$$

$$= \frac{(2-U)m^2 + 2Um + U}{(m-1)m(m+1)(m+U)} = 0, \qquad (4.48)$$

we have roots

$$m = \frac{-U \pm \sqrt{2U^2 - 2U}}{2-U} \text{ when } U \geq 1.$$

For $m \geq 2$, the denominator of (4.48) is always positive. For the numerator, we define $f(m) = (2-U)m^2 + 2Um + U$ and discuss several scenarios.

- When $0 < U < 1$, $f(m)$ is a U-shape curve above 0. Thus $\partial \log(Q)/\partial m > 0$ for $m \geq 2$, i.e., Q is increasing with m and the minimum is achieved at $m^o = 2$.

- When $1 < U < 2$, $f(m)$ is a U-shape curve centered at $-U/(2-U) < 0$. Thus we have $f(m) > f(2) = 8 + U > 0$ for $m \geq 2$. As a result, Q is also increasing over $m \geq 2$ and the minimum is achieved at $m^o = 2$.

- When $U > 2$, $f(m)$ is a reversed U-shape curve with two roots

$$m = \frac{-U \pm \sqrt{2U^2 - 2U}}{2 - U}.$$

Lemma 1

$$\frac{-U + \sqrt{2U^2 - 2U}}{2 - U} \leq 2 \leq \frac{-U - \sqrt{2U^2 - 2U}}{2 - U}.$$

Proof. If Lemma 1 does not hold, then solving inequality

$$\frac{-U + \sqrt{2U^2 - 2U}}{2 - U} \geq 2$$

yields that $-8 \leq U \leq 2$, which contradicts with the requirement of $U \geq 2$. Similarly, solving

$$\frac{-U - \sqrt{2U^2 - 2U}}{2 - U} \leq 2$$

given $U > 2$ also yields an empty set. Thus we finish the proof of Lemma 1. $\qquad \square$

Lemma 1 suggests that $f(m)$ or $\partial \log(Q)/\partial m$ changes from positive to negative at

$$m^* = \frac{-U - \sqrt{2U^2 - 2U}}{2 - U},$$

i.e., Q increases on the left of m^* and decreases on the right of m^*. Because $2 \leq m^* \leq m_{max}$, the maximum of Q is achieved at m^*, and the minimum of Q is achieved at one of the two extreme values of m (2 or m_{max}).

Appendix A.2. Proof of Theorem 1

First we derive the expression of n_0, which can be obtained from (4.20) by setting all $\delta_j = 1$ and $\delta_{jj'} = 1$,

$$n_0 = \frac{\sigma^2 \left(\sum_{j=1}^{m} \sum_{j'=1}^{m} \rho_{jj'} \right) \left(z_{1-\alpha/2} + z_{1-\gamma} \right)^2}{\beta_2^2 m^2 \sigma_r^2}.$$

Thus

$$\frac{n}{n_0/\delta_m} = \frac{m^2\delta_m}{\left(\sum_{j=1}^m \delta_j\right)^2} \cdot \frac{\sum_{j=1}^m \sum_{j'=1}^m \delta_{jj'}\rho_{jj'}}{\sum_{j=1}^m \sum_{j'=1}^m \rho_{jj'}}.$$

Defining $\bar{\delta} = \sum_{j=1}^m \delta_j/m$ and using the fact that $\rho_{jj} = 1$, we have

$$\frac{n}{n_0/\delta_m} = \frac{m^2\delta_m}{m^2\bar{\delta}^2} \cdot \frac{m\bar{\delta} + 2\sum_{j=1}^{m-1}\sum_{j'=j+1}^m \delta_{jj'}\rho_{jj'}}{m + 2\sum_{j=1}^{m-1}\sum_{j'=j+1}^m \rho_{jj'}}.$$

Lemma 2 *Under conditions 1 and 2,*

$$\sum_{j=1}^{m-1}\sum_{j'=j+1}^m \delta_{jj'}\rho_{jj'} \le \bar{\delta}\sum_{j=1}^{m-1}\sum_{j'=j+1}^m \rho_{jj'}.$$

Proof. The inequality is equivalent to

$$\frac{\sum_{j=1}^{m-1}\sum_{j'=j+1}^m \delta_{jj'}\rho_{jj'}}{\sum_{j=1}^{m-1}\sum_{j'=j+1}^m \rho_{jj'}} \le \bar{\delta}.$$

Because $\rho_{jj'} \ge 0$ and $\delta_{jj'} \le \delta_{j'}$, the left−hand side is smaller than

$$\frac{\sum_{j=1}^{m-1}\sum_{j'=j+1}^m \delta_j\rho_{jj'}}{\sum_{j=1}^{m-1}\sum_{j'=j+1}^m \rho_{jj'}} = \frac{\sum_{j'=2}^m \left(\delta_j \sum_{j=1}^{j'-1}\rho_{jj'}\right)}{\sum_{j'=2}^m \sum_{j=1}^{j'-1}\rho_{jj'}} = \sum_{j'=2}^m \delta_{j'}w_{j'},$$

which is a weighted average of $\{\delta_{j'} : j' = 2,\ldots,m\}$, with weights

$$w_{j'} = \frac{\sum_{j=1}^{j'-1}\rho_{jj'}}{\sum_{l=2}^m \sum_{j=1}^{l-1}\rho_{jl}}.$$

With $\rho_{jj'} \ge 0$ from Condition 2, we have $w_2 \le w_3 \le \cdots \le w_m$. Furthermore, Condition 1 indicates that $\delta_1 \ge \delta_2 \ge \cdots \ge \delta_m$. In the weighted average, the weights decrease with the values of the elements. Thus

$$\sum_{j'=2}^m \delta_{j'}w_{j'} \le \bar{\delta}_{(-1)} \le \bar{\delta}.$$

Here $\bar{\delta}_{(-1)} = \sum_{j'=2}^m \delta_{j'}/(m-2)$ is the unweighted average of $\{\delta_{j'} : j' = 2,\ldots,m\}$. We have the last "$\le$" sign because $\bar{\delta}$ includes an additional element (δ_1) which is no less than any of the elements in $\delta_{(-1)}$. Thus we complete the proof of the lemma. $\square$

Based on the lemma, we have

$$\frac{n}{n_0/\delta_m} \le \frac{\delta_m}{\bar{\delta}^2} \cdot \frac{m\bar{\delta} + 2\bar{\delta}\sum_{j=1}^{m-1}\sum_{j'=j+1}^m \rho_{jj'}}{m + 2\sum_{j=1}^{m-1}\sum_{j'=j+1}^m \rho_{jj'}} = \frac{\delta_m\bar{\delta}}{\bar{\delta}^2} = \delta_m/\bar{\delta} \le 1.$$

The last "$\leq$" sign comes from the fact that δ_m is the smallest element in $\{\delta_j : j = 1, \ldots, m\}$, so $\delta_m \leq \bar{\delta}$.

It is obvious from the foregoing derivation that the equality sign in (4.21) only holds when $\delta_1 = \cdots = \delta_m = 1$ and $\rho_{jj'} = 0 (j \neq j')$.

Bibliography

[1] N.M. Laird and J. H. Ware. Random effects models for longitudinal data. *Biometrics*, 38:963–974, 1982.

[2] R.I. Jennrich and M.D. Schlucter. Unbalanced repeated-measures models with structured covariance matric. *Biometrics*, 42:805–820, 1986.

[3] H. D. Patterson and R. Thompson. Recovery of inter-block information when block sizes are unequal. *Biometrika*, 58(3):545–554, 1971.

[4] D.A. Harville. Maximum likelihood approaches to variance component estimation and to related problems. *Journal of the American Statistical Association*, 72(358):320–338, 1977.

[5] K. Liang and L. L. Zeger. Longitudinal data analysis using generalized linear models. *Biometrika*, 73:45–51, 1986.

[6] A. W. Kimball. Errors of the third kind in statistical consulting. *Journal of the American Statistician Association*, 57:133–142, 1957.

[7] M. Crowder. On the use of a working correlation matrix in using generalised linear models for repeated measures. *Biometrika*, 82:407–410, 1995.

[8] B.W. McDonald. Estimating logistic regression parameters for bivariate binary data. *Journal of the Royal Statistical Society, Series. B*, 55:391–397, 1993.

[9] W. J. Shih. Sample size and power calculations for periodontal and other studies with clustered samples using the method of generalized estimating equations. *Biometrical Journal*, 39(8):899–908, 1997.

[10] J. Rochon. Application of GEE procedures for sample size calculations in repeated measures experiments. *Statistics in Medicine*, 17(14):1643–1658, 1998.

[11] W. Pan. Sample size and power calculations with correlated binary data. *Controlled Clinical Trials*, 22(3):211–227, 2001.

[12] A. Liu, W. J. Shih, and E. Gehan. Sample size and power determination for clustered repeated measurements. *Statistics in Medicine*, 21(12):1787–1801, 2002.

[13] S.H. Jung and C. Ahn. Sample size estimation for GEE method for comparing slopes in repeated measurements data. *Statistics in Medicine*, 22(8):1305–1315, 2003.

[14] H. Kim, J. M. Williamson, and C. M. Lyles. Sample-size calculations for studies with correlated ordinal outcomes. *Statistics in Medicine*, 24(19):2977–2987, 2005.

[15] G. Liu and K. Liang. Sample size calculations for studies with correlated observations. *Biometrics*, 53(3):937–947, 1997.

[16] Steven G. Self and Robert H. Mauritsen. Power/sample size calculations for generalized linear models. *Biometrics*, 44:79–86, 1988.

[17] P. S. Albert. Longitudinal data analysis (repeated measures) in clinical trials. *Statistics in Medicine*, 18(13):1707–1732, 1999.

[18] J. E. Overall, G. Shobaki, C. Shivakumar, and J. Steele. Adjusting sample size for anticipated dropouts in clinical trials. *Psychopharmacology bulletin*, 34(1):25–33, 1998.

[19] Donald B. Rubin. Inference and missing data. *Biometrika*, 63:581–590, 1976.

[20] A. Munoz, V. Carey, J. P. Schouten, M. Segal, and B. Rosner. A parametric family of correlation structures for the analysis of longitudinal data. *Biometrics*, 48(3):733–742, 1992.

[21] J. D. Dawson. Sample size calculations based on slopes and other summary statistics. *Biometrics*, 54(1):323–330, 1998.

[22] A.J. Vicker. How many repeated measures in repeated measures designs? Statistical issues for comparative trials. *BMC Medical Research Methodology*, 3(22), 2003.

[23] P. Liu and S. Dahlberg. Design and analysis of multiarm clinical trials with survival endpoints. *Controlled Clinical Trials*, 16(2):119–130, 1995.

[24] R. W. Makuch and R. M. Simon. Sample size requirements for comparing time-to-failure among k treatment groups. *Journal of Chronic Diseases*, 35(11):861–867, 1982.

[25] S.H. Jung and C. Ahn. K-sample test and sample size calculation for comparing slopes in data with repeated measurements. *Biometrical Journal*, 46(5):554–564, 2004.

[26] G.W. Snedecor and W.G. Cochran. *Statiscal Methods (8th ed.)*. Ames: The Iowa State University Press, 1989.

[27] J.L. Fleiss. *The Design and Analysis of Clinical Experiments*. New York: Wiley, 1986.

[28] C. Ahn and S.H. Jung. Efficiency of general estimating equations estimators of slopes in repeated measurements: Adding subjects or adding measurements? *Drug Information Journal*, 37(3):309–316, 2003.

[29] C. Ahn and S.H. Jung. Efficiency of the slope estimator in repeated measurements. *Drug Information Journal*, 38(2):143–148, 2004.

[30] D.A. Bloch. Sample size requirements and the cost of a randomized clinical trial with repeated measurements. *Statistics in Medicine*, 5(6):663–667, 1986.

[31] D. Lai, T.M. King, L.A. Moyé, and Q. Wei. Sample size for biomarker studies: More subjects or more measurements per subject? *Annals of Epidemiology*, 13(3):204–208, 2003.

[32] K. J. Lui and W.G. Cumberland. Sample size requirement for repeated measurements in continuous data. *Statistics in Medicine*, 11(5):633–641, 1992.

[33] B. Winkens, H.J.A. Schouten, G.J.P. Van Breukelen, and M.P.F. Berger. Optimal number of repeated measures and group sizes in clinical trials with linearly divergent treatment effects. *Contemporary Clinical Trials*, 27(1):57–69, 2006.

[34] B. Winkens, H.J.A. Schouten, G.J.P. Van Breukelen, and M.P.F. Berger. Optimal designs for clinical trials with second-order polynomial treatment effects. *Statistical Methods in Medical Research*, 16(6):523–537, 2007.

[35] S. Zhang and C. Ahn. Adding subjects or adding measurements in repeated measurement studies under financial constraints. *Statistics in Biopharmaceutical Research*, 3(1):54–64, 2011.

[36] R.B. Davies. Algorithm AS 155: The distribution of a linear combination of χ^2 random variables. *Journal of the Royal Statistical Society. Series C (Applied Statistics)*, 29:323–333, 1980.

[37] P.J. Diggle, P. Heagerty, K.Y. Liang, and S.L. Zeger. *Analysis of Longitudinal Data (2nd ed.)*. Oxford University Press, 2002.

[38] H. Liu and T. Wu. Sample size calculation and power analysis of time-averaged difference. *Journal of Modern Applied Statistical Methods*, 4(2):434–445, 2005.

[39] T. Heiberg, T.K. Kvien, Ø. Dale, P. Mowinckel, G.J. Aanerud, A.B. Songe-MØller, T. Uhlig, and K.B. Hagen. Daily health status registration (patient diary) in patients with rheumatoid arthritis: A comparison between personal digital assistant and paper-pencil format. *Arthritis Care and Research*, 57(3):454–460, 2007.

[40] P. Mowinckel, K.B. Hagen, T. Heiberg, and T.K. Kvien. Repeated measures in rheumatoid arthritis reduced the required sample size in a two-armed clinical trial. *Journal of Clinical Epidemiology*, 61(9):940–944, 2008.

[41] A. Donner and N. Klar. *Design and Analysis of Cluster Randomization Trials in Health Research*. Oxford University Press, 2000.

[42] S. Zhang and C. Ahn. How many measurements for time-averaged differences in repeated measurement studies. *Contemporary Clinical Trials*, 32(3):412–417, 2011.

[43] S. Zhang and C. Ahn. Effects of correlation and missing data on sample size estimation in longitudinal clinical trials. *Pharmaceutical Statistics*, 9(1):2–9, 2010.

[44] S. Zhang and C. Ahn. Sample size calculation for comparing time-averaged responses in K-group repeated-measurement studies. *Computational Statistics and Data Analysis*, 58(1):283–291, 2013.

[45] K.J. Lui. Sample sizes for repeated measurements in dichotomous data. *Statistics in Medicine*, 10(3):463–472, 1991.

[46] S.H. Jung, S.H. Kang, and C. Ahn. Sample size calculations for clustered binary data. *Statistics in Medicine*, 20(13):1971–1982, 2001.

[47] S.R. Lipsitz and G.M. Fitzmaurice. Sample size for repeated measures studies with binary responses. *Statistics in Medicine*, 13(12):1233–1239, 1994.

[48] S.H. Jung and C.W. Ahn. Sample size for a two-group comparison of repeated binary measurements using GEE. *Statistics in Medicine*, 24(17):2583–2596, 2005.

[49] J.D. Reveille, M. Fischbach, T. McNearney, A.W. Friedman, M.B. Aguilar, J. Lisse, M.J. Fritzler, C. Ahn, and F.C. Arnett. Systemic sclerosis in 3 US ethnic groups: A comparison of clinical, sociodemographic, serologic, and immunogenetic determinants. *Seminars in Arthritis and Rheumatism*, 30(5):332–346, 2001.

[50] H.I. Patel and E. Rowe. Sample size for comparing linear growth curves. *Journal of Biopharmaceutical Statistics*, 9(2):339–50, 1999.

[51] K. Ogungbenro and L. Aarons. Sample size/power calculations for population pharmacodynamic experiments involving repeated-count measurements. *Journal of Biopharmaceutical Statistics*, 20(5):1026–1042, 2010.

[52] A.M. Spleen, B.C. Kluhsman, A.D. Clark, M.B. Dignan, E.J. Lengerich, and The ACTION Health Cancer Task Force. An increase in HPV-related knowledge and vaccination intent among parental and non-parental caregivers of adolescent girls, age 9-17 years, in Appalachian Pennsylvania. *Journal of Cancer Education*, 27(2):312–319, 2012.

[53] M.C. Rossi, C. Perozzi, C. Consorti, T. Almonti, P. Foglini, N. Giostra, P. Nanni, S. Talevi, D. Bartolomei, and G. Vespasiani. An interactive diary for diet management (DAI): A new telemedicine system able to promote body weight reduction, nutritional education, and consumption of fresh local produce. *Diabetes Technology and Therapeutics*, 12(8):641–647, 2010.

[54] A. Wajnberg, K.H. Wang, M. Aniff, and H.V. Kunins. Hospitalizations and skilled nursing facility admissions before and after the implementation of a home-based primary care program. *Journal of the American Geriatrics Society*, 58(6):1144–1147, 2010.

[55] E.J. Knudtson, L.B. Lorenz, V.J. Skaggs, J.D. Peck, J.R. Goodman, and A.A. Elimian. The effect of digital cervical examination on group B streptococcal culture. *American Journal of Obstetrics and Gynecology*, 202(1):58.e1–58.e4, 2010.

[56] T. Zieschang, I. Dutzi, E. Müller, U. Hestermann, K. Grünendahl, A.K. Braun, D. Hüger, D. Kopf, N. Specht-Leible, and P. Oster. Improving care for patients with dementia hospitalized for acute somatic illness in a specialized care unit: A feasibility study. *International Psychogeriatrics*, 22(1):139–146, 2010.

[57] Q. McNemar. Note on the sampling error of the difference between correlated proportions or percentages. *Psychometrika*, 12(2):153–157, 1947.

[58] O. S. Miettinen. The matched pairs design in the case of all-or-none responses. *Biometrics*, 24(2):339–352, 1968.

[59] R. J. Connor. Sample size for testing differences in proportions for the paired-sample design. *Biometrics*, 43(1):207–211, 1987.

[60] M.A. Shork and G.W. Williams. Number of observations required for the comparison of two correlated proportions. *Communications in Statistics - Simulation and Computation*, 9(4):349–357, 1980.

[61] J.M. Lachin. Power and sample size evaluation for the McNemar test with application to matched case-control studies. *Statistics in Medicine*, 11(9):1239–1251, 1992.

[62] Y. Lu and J.A. Bean. On the sample size for one-sided equivalence of sensitivities based upon McNemar's test. *Statistics in Medicine*, 14(16):1831–1839, 1995.

[63] S. Zhang, J. Cao, and C. Ahn. A GEE approach to determine sample size for pre- and post-intervention experiments with dropout. *Computational Statistics and Data Analysis*, 69:114–121, 2014.

[64] X.M. Tu, J. Kowalski, J. Zhang, K.G. Lynch, and P. Crits-Christoph. Power analyses for longitudinal trials and other clustered designs. *Statistics in Medicine*, 23(18):2799–2815, 2004.

[65] J. Geweke. Bayesian inference in econometric models using Monte Carlo integration. *Econometrica*, 57:1317–1339, 1989.

[66] K.E. Muller, L.M. LaVange, S.L. Ramey, and C.T. Rame. Power calculations for general linear multivariate models including repeated measures applications. *Journal of American Statistical Association*, 87:1209–1226, 1992.

[67] S. Zhang and C. Ahn. Sample size calculation for time-averaged differences in the presence of missing data. *Contemporary Clinical Trials*, 33(3):550–556, 2012.

[68] M.P. Feng, T.M. Braun, and C. McCulloch. Small sample inference for clustered data. In D.Y. Lin and P.J. Heagerty, editors, *Proceedings of the Second Seattle Symposium in Biostatistics*, pages 71–87. Springer, 2004.

[69] L.A. Mancl and T.A. DeRouen. A covariance estimator for GEE with improved small-sample properties. *Biometrics*, 57(1):126–134, 2001.

[70] W. Pan and M.M. Wall. Small-sample adjustments in using the sandwich variance estimator in generalized estimating equations. *Statistics in Medicine*, 21(10):1429–1441, 2002.

[71] X. Guo, W. Pan, J.E. Connett, P.J. Hannan, and S.A. French. Small-sample performance of the robust score test and its modifications in generalized estimating equations. *Statistics in Medicine*, 24(22):3479–3495, 2005.

[72] S. Teerenstra, B. Lu, J. S. Preisser, T. Van Achterberg, and G.F. Borm. Sample size considerations for GEE analyses of three-level cluster randomized trials. *Biometrics*, 66(4):1230–7, 2010.

[73] G. Kauermann and R.J. Carroll. A note on the efficiency of sandwich covariance matrix estimation. *Journal of the American Statistical Association*, 96(456):1387–1396, 2001.

[74] M. Heo and A.C. Leon. Statistical power and sample size requirements for three-level hierarchical cluster randomized trials. *Biometrics*, 64(4):1256–1262, 2008.

[75] W.J. Shih and P.L. Zhao. Design for sample size re-estimation with interim data for double-blind clinical trials with binary outcomes. *Statistics in Medicine*, 16(17):1913–1923, 1997.

[76] J. Wittes and E. Brittain. The role of internal pilot studies in increasing the efficiency of clinical trials. *Statistics in Medicine*, 9(1-2):65–72, 1990.

[77] A.L. Gould. Interim analyses for monitoring clinical trials that do not materially affect the type I error rate. *Statistics in Medicine*, 11(1):55–66, 1992.

5

Sample Size Determination for Correlated Outcomes from Two-Level Randomized Clinical Trials

5.1 Introduction

Two-level data structures arise frequently in randomized clinical trials. For example, in cross-sectional studies which enroll clusters or groups of subjects and evaluate them at a single point in time, the subjects nested within the groups would be considered the first level and the organizational entity that forms the groups (e.g., households or clinics) are the second level. This two-level structure also arises in longitudinal trials in which repeated measurements are obtained on study subjects during follow-up. In this case, the repeated measures are the first level data units and the subjects are the second level data units. Thus, depending on the study design, the nature of the first and second level units may differ, and randomization to different treatments may occur at either level. In determining sample sizes for such two-level clinical trials, the hierarchical data structure must be considered since both the first and second level units contribute to the total variation in the observed outcomes. In addition, level 1 data units from the same level 2 unit tend to be positively correlated. Furthermore, in some cases, additional between-subject variability in the longitudinal trends of the outcome needs to be taken into account in sample size determination. A vast body of literature addresses statistical power and sample size determinations for two-level data structures (e.g., [1–4]). Chapter 2 also includes several sample size approaches for two-level randomized trials for the special case in which randomization occurs at level 2 (cluster level) in a 1:1 balanced ratio across treatment arms. In this chapter we explore further the sample size issues arising in two-level cross-sectional and longitudinal randomized trials. We summarize and present closed-form sample size approaches that are based on mixed-effects models for parallel and factorial randomized trials. In addition, we consider both balanced and unbalanced randomization designs, and also evaluate and compare sample size requirements for level 2 versus level 1 randomization schemes.

5.2 Statistical Models for Continuous Outcomes

For outcomes which are continuous variables, we consider the following mixed-effects linear model for two-level data structures [5]:

$$Y_{ij} = h\left(X_{ij}, T_{ij}; \theta\right) + \nu_i T_{ij} + u_i + e_{ij}. \tag{5.1}$$

where Y_{ij} denotes the observed outcome on the jth level 1 unit in the ith level 2 unit where $i = 1, 2, \ldots, N_2$, and $j = 1, 2, \ldots, N_1$. As discussed further below, N_2 will either be defined as the number of level 2 units in each treatment arm or the total number of level two units, depending on the study design. Likewise, N_1 will either be defined as the number of level one units in each intervention arm per level two unit or the total number of level one units per level two unit. Regardless, if a study design is balanced, the total number of observations will be equal to $2N_1 N_2$. The variable X_{ij} is an indicator variable for treatment assignment ($X_{ij} = 1$ for experimental treatment and 0 for control). For longitudinal clinical trials, T_{ij} denotes the time point at which Y_{ij} is measured and ν_i denotes the subject-specific effect of time on the outcome (i.e., slope). The fixed effects are represented by the function h with parameter vector θ, the exact specification depending on the study design and the parameter of interest. We note that for cross-sectional studies that do not have a time component, the fixed effects $h(X, T; \theta)$ reduce to $h(X; \theta)$, and the term $\nu_i T_{ij}$ is removed from model (5.1).

The u_i and e_{ij} terms in model (5.1) are random effects. The error term e_{ij} is assumed to follow a $N\left(0, \sigma_e^2\right)$ distribution where the variance term, σ_e^2, represents the variance of the level one random effects. The level 2 random components (i.e., subject-specific intercepts) are distributed as $u_i \sim N\left(0, \sigma_u^2\right)$, where σ_u^2 denotes the variance of the level two random effects. If the slope of the outcome over time, ν_i, is considered random, then we assume $\nu_i \sim N\left(0, \sigma_\tau^2\right)$, where σ_τ^2 is the variance in the slopes across the level two data units. It is further assumed that $u_i \perp e_{ij} \perp \nu_i$, i.e., these three random components are mutually independent. In addition, *conditional independence* is assumed for all u_i, ν_i and e_{ij}, whereas the u_i are *unconditionally* independent. That is, both u_i and ν_i are independent conditional on u_i, and e_{ij} are independent over j conditional on u_i and ν_i. Throughout this chapter, all of these variances of the random effects are assumed to be known.

Under model (5.1), it can be shown that the elements of the mean vector for the outcome are equal to

$$E(Y_{ij}) = h\left(X_{ij}, T_{ij}; \theta\right)$$

and the elements of the covariance matrix are

$$Cov\left(Y_{ij}, Y_{i'j'}\right) = 1(i = i' \cap j = j')\sigma_e^2 + 1(i = i')\left(T_j T_{j'}\sigma_\tau^2 + \sigma_u^2\right),$$

where $1(.)$ is an indicator function and T_{ij} is assumed to be equal to T_j for all i. It follows that

$$Var\,(Y_{ij}) = Cov\,(Y_{ij}, Y_{ij}) = \sigma^2 + T_j^2 \sigma_\tau^2,$$

where

$$\sigma^2 \equiv \sigma_e^2 + \sigma_u^2,$$

which is the variance of Y when the fixed slope model is assumed ($\sigma_\tau^2 = 0$). Therefore, the correlations among the level 1 data, i.e., among observations from different level 1 units within the same level 2 unit, can be expressed for $j \neq j'$ as follows:

$$Corr\,(Y_{ij}, Y_{ij'}) = \frac{\sigma_u^2}{\sqrt{\sigma^2 + T_j^2 \sigma_\tau^2}\sqrt{\sigma^2 + T_j'^2 \sigma_\tau^2}}.$$

Under the fixed slope model, i.e., when $\sigma_\tau^2 = 0$, the correlations reduce to the following,

$$\rho = \sigma_u^2 / \sigma^2, \tag{5.2}$$

which is often referred to as the intra class correlations (ICC) [6].

5.3   Testing Main Effects

We first consider the two-level randomized trial in which clusters of subjects are enrolled in the study and a single measurement is available for each subject, whether the measurement is the study outcome assessed at a single point in time in a cross-sectional design, a within-subject pre-post treatment difference, or the average of the subject's repeated measurements obtained over several time points. In such two-level trials, subjects are considered the first level data units; the hospitals, research centers, primary care clinics or households that form the clusters of subjects are considered the second level units. Furthermore, randomization to interventions or treatments can occur at either the first or second level. For the former, the randomizations are assigned at the cluster level and thus the subjects within the cluster receive identical treatments (Section 5.3.1). For the latter, the randomizations are be assigned at the subject level within clusters and thus the subjects within the cluster may receive different treatments (Section 5.3.2). We compare the sample sizes required for testing main effects with the second and first level randomization approaches in Section 5.3.3, which is followed by an example in Section 5.3.4.

5.3.1   Randomization at the Second Level

For the design in which randomization occurs at the second level (e.g. clinic), we further define $X_i = 0$ if the ith level 2 unit is assigned to a control intervention and $X_i = 1$ if assigned to an experimental intervention. This implies that $X_{ij} = X_i$ for all j within i. Because the outcome is measured at a single point in time on each subject, the fixed-effect term $h(X_{ij}, T_{ij}; \theta)$ in model (5.1) is reduced to

$$h(X_{ij}, T_{ij}; \theta) = \beta_0 + \delta_{(2)} X_{ij},$$

resulting in the following model:

$$Y_{ij} = \beta_0 + \delta_{(2)} X_{ij} + u_i + e_{ij}, \tag{5.3}$$

where β_0 denotes the overall fixed intercept term and the parameter $\delta_{(2)}$ represents the main treatment effect of interest. The null hypothesis to be tested is $H_0 : \delta_{(2)} = 0$. Let us denote the level 2 sample sizes for the control $(X = 0)$ and intervention $(X = 1)$ arms by $N_2^{(0)}$ and $N_2^{(1)}$ while N_1 represents the level 1 sample size per level two unit.

5.3.1.1   Balanced Allocations

For a balanced design, it is assumed that $\Sigma_i X_i = N_2^{(1)} = N_2^{(0)}$ and $i = 1, 2, \ldots, 2N_2^{(0)}$. That is, $N_2 = N_2^{(0)} = N_2^{(1)}$ corresponds to the level 2 sample size *per treatment arm*. To test the significance of $\delta_{(2)}$ in model (5.3), the following test statistic can be used:

$$D_2 = \frac{\sqrt{N_2 N_1} \left(\bar{Y}_1 - \bar{Y}_0 \right)}{\sigma \sqrt{2f}}.$$

Here, $\bar{Y}_g = \frac{1}{N_2 N_1} \sum_{i=1}^{N_2} \sum_{j=1}^{N_1} Y_{ij} (g = 0, 1)$ is the mean of the outcome Y for the gth arm, for which $X_i = g$, and

$$f = 1 + (N_1 - 1)\rho \tag{5.4}$$

is referred to as a *variance inflation factor* or *design effect* [7, 8] and does not depend on N_2. Note that $\hat{\delta}_{(2)} = \bar{Y}_1 - \bar{Y}_0$ is an unbiased estimate of $\delta_{(2)}$ and $Var\left(\bar{Y}_g \right) = f\sigma^2/(N_2 N_1)$ yielding

$$Var\left(\hat{\delta}_{(2)} \right) = \frac{2f\sigma^2}{N_2 N_1}.$$

It can be shown that the test statistic D_2 is normally distributed with mean $\delta_{(2)}$ and variance 1. The power of the test statistic D_2, denoted by $\varphi_{(2)}$, can therefore be written as follows:

$$\varphi_{(2)} = \Phi\left\{ \Delta_{(2)} \sqrt{N_2 N_1 / 2f} - \Phi^{-1}(1 - \alpha/2) \right\}.$$

The parameter $\Delta_{(2)} = |\delta_{(2)}/\sigma|$ is referred to as a standardized effect size or Cohen's d [9] which corresponds to the mean difference in outcome Y expressed in units of standard deviation σ. The cumulative distribution function (CDF) and the inverse CDF of a standard normal distribution are denoted by Φ and Φ^{-1}, respectively. A two-sided significance level is denoted by α. It follows that the required number of level 2 units per treatment arm, N_2, for a desired statistical power $\varphi_{(2)} = \phi$ can be expressed as follows [10]:

$$N_2 = \frac{2f z_{\alpha,\phi}^2}{N_1 \Delta_{(2)}^2}, \qquad (5.5)$$

where

$$z_{\alpha,\phi} = \Phi^{-1}\left(1 - \alpha/2\right) + \Phi^{-1}\left(\phi\right). \qquad (5.6)$$

It should be noted that Equation (5.5) is identical to Equation (2.10) except for a difference in notation. To determine the required total number of level 1 units per level 2 unit, N_1, for a given N_2, equation (5.5) can be used to solve for N_1:

$$N_1 = \frac{2\left(1 - \rho\right) z_{\alpha,\phi}^2}{N_2 \Delta_{(2)}^2 - 2\rho z_{\alpha,\phi}^2}.$$

We note that N_1 cannot be determined for a given N_2 if ρ in particular is not small enough to satisfy $\rho < N_2 \Delta_{(2)}^2 / 2z_{\alpha,\phi}^2$ so that the denominator of the right hand side of the above equation must be positive. *This requirement for a positive denominator applies throughout this chapter and beyond.*

5.3.1.2 Unbalanced Allocations

Suppose that the random allocation needs to be unbalanced so that $\Sigma_i X_i = \lambda_2 N_2^{(0)}$ for $\lambda_2 > 0$ and $i = 1, 2, \ldots, (1 + \lambda_2) N_2^{(0)}$. The ratio λ_2 represents the ratio of the number of level 2 units in the intervention arm to the corresponding number in the control arm, i.e., $\lambda_2 = N_2^{(1)} / N_2^{(0)}$. If $\lambda_2 > 1$, the sample size in the intervention arm is larger, which is usually the case in practice. Then, the sample size $N_2^{(0)}$ in the control arm can be expressed as

$$N_2^{(0)} = \frac{f\left(1 + 1/\lambda_2\right) z_{\alpha,\phi}^2}{N_1 \Delta_{(2)}^2}. \qquad (5.7)$$

And the sample size of the level 1 units given N_2 as

$$N_1 = \frac{\left(1 - \rho\right)\left(1 + 1/\lambda_2\right) z_{\alpha,\phi}^2}{N_2^{(0)} \Delta_{(2)}^2 - \left(1 + 1/\lambda_2\right) \rho z_{\alpha,\phi}^2}.$$

If the sample sizes N_1 needs to vary across the level 2 units, that is, $j = 1, 2, \ldots, n_i$, then N_1 in (5.5) and (5.7) can be replaced by $\tilde{N}_1 = \sum_{i=1}^{N_2^{(0)} + N_2^{(1)}} n_i / (N_2^{(0)} + N_2^{(1)})$ to yield an approximate sample size, even if derivation of a more accurate exact formula with varying cluster sizes is possible. An alternative approach for handling unequal cluster sizes in the sample size determination of two-level cluster randomized trials is discussed in Section 2.4.2.

TABLE 5.1
Sample size and power for detecting a main effect $\delta_{(2)}$ in model (5.3) when randomizations occur at the second level (two-sided significance level $\alpha = 0.05$)

λ_2	$\Delta_{(2)}$	ρ	N_1	$N_2^{(0)}$	Total N	$\varphi_{(2)}$
1	0.4	0.1	10	19	380	0.807
			20	15	600	0.820
		0.2	10	28	560	0.807
			20	24	960	0.807
	0.5	0.1	10	12	240	0.802
			20	10	400	0.836
		0.2	10	18	360	0.809
			20	16	640	0.823
1.5	0.4	0.1	10	16	400	0.811
			20	12	600	0.807
		0.2	10	23	575	0.801
			20	20	1000	0.807
	0.5	0.1	10	10	250	0.802
			20	8	400	0.820
		0.2	10	15	375	0.809
			20	13	650	0.813

Note: λ_2 represents the ratio of the number of level 2 units in the intervention arm to the corresponding number in the control arm; $\Delta_{(2)} = |\delta_{(2)}/\sigma|$; ρ (5.2) is the correlations among the level 1 data, often referred to as intra class correlations (ICC); N_1 is the number of level 1 units per level 2 units; $N_2^{(0)}$ is the level 2 sample size in the *control* arm; Total $N = (1 + \lambda_2)N_1 N_2^{(0)}$ is the required total number of observations; $\varphi_{(2)}$ is the statistical power to test $H_0 : \delta_{(2)} = 0$.

5.3.1.3 Summary

The sample size N_2 in (5.5) or $N_2^{(0)}$ in (5.7) required for a desired magnitude of statistical power decreases with increasing N_1, and $\Delta_{(2)}$, and increases with increasing ρ. However, with all the other design parameters fixed, the total sample size $(1+\lambda_2)N_2^{(0)}$ is smallest for balanced designs ($\lambda_2 = 1$) compared to any unbalanced design ($\lambda_2 \neq 1$). *In fact, this property of minimally required total sample size of a balanced design holds for any levels of randomization and also for testing any effects, and thus will not be repeated throughout this chapter and beyond.* Table 5.1 displays examples of statistical power for testing the null hypothesis $H_0 : \delta_{(2)} = 0$.

5.3.2   Randomization at the First Level

When randomization occurs at level 1 (e.g., subjects), the fixed-effect term $h(X_{ij}, T_{ij}; \theta)$ in model (5.1) is reduced to

$$h(X_{ij}, T_{ij}; \theta) = \beta_0 + \delta_{(1)} X_{ij},$$

resulting in the model

$$Y_{ij} = \beta_0 + \delta_{(1)} X_{ij} + u_i + e_{ij}, \tag{5.8}$$

where the parameter $\delta_{(1)}$ represents the intervention effect. The null hypothesis to be tested is $H_0 : \delta_{(1)} = 0$. Let us denote the level 1 sample sizes for the control ($X = 0$) and intervention ($X = 1$) arms by $N_1^{(0)}$ and $N_1^{(1)}$ while N_2 represents the *total* level 2 sample size.

5.3.2.1 Balanced Allocations

The intervention assignment indicator variable $X_{ij} = 0$ if the jth level 1 unit in the ith level 2 unit is assigned to a control intervention and $X_{ij} = 1$ if assigned to an experimental intervention. For a balanced design, it is assumed that $\Sigma_j X_{ij} = N_1^{(1)}$ for all i so that $i = 1, 2, \ldots, N_2$, and $j = 1, \ldots, 2N_1^{(0)}$. That is, $N_1 = N_1^{(0)} = N_1^{(1)}$ corresponds to the level 1 sample size *per treatment arm* within level 2 units.

To test the significance of $\delta_{(1)}$ in model (5.8), the following test statistic can be used:

$$D_1 = \frac{\sqrt{N_2 N_1} \left(\bar{Y}_1 - \bar{Y}_0 \right)}{\sigma \sqrt{2 \left(1 - \rho \right)}}.$$

Here, $\bar{Y}_g = \frac{1}{N_2 N_1} \sum_{i=1}^{N_2} \sum_{j=1}^{N_1} Y_{ij} (g = 0, 1)$ is the mean of the outcome Y for the gth arm, for which $X_j = g$, and the design effect is simply $1 - \rho$. We note that $\hat{\delta}_{(1)} = \bar{Y}_1 - \bar{Y}_0$ is an unbiased estimate of $\delta_{(1)}$ and

$$Var(\hat{\delta}_{(1)}) = \frac{2 \left(1 - \rho \right) \sigma^2}{N_2 N_1}.$$

It can be shown that the test statistic D_1 is normally distributed with mean $\delta_{(2)}$ and variance 1. The power of the test statistic D_1 can therefore be written in terms of $\Delta_{(1)} = |\delta_{(1)}/\sigma|$ as follows:

$$\varphi_{(1)} = \Phi \left\{ \Delta_{(1)} \sqrt{N_2 N_1 / 2 \left(1 - \rho \right)} - \Phi^{-1}(1 - \alpha/2) \right\}.$$

Therefore, the required total number of level 2 units, N_2, for a desired statistical power $\varphi_{(1)} = \phi$ can be expressed as [11]

$$N_2 = \frac{2 \left(1 - \rho \right) z_{\alpha, \phi}^2}{N_1 \Delta_{(1)}^2}. \tag{5.9}$$

It should be noted that the sample size when randomization occurs at the first level depends only on the residual variance $\sigma_e^2 = (1 - \rho)\sigma^2$ but not on σ_u^2. Furthermore, determination of N_1 for a given N_2 is straightforward since the design effect $1 - \rho$ under first level randomization does not depend on N_1, which is the number of level 1 units in each treatment arm per level 2 unit. In other words, the statistical power $\varphi_{(1)}$ is invariant over the product $N_1 N_2$, which we call *invariance over the product of $N_1 N_2$ property*. Due to this property, N_2 and N_1 can be simply switched in (5.9) to solve for N_1:

$$N_1 = \frac{2(1-\rho)z_{\alpha,\phi}^2}{N_2\Delta_{(1)}^2}.$$

5.3.2.2 Unbalanced Allocations

Suppose that the random allocation at the second level needs to be unbalanced such that $\Sigma_j X_j = \lambda_1 N_1^{(0)}$ for $\lambda_1 > 0$ and $j = 1, 2, \ldots, (1 + \lambda_1)N_1^{(0)}$, where $N_1^{(0)}$ is the number of subjects within each cluster assigned to the *control* arm. The ratio λ_1 represents the ratio of the number of the first level units in each second level unit that are assigned to the intervention arm to the corresponding number assigned to the control arm, i.e., $\lambda_1 = N_1^{(1)}/N_1^{(0)}$. The sample size N_2 and $N_1^{(0)}$ can be expressed respectively as

$$N_2 = \frac{(1 + 1/\lambda_1)(1 - \rho)z_{\alpha,\phi}^2}{N_1^{(0)}\Delta_{(1)}^2} \qquad (5.10)$$

and

$$N_1^{(0)} = \frac{(1 + 1/\lambda_1)(1 - \rho)z_{\alpha,\phi}^2}{N_2\Delta_{(1)}^2}.$$

Again, if N_1 needs to vary across the level 2 units, that is, $j = 1, 2, .., n_i$, then N_1 in (5.9) and $N_1^{(0)}$ in (5.10) can be replaced by $\tilde{N}_1 = \sum_{i=1}^{N_2} n_i/N_2$. While this approximation strategy is useful, exact sample size formulas are presented in Section (2.4.2).

5.3.2.3 Summary

The sample size N_2 in (5.9) or in (5.10) required for a desired magnitude of statistical power decreases with increasing $N_1^{(0)}, \Delta_{(1)}$, and ρ unlike the case with the second-level randomization. Table 5.2 displays examples of statistical power for testing the null hypothesis $H_0 : \delta_{(1)} = 0$, and shows that the aforementioned invariance over product property (see section 5.3.2.1) holds regardless of whether or not λ_1 is equal to 1.

5.3.3 Comparison of Sample Sizes

Either for balanced designs or for an unbalanced design with $\lambda_1 = \lambda_2$, the ratio of the total number of *observations*, that is $2N_2N_1$, required for trials with second level randomization to that for first level randomization can be

TABLE 5.2

Sample size and power for detecting a main effect $\delta_{(1)}$ in model (5.8) when randomizations occur at the first level (two-sided significance level $\alpha = 0.05$)

λ_1	$\Delta_{(1)}$	ρ	$N_1^{(0)}$	N_2	Total N	$\varphi_{(1)}$
1	0.4	0.1	10	10	200	0.846
			20	5	200	0.846
		0.2	10	8	160	0.807
			20	4	160	0.807
	0.5	0.1	10	6	120	0.823
			20	3	120	0.823
		0.2	10	6	120	0.865
			20	3	120	0.865
1.5	0.4	0.1	10	8	200	0.832
			20	4	200	0.832
		0.2	10	8	200	0.873
			20	4	200	0.873
	0.5	0.1	10	6	150	0.885
			20	3	150	0.885
		0.2	10	6	150	0.918
			20	3	150	0.918

Note: λ_1 represents the ratio of the number of level 1 units in the intervention arm to the corresponding number in the control arm within level 2 units; $\Delta_{(1)} = |\delta_{(1)}/\sigma|$; ρ (5.2) is the correlations among the level 1 data, often referred to as intra class correlations (ICC); $N_1^{(0)}$ is the level 1 sample size in the *control* arm within level 2 units; N_2 is the total level 2 sample size; Total $N = (1 + \lambda_1)N_1^{(0)}N_2$ is the required total number of observations; $\varphi_{(1)}$ is the statistical power to test $H_0 : \delta_{(1)} = 0$.

derived from the above equations to yield:

$$R_{2.1} = f_1/(1 - \rho) = 1 + N_1\rho/(1 - \rho) \geq 1. \tag{5.11}$$

This ratio is identical to the ratio of the design effects between the first and second level randomizations. Therefore, unless $\rho = 0$, the trials using second level randomizations require more observations for a given effect size (Δ), statistical power (φ), two-sided significance level (α), and correlation (ρ). In addition, as indicated in (5.3.9), the ratio of the required sample sizes increases with increasing values of N_1 and ρ. This is also evident by comparing the results in Tables 5.1 and 5.2.

5.3.4 Example

The prevalence of type 2 diabetes in the United Kingdom is four- to six-fold higher in south Asians (UK decennial census categories Indian, Pakistani,

Bangladeshi, and other Asians) than in white Europeans. Furthermore, south Asians show higher rates of disease related to morbidity and mortality, and health-care delivery in this population is more challenging because of cultural differences and communication difficulties. Bellary et al. [12] conducted the United Kingdom Asian Diabetes Study (UKADS) to assess a community-based intervention for reducing cardiovascular risk in south Asian patients with type 2 diabetes. The intervention was tailored to the needs of the south Asian community and consisted of additional time with a practice nurse, Asian link workers to facilitate communication and improve compliance, and input from diabetes-specialist nurses. The UKADS study hypothesis was that this enhanced clinical management approach for diabetes would improve cardio-vascular risk profile in patients of south Asian origin with established type 2 diabetes. In this cluster randomized trial, inner-city practices with a very high proportion (more than 80%) of south Asian patients in the UK were ran-domly assigned to the intervention or control (standard care). The primary outcomes were changes in blood pressure, total cholesterol, and glycaemic control (haemoglobin A1c) after 2 years.

The design of this study is a balanced two-level cluster randomized trial, with patients as the first level and practice as the second level, and random-ization occurring at the second level. To estimate the required sample size for this trial, the within-practice intraclass correlations (ICC) for each outcome were obtained from the investigators' pilot data or, if unavailable, assumed to be equal to ICC = 0.05 based on published estimates for primary care stud-ies. Minimum detectable effect sizes were chosen to be similar to intervention effects observed in the pilot study. Specifically, the target between group dif-ferences in changes were systolic blood pressure of 7 mm Hg (SD = 21. 25, ICC = 0·035), total cholesterol of 0.45 mmol/L (SD = 1.1; ICC = 0·05), and haemoglobin A1c of 0.75 (SD = 2.1; ICC = 0.05); the corresponding effects sizes $\Delta_{(2)}$ were $7/21.25 = 0.33$, $0.45/1.1 = 0.41$, and $0.75/2.1 = 0.36$, respec-tively. In regard to required sample sizes, applying the methods in 5.3.1 for testing main effects $\delta_{(2)}$ for two-level trials with second level randomization, it can be determined that a total of 16–18 clusters (8–9 clusters per arm) with 80–100 patients in each are needed for a minimum of 80% power to evaluate each of the three outcomes with a two-sided Type I error rate of 5%.

5.4   Two-Level Longitudinal Designs: Testing Slope Differences

For two-level longitudinal designs in which subjects are randomly assigned to treatments and then followed up at a pre-specified number of visits, the subjects are the second level data units and the repeated measures are the first level data units. With this design, it is often of primary interest to test between intervention arms the difference in the slopes or trends in the outcome over time. There are two ways to model slope variations across subjects: fixed and random slope effect models. For the former, subjects are assumed to have a common slope (Section 5.4.1) and for the latter, each subject is assumed to have his/her own trend in outcome overtime (Section 5.4.2). In this case, the variability in outcomes is greater than those in the fixed slope models, especially at larger time points. We compare the required sample sizes with fixed and random slope models in Section 5.4.3 and discuss the testing of main effects at the end of study in Section 5.4.4. We note that Chapters 3 and 4 also include additional discussion of sample size approaches for the comparisons of slopes in two-level longitudinal designs under different correlation structures and a fixed slope assumption.

5.4.1 Fixed Slope Model

When the subject-specific slopes of the outcome over time are modeled as fixed as opposed to random values, the fixed effect term $h(X, T; \theta)$ in model (5.1) can be specified as

$$h(X, T; \theta) = \beta_0 + \xi X + \tau T + \delta_{(f)} XT,$$

and the variance of the random slopes is set as $\sigma_\tau^2 = 0$. This results in the following model for the outcome:

$$Y_{ij} = \beta_0 + \xi X_{ij} + \tau T_{ij} + \delta_{(f)} X_{ij} T_{ij} + u_i + e_{ij}. \tag{5.12}$$

Let us denote the level 2 sample sizes for the control ($X = 0$) and intervention ($X = 1$) arms by $N_2^{(0)}$ and $N_2^{(1)}$ while N_1 represents the level 1 sample size per level 2 units.

5.4.1.1 Balanced Allocations

Under a balanced design such that $\Sigma_i X_i = N_2^{(1)}$, the second level unit is indexed by $i = 1, 2, \ldots, 2N_2^{(0)}$ That is, $N_2 = N_2^{(0)} = N_2^{(1)}$ corresponds to the level 2 sample size *per treatment arm*. The first level units nested within each i are indexed by $j = 1, \ldots, N_1$. Here, it is assumed that $T_{ij} = T_j$ for all i, and that the time increases from 0 (baseline) to $T_{end} = N_1 - 1$ (the last time point) by unit increments. However, these assumptions are not required in general.

The parameter ξ represents the intervention effect at baseline, which is usually assumed to equal zero in randomized trials. The parameter τ represents the slope of the time effect, that is, the change in outcome over time. The overall fixed intercept is denoted by β_0. The main parameter of interest is the intervention by time effect, $\delta_{(f)}$, or the difference in slopes between arms. The null hypothesis to be tested is $H_0 : \delta_{(f)} = 0$. Figure 5.1 depicts a geometrical relationship among the fixed parameters in model (5.12).

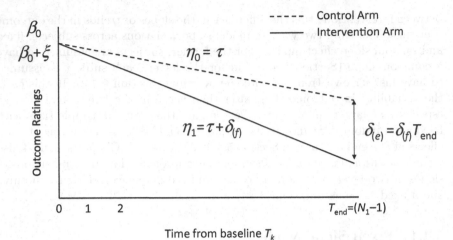

FIGURE 5.1
Geometrical representations of fixed parameters in model (5.12) for a parallel-arm longitudinal cluster randomized trial.

To test the significance of $\delta_{(f)}$ in model (5.12), the following test statistic can be used:

$$D_f = \frac{\sqrt{N_2 N_1 Var_p(T)}\,(\hat{\eta}_1 - \hat{\eta}_0)}{\sigma\sqrt{2(1-\rho)}},$$

where $\hat{\eta}_g (g = 0, 1)$ is an ordinary least square estimate (OLS) of the slope for the outcome Y in the gth arm, in which $X_i = g$, so that $\hat{\delta}_{(f)} = \hat{\eta}_1 - \hat{\eta}_0$, which is an unbiased estimate of $\delta_{(f)}$. When $\sigma_\tau^2 = 0$, this OLS is identical to a maximum likelihood estimate [13]. Specifically, for the gth arm,

$$
\begin{aligned}
\hat{\eta}_g &= \sum_{i=1}^{N_2} \sum_{j=1}^{N_1} (T_j - \bar{T})(Y_{ij} - \bar{Y}_g) \Big/ \sum_{i=1}^{N_2} \sum_{j=1}^{N_1} (T_j - \bar{T})^2 \\
&= \sum_{i=1}^{N_2} \sum_{j=1}^{N_1} (T_j - \bar{T})(Y_{ij} - \bar{Y}_g) \Big/ N_2 N_1 Var_p(T)
\end{aligned}
\tag{5.13}
$$

where: 1) $\bar{Y}_g (g = 0, 1)$ is the overall mean of the outcome Y for the gth arm; 2) $\bar{T} = \sum_{j=1}^{N_1} T_j/N_1$ is the "mean" time point; and 3) $Var_p(T) = \sum_{j=1}^{N_1} (T_j - \bar{T})^2/N_1$ is the "population variance" of the time variable T. The variance of $\hat{\delta}_{(f)}$ can be obtained as

$$Var\left(\hat{\delta}_{(f)}\right) = Var\left(\hat{\eta}_1 - \hat{\eta}_0\right) = \frac{2(1-\rho)\sigma^2}{N_2 N_1 Var_p(T)}.$$

It should be noted that the variance of $\hat{\delta}_{(f)}$ depends only on the residual variance $\sigma_e^2 = (1-\rho)\sigma^2$ but not on σ_u^2. Therefore, *for a given total variance σ^2,* the variance of $\hat{\delta}_{(f)}$ decreases with decreasing σ_e^2 or increasing ρ, the correlation among the first level units.

The power of the test statistic D_f can be written as follows:

$$\varphi_{(f)} = \Phi\left\{\Delta_{(f)}\sqrt{N_2 N_1 Var_p(T)/2(1-\rho)} - \Phi^{-1}(1-\alpha/2)\right\},$$

where the slope difference is expressed in terms of a standardized effect size, that is, $\Delta_{(f)} = |\delta_{(f)}/\sigma|$. It follows that the second level unit sample size N_2 per arm for a desired statistical power $\varphi_f = \phi$ can be obtained as [14]

$$N_2 = \frac{2(1-\rho)z_{\alpha,\phi}^2}{N_1 Var_p(T)\Delta_{(f)}^2}. \tag{5.14}$$

The determination of size sample size N_1 for given N_2 should be an iterative solution of the following equation since $Var_p(T)$ is a function of N_1:

$$N_1 = \frac{2(1-\rho)z_{\alpha,\phi}^2}{N_2 Var_p(T)\Delta_{(f)}^2}.$$

5.4.1.2 Unbalanced Allocations

Again, suppose that the random allocation needs to be unbalanced where $\Sigma_i X_i = \lambda_2 N_2^{(0)}$ for $\lambda_2 = N_2^{(1)}/N_2^{(0)} > 0$ and $i = 1, 2, \ldots, (1+\lambda_2)N_2^{(0)}$. Then, the sample size $N_2^{(0)}$ in the control arm can be expressed as

$$N_2^{(0)} = \frac{(1-\rho)(1+1/\lambda_2)z_{\alpha,\phi}^2}{N_1 Var_p(T)\Delta_{(f)}^2}. \tag{5.15}$$

And using an iterative approach to solve for N_1, we have

$$N_1 = \frac{(1-\rho)(1+1/\lambda_2)z_{\alpha,\phi}^2}{N_2^{(0)}Var_p(T)\Delta_{(f)}^2}.$$

5.4.1.3 Summary

The sample size N_2 in (5.14) or $N_2^{(0)}$ in (5.15) required for a desired magnitude of statistical power decreases with increasing $N_1, \Delta_{(f)}$, and ρ. The effect of N_1 on decreasing N_2 could also be accelerated by increasing $Var_p(T)$ if the effect size $\Delta_{(f)}$ is held fixed. Table 5.3 displays examples of statistical power for testing the null hypothesis $H_0 : \delta_{(f)} = 0$.

5.4.2 Random Slope Model

When the subject-specific slopes of the outcome measurements over time are assumed to be random, the fixed-effect term $h(X, T; \theta)$ in model (5.1) is specified as

TABLE 5.3
Sample size and power for detecting an effect $\delta_{(f)}$ on slope differences in a fixed-slope model (5.12) with $r_\tau = 0$ when randomizations occur at the second level (two-sided significance level $\alpha = 0.05$)

λ_2	ρ	$\Delta_{(f)}$	N_1	$N_2^{(0)}$	Total N	$\varphi_{(f)}$
1	0.1	0.4/4	5	142	1420	0.802
		0.4/8	9	95	1710	0.803
	0.2	0.4/4	5	126	1260	0.801
		0.4/8	9	84	1512	0.801
	0.1	0.5/4	5	91	910	0.803
		0.5/8	9	61	1098	0.805
	0.2	0.5/4	5	81	810	0.803
		0.5/8	9	54	972	0.803
1.5	0.1	0.4/4	5	118	1475	0.801
		0.4/8	9	79	1178	0.803
	0.2	0.4/4	5	105	1313	0.801
		0.4/8	9	70	1575	0.801
	0.1	0.5/4	5	76	950	0.803
		0.5/8	9	51	1148	0.806
	0.2	0.5/4	5	67	838	0.800
		0.5/8	9	45	1013	0.803

Note: λ_2 represents the ratio of the number of level 2 units in the intervention arm to the corresponding number in the control arm; $\Delta_{(f)} = \left|\delta_{(f)}/\sigma\right|$; ρ (5.2) is the correlations among the level 1 data, often referred to as intra class correlations (ICC); N_1 is the number of level 1 units per level 2 units; $N_2^{(0)}$ is the level 2 sample size in the *control* arm; Total $N = (1 + \lambda_2)N_1 N_2^{(0)}$ is the required total number of observations; $\varphi_{(f)}$ is the statistical power to test $H_0 : \delta_{(f)} = 0$.

$$h(X, T; \theta) = \beta_0 + \xi X + \tau T + \delta_{(r)} XT,$$

and the subject specific slopes are assumed to be distributed as $\nu_i \sim N\left(0, \sigma_\tau^2\right)$. These specifications result in the following model for the outcome variable:

$$Y_{ij} = \beta_0 + \xi X_{ij} + \tau T_{ij} + \delta_{(r)} X_{ij} T_{ij} + \nu_i T_{ij} + u_i + e_{ijk}. \qquad (5.16)$$

Again, let us denote the level 2 sample sizes for the control ($X = 0$) and intervention ($X = 1$) arms by $N_2^{(0)}$ and $N_2^{(1)}$ while N_1 represents the level 1 sample size per level 2 units. For balanced designs, $N_2 = N_2^{(0)} = N_2^{(1)}$.

5.4.2.1 Balanced Allocations

To test the null hypothesis $H_0 : \delta_{(r)} = 0$, the following test statistic can be used:

$$D_r = \frac{(\hat{\eta}_1 - \hat{\eta}_0)\sqrt{N_2 N_1 Var_p(T)}}{\sqrt{2\{(1 - \rho)\sigma^2 + N_1 Var_p(T)\sigma_\tau^2\}}},$$

where $\hat{\eta}_g (g = 0, 1)$ is an ordinary least square (OLS) estimate of the slope for the outcome Y as in Equation (5.13) so that $\hat{\delta}_{(r)} = \hat{\eta}_1 - \hat{\eta}_0$, which is an unbiased estimate of $\delta_{(r)}$. For the random slope with $\nu_i \sim N(0, \sigma_\tau^2)$ this OLS is not necessarily identical to the maximum likelihood estimate (MLE). However, the test statistic can be constructed based on the OLS estimate. In addition, the variance of $\hat{\delta}_{(r)}$ can be obtained as follows:

$$Var\left(\hat{\delta}_{(r)}\right) = Var\left(\hat{\eta}_1 - \hat{\eta}_0\right) = \frac{2\left\{(1-\rho)\sigma^2 + N_1 Var_p(T)\sigma_\tau^2\right\}}{N_2 N_1 Var_p(T)},$$

which is again not a function of ρ_2. The power of the test statistic D can therefore be expressed as follows:

$$\varphi_{(r)} - \Phi\left\{\Delta_{(r)}\sqrt{\frac{N_2 N_1 Var_p(T)}{2\left\{(1-\rho) + r_\tau N_1 Var_p(T)\right\}}} - \Phi^{-1}(1-\alpha/2)\right\},$$

where $\Delta_{(r)} = |\delta_{(r)}/\sigma|$ and

$$r_\tau = \sigma_\tau^2/(\sigma_2^2 + \sigma_e^2) = \sigma_\tau^2/\sigma^2 \tag{5.17}$$

is the ratio of the random slope variance to the sum of the other variance terms. It follows that the required sample size per arm for the second level unit N_2 can be obtained as

$$N_2 = \frac{2\left\{(1-\rho) + r_\tau N_1 Var_p(T)\right\} z_{\alpha,\phi}^2}{N_1 Var_p(T)\Delta_{(r)}^2}. \tag{5.18}$$

The sample size for the level 1 data, N_1, needs to be determined in an iterative manner because $Var_p(T)$ is a function of N_1. The iterative solution for N_1 must satisfy the following equation:

$$N_1 = \frac{(1-\rho)}{Var_p(T)\left\{N_2 N_1 \Delta_{(r)}^2 \Big/ \left(2z_{\alpha,\phi}^2\right) - r_\tau\right\}}.$$

5.4.2.2   Unbalanced Allocations

Again, suppose that the random allocation needs to be unbalanced such that $\Sigma_i X_i = \lambda_2 N_2^{(0)}$ for $\lambda_2 = N_2^{(1)}/N_2^{(0)} > 0$ and $i = 1, 2, \ldots, (1 + \lambda_2)N_2^{(0)}$. Then, the sample size $N_2^{(0)}$ in the control arm can be expressed as

$$N_2^{(0)} = \frac{(1 + 1/\lambda_2)\left\{(1-\rho) + r_\tau N_1 Var_p(T)\right\} z_{\alpha,\phi}^2}{N_1 Var_p(T)\Delta_{(r)}^2}. \tag{5.19}$$

Subsequently, N_1 can be determined iteratively using the following equation:

$$N_1 = \frac{(1-\rho)}{Var_p(T)\left\{N_2^{(0)} \Delta_{(r)}^2 \Big/ \left\{(1 + 1/\lambda_2) z_{\alpha,\phi}^2\right\} - r_\tau\right\}}$$

TABLE 5.4
Sample size and power for detecting an effect $\delta_{(f)}$ on slope differences in a random-slope model (5.4.5) with $r_\tau = 0.1$ when randomizations occur at the second level (two-sided significance level $\alpha = 0.05$)

λ_2	ρ	$\Delta_{(r)}$	N_1	$N_2^{(0)}$	Total N	$\varphi_{(r)}$
1	0.1	0.4/4	5	299	2990	0.801
		0.4/8	9	723	13104	0.800
	0.2	0.4/4	5	283	2830	0.801
		0.4/8	9	712	12816	0.800
	0.1	0.5/4	5	191	1910	0.800
		0.5/8	9	463	8834	0.801
	0.2	0.5/4	5	181	1810	0.800
		0.5/8	9	456	8208	0.800
1.5	0.1	0.4/4	5	249	3113	0.801
		0.4/8	9	602	13545	0.800
	0.2	0.4/4	5	236	2950	0.801
		0.4/8	9	594	13365	0.801
	0.1	0.5/4	5	160	2000	0.802
		0.5/8	9	386	8685	0.801
	0.2	0.5/4	5	151	1888	0.801
		0.5/8	9	380	8550	0.800

Note: λ_2 represents the ratio of the number of level 2 units in the intervention arm to the corresponding number in the control arm; $\Delta_{(r)} = \left|\delta_{(r)}/\sigma\right|$; ρ (5.2) is the correlations among the level 1 data, often referred to as intra class correlations (ICC); N_1 is the number of level 1 units per level 2 unit; $N_2^{(0)}$ is the level 2 sample size in the *control* arm; Total $N = (1 + \lambda_2)N_1 N_2^{(0)}$ is the required total number of observations; $\varphi_{(r)}$ is the statistical power to test $H_0 : \delta_{(r)} = 0$.

5.4.2.3   Summary

The sample size N_2 in (5.18) or $N_2^{(0)}$ in (5.19) required for a desired magnitude of statistical power decreases with increasing N_1, and $\Delta_{(f)}$, and ρ, and increases with increasing r_τ. Table 5.4 displays examples of statistical power for testing the null hypothesis $H_0 : \delta_{(r)} = 0$. We note that additional sample size formulas assuming different correlation structures are presented in Section (4.3).

5.4.3   Comparison of Sample Sizes

The ratio $R_{r.f}(N_2)$ of sample size N_2 required with random slopes to that required for fixed slopes for balanced trials can be expressed as follows using Equations (5.14) and (5.18) for fixed Δ, φ, α, and ρ:

$$R_{r.f}(N_2; r_\tau, N_1, \rho) \equiv \frac{N_2 \left| r_\tau > 0 \right.}{N_2 \left| r_\tau = 0 \right.} = 1 + r_\tau N_1 Var_p(T)/(1 - \rho). \qquad (5.20)$$

This ratio holds even for unbalanced trials if $\lambda_1 = \lambda_2$. Under the assumption that the value of T increases from 0 to $T_{end} = N_1 - 1$ by unit time increments, the population variance of T reduces to $Var_p(T) = (N_1-1)(N_1+1)/12$ which yields

$$R_{\mathrm{r.f}}(N_2) = 1 + r_\tau N_1(N_1^2 - 1)/\{12(1 - \rho)\}. \tag{5.21}$$

Thus, $R_{r.f}(N_2)$ is an increasing function of r_τ, ρ, and N_1. For example, the effect of N_1 on the sample size ratio $R_{r.f}(N_2)$ is greatest for larger r_τ because the variance of the outcome increases quadratically with N_1 and the magnitude of the increase in the variance is larger for larger r_τ. The effect of r_τ is in fact enormous in that $R_{r.f}(N_2)$ is as high as 7.7 even for $r_\tau = 0.1$ for $N_1 = 9$, and $\rho = 0.1$, as also shown when comparing the results in Table 5.3 ($r_\tau = 0$) and in Table 5.4 ($r_\tau = 0.1$).

5.4.4 Testing Treatment Effects at the End of Follow-Up

For longitudinal trials, it may be of interest to test a treatment effect at a certain point in time, usually the end of follow-up (i.e., when $T_j = T_{end}$). Sample size requirements for testing such a local intervention effect at the end of the study can be derived from model (5.12) assuming that subject-specific slopes are fixed. Specifically, a contrast representing the intervention effect can be constructed on a shifted scale of the time variable $T'_j = T_j - T_{end}$ (increasing from $-T_{end}$ ($=1 - N_1$) to 0 by 1) as follows:

$$Y_{ij} = \beta'_0 + \delta_{(e)} X_i + \tau' l'_j + \delta_{(f)} X_i l'_j + u_i + e_{ij}, \tag{5.22}$$

where $\beta'_0 = \beta_0 + \tau T_{end}$ and $\delta_{(e)} = \xi + \delta_{(f)} T_{end}$. Since $T_j = T_{end}$ is equivalent to $T'_j = 0$, the re-parameterized intervention effect $\delta_{(e)}$ in model (5.22) represents the intervention effect at the end of the study, and is geometrically depicted in Figure 5.1. Here, however, we assume for the purpose of deriving the power function that the parameter ξ in model (5.12) representing the intervention effect at baseline is zero, which is plausible due to randomization.

5.4.4.1 Balanced Allocations

The null hypothesis to be tested is $H_0 : \delta_{(e)} = 0$. Randomized treatment assignment implies the mean difference in outcome Y at baseline between the two arms is zero, that is, $\xi = 0$ in model (5.12). Under this assumption, the null hypothesis can be tested by

$$D_e = \frac{\sqrt{N_2 N_1}\hat{\delta}_{(e)}}{\sigma\sqrt{2fC_{(2)}}},$$

where $f = 1 + (N_1 - 1)\rho$, $\hat{\delta}_{(e)} = (\bar{Y}_1 - \bar{Y}_0) - \hat{\delta}_{(f)}\bar{T}'$, $\bar{T}' = \sum_{j=1}^{N_1} T'_j/N_1$, $\hat{\delta}_{(f)} = \hat{\eta}_1 - \hat{\eta}_0$, and $C_{(2)} = 1 + (1 - \rho)CV^{-2}(T')/f$; $CV(T') = SD_p(T')/\bar{T}'$ is the coefficient of variation (CV) of the time variable T', and $SD_p(T')$ is a "population" standard deviation, the square root of $Var_p(T')$. The variance of $\hat{\delta}_{(e)}$, an unbiased estimate of $\delta_{(e)}$, can be computed as

$$Var(\hat{\delta}_{(e)}) = Var\left(\bar{Y}_1 - \bar{Y}_0\right) + \bar{T}'^2 Var(\hat{\delta}_{(f)}) = \frac{2fC_{(2)}\sigma^2}{N_2 N_1}.$$

Accordingly, the statistical power of the test statistic D_e can be written as follows:

$$\varphi_{(e)} = \Phi\left\{\Delta_{(e)}\sqrt{\frac{N_2 N_1}{2fC_{(2)}}} - \Phi^{-1}(1-\alpha/2)\right\},$$

where $\Delta_{(e)} = |\delta_{(e)}/\sigma|$ is a standardized effect size.

The second level unit sample size N_2 *per arm* (i.e., the number of subjects per treatment arm) for a desired statistical power $\varphi_{(e)} = \phi$ can be obtained as

$$N_2 = \frac{2fC_{(2)}z_{\alpha,\phi}^2}{N_1\Delta_{(e)}^2}. \tag{5.23}$$

The sample size N_1 for the level 1 data should, however, be an iterative solution of the following equation since $CV(T')$ is a function of N_1:

$$N_1 = \frac{2\left(1-\rho\right)\left(1+CV^{-2}(T')\right)z_{\alpha,\phi}^2}{N_2\Delta_{(e)}^2 - 2\rho z_{\alpha,\phi}^2}.$$

Note that elongation of the time intervals will not affect the sample size (5.23) so that the required sample sizes with time intervals $t = T_k - T_{k-1}$ for all k will be the same as that with time intervals ωt for any $\omega > 0$.

5.4.4.2   Unbalanced Allocations

Again, suppose that the random allocation needs to be unbalanced such that $\Sigma_i X_i = \lambda_2 N_2^{(0)}$ for $\lambda_2 > 0$ and $i = 1, 2, \ldots, (1+\lambda_2)N_2^{(0)}$ Then, the sample size $N_2^{(0)}$ in the control arm can be expressed as

$$N_2^{(0)} = \frac{fC\left(1+1/\lambda_2\right)z_{\alpha,\phi}^2}{N_1\Delta_{(e)}^2}. \tag{5.24}$$

And subsequently N_1 for a given N_2 can be determined by iteratively solving the following equation:

$$N_1 = \frac{(1+1/\lambda_2)\left\{1-\rho+(1-\rho)CV^{-2}(T')\right\}z_{\alpha,\phi}^2}{N_2\Delta_{(e)}^2 - \rho\left(1+1/\lambda_2\right)z_{\alpha,\phi}^2}.$$

5.4.4.3   Summary

The sample size N_2 in (5.23) or $N_2^{(0)}$ in (5.24) required for a desired magnitude of statistical power decreases with increasing N_1, and $\Delta_{(e)}$, and increases with

TABLE 5.5

Sample size and power for detecting a main effect $\delta_{(e)}$ at the end of study in a fixed-slope model (5.22) when randomizations occur at the second level (two-sided significance level $\alpha = 0.05$)

λ_2	$\Delta_{(e)}$	ρ	N_1	$N_2^{(0)}$	$\varphi_{(e)}$
1	0.4	0.1	5	63	0.801
			9	44	0.807
		0.2	5	67	0.802
			9	50	0.806
	0.5	0.1	5	41	0.808
			9	28	0.805
		0.2	5	43	0.803
			9	32	0.806
1.5	0.4	0.1	5	53	0.804
			9	36	0.800
		0.2	5	56	0.803
			9	42	0.809
	0.5	0.1	5	34	0.806
			9	24	0.816
		0.2	5	36	0.805
			9	27	0.811

Note: λ_2 represents the ratio of the number of level 2 units in the intervention arm to the corresponding number in the control arm; $\Delta_{(e)} = \left| \delta_{(e)}/\sigma \right|$; ρ (5.2) is the correlations among the level 1 data, often referred to as intra class correlations (ICC); N_1 is the number of level 1 units per level 2 units; $N_2^{(0)}$ is the level 2 sample size in the *control* arm; $\varphi_{(e)}$ is the statistical power to test $H_0 : \delta_{(e)} = 0$.

increasing ρ. The sample size N_2 in (5.23) or $N_2^{(0)}$ in (5.24) is smaller for the same statistical power than that required for a test such as a t-test under a parallel group design with single level data. Table 5.5 displays examples of statistical power for testing the null hypothesis $H_0 : \delta_{(e)} = 0$.

5.5 Cross-Sectional Factorial Designs: Interactions between Treatments

In some circumstances, clinical trials use a 2×2 factorial design with two experimental treatments, X and Z, each with two-levels (0 and 1). This results in four treatment arms: X alone, Z alone, both X and Z, and neither X nor Z (control). A key question of interest may be whether the effect of the combination of the two experimental treatments is larger than what would be expected if the treatment effects were purely additive, that is, whether there is evidence of an interaction effect between X and Z on the study outcome. Fleiss [15] derived sample size determinations for such trials with single level data. In this section, we extend Fleiss' derivation to factorial designs with a two-level data structure. As we will show, the sample sizes required for testing an interaction effect are four times larger than that required for testing a main effect, whether for one or two level data structures. Randomization can occur at the second level (Section 5.5.1) or at the first level (Section 5.5.2). The comparison of sample sizes between different levels of randomization is discussed in Section 5.5.3.

5.5.1   Randomization at the Second Level

When the unit of randomization is the second level unit for both X and Z, a mixed-effects linear regression model can be constructed as follows:

$$Y_{ij} = \beta_0 + \delta_{X(2)}X_{ij} + \delta_{Z(2)}Z_{ij} + \delta_{XZ(2)}X_{ij}Z_{ij} + u_i + e_{ij}. \tag{5.25}$$

The intervention assignment indicator variable $X_i = 0$ if the ith level 2 unit is assigned to a control intervention and $X_i = 1$ if assigned to an experimental intervention; therefore $X_{ij} = X_i$ for all j. Likewise, we define another intervention assignment indicator variable $Z_i = 0$ if the ith level 2 unit is assigned to a control intervention and $Z_i = 1$ if assigned to an experimental intervention; therefore $Z_{ij} = Z_i$ for all j.

5.5.1.1   Balanced Allocations

For a balanced design it is assumed that $\Sigma_i X_i Z_i = N_2^{(1,1)}$ so that $i = 1, 2, \ldots, 4N_2^{(1,1)}$ and $j = 1, \ldots, N_1$. Specifically, $N_2^{(1,1)}$ represents the level 2 sample size assigned to $(X = 1, Z = 1)$ group, or the $(1,1)$ cell in a 2-by-2 factorial table of X and Z, and N_1 is the level 1 sample size per level 2 units. In a balanced design, the sample sizes of the other cells are identical to $N_2^{(1,1)}$, i.e., $N_2^{(1,1)} = N_2^{(1,0)} = N_2^{(0,1)} = N_2^{(0,0)}$. Here, the null hypothesis to be tested is $H_0 : \delta_{XZ(2)}x = 0$ which can be evaluated using the following test statistic test:

$$D_{XZ(2)} = \frac{\sqrt{N_2^{(0,0)}N_1}\left\{\left(\bar{Y}^{(1,1)} - \bar{Y}^{(1,0)}\right) - \left(\bar{Y}^{(0,1)} - \bar{Y}^{(0,0)}\right)\right\}}{\sigma\sqrt{4f}},$$

where $\bar{Y}^{(x,z)}$ is the mean of Y in $(X = x, Z = z)$ cells. We note that $\hat{\delta}_{XZ(2)} = \left(\bar{Y}^{(1,1)} - \bar{Y}^{(1,0)}\right) - \left(\bar{Y}^{(0,1)} - \bar{Y}^{(0,0)}\right)$ is an unbiased estimate of $\delta_{XZ(2)}$ and

$$Var\left(\hat{\delta}_{XZ(2)}\right) = \frac{4f\sigma^2}{N_2^{(0,0)}N_1}.$$

The corresponding statistical power can be expressed as

$$\varphi_{xz(2)} = \Phi\left\{\Delta_{XZ(2)}\sqrt{N_2^{(0,0)}N_1\big/4f} - \Phi^{-1}(1 - \alpha/2)\right\}.$$

where $\Delta_{XZ(2)} = \left|\delta_{XZ(2)}/\sigma\right|$. Accordingly, the corresponding required sample size is [16]:

$$N_2^{(0,0)} = \frac{4fz_{\alpha,\phi}^2}{N_1\Delta_{XZ(2)}^2}. \tag{5.26}$$

It is notable that when Equations (5.5) and (5.26) are compared, it is clear that $N_2^{(0,0)} = 2N_2^{(0)}$, where $N_2^{(0)}$ is the required sample size for the control arm ($X = 0$) for testing $H_0 : \delta_{(2)} = 0$ in model (5.3). If follows that the total number of measurements, that is, $4N_2^{(0,0)}N_1$ required for testing $H_0 : \delta_{XZ(2)} = 0$ in model (5.25) is four times larger than the total number of measurements, $2N_2^{(0)}N_1$, required for testing $H_0 : \delta_{(2)} = 0$. In other words, the required sample sizes for testing an interaction is four times larger than that required for testing a main effect. Sample sizes of the first level units per cell can be determined by solving (5.26) for N_1, that is,

$$N_1 = \frac{4z_{\alpha,\phi}^2}{N_2^{(0,0)}\Delta_{XZ(2)}^2 - 4(1 - \rho)z_{\alpha,\phi}^2}.$$

5.5.1.2 Unbalanced Allocations

Under an unbalanced design where $N_2^{(0,1)} = \lambda_{01}N_2^{(0,0)}, N_2^{(1,0)} = \lambda_{10}N_2^{(0,0)}$, and $N_2^{(1,1)} = \lambda_{11}N_2^{(0,0)}$, the sample size $N_2^{(0,0)}$ can be determined as

$$N_2^{(0,0)} = \frac{f\left(1 + 1/\lambda_{11} + 1/\lambda_{10} + 1/\lambda_{01}\right)z_{\alpha,\phi}^2}{N_1\Delta_{XZ(2)}^2} \tag{5.27}$$

Even under unbalanced allocations, the total number measurements required for testing $H_0 : \delta_{XZ(2)} = 0$ in model (5.25) is either approximately or exactly four times larger than the total number of measurements required for testing $H_0 : \delta_{XZ(2)} = 0$ in model (5.3) depending on the values of λ_2 in Section 5.3.1.2 and the values of $\lambda = (\lambda_{01}\lambda_{10}, \lambda_{11})$ in this section.

5.5.1.3 Summary

The sample size $N_2^{(0,0)}$ in (5.26) or in (5.27) required for a desired magnitude of statistical power decreases with increasing N_1, and $\Delta_{XZ(2)}$, and increases with increasing ρ. Table 5.6 displays examples of statistical power for testing the null hypothesis $H_0 : \delta_{XZ(2)} = 0$ and can be compared with Table 5.1 to confirm that the required samples size for testing the interaction is four times that required for testing the main effect.

TABLE 5.6
Sample size and power for detecting a two-way interaction XZ effect $\delta_{XZ(2)}$ in model (5.25) for a 2-by-2 factorial design when randomizations occur at the second level (two-sided significance level $\alpha = 0.05$)

$\lambda_{01}\lambda_{10}, \lambda_{11}$	$\Delta_{XZ(2)}$	ρ	N_1	$N_2^{(0,0)}$	Total N	$\varphi_{XZ(2)}$
1,1,1	0.4	0.1	10	38	1520	0.807
			20	29	2320	0.807
		0.2	10	55	2200	0.800
			20	48	3840	0.807
	0.5	0.1	10	24	960	0.802
			20	19	1520	0.816
		0.2	10	36	1440	0.809
			20	31	2480	0.811
1.5,1.5,1.5	0.4	0.1	10	28	1540	0.801
			20	22	2420	0.812
		0.2	10	42	2310	0.807
			20	36	3960	0.807
	0.5	0.1	10	18	990	0.802
			20	14	1540	0.810
		0.2	10	27	1485	0.809
			20	23	2530	0.807

Note: $N_2^{(0,1)} = \lambda_{01} N_2^{(0,0)}$; $N_2^{(1,0)} = \lambda_{10} N_2^{(0,0)}$; $N_2^{(1,1)} = \lambda_{11} N_2^{(0,0)}$; $\Delta_{XZ(2)} = |\delta_{XZ(2)}/\sigma|$; ρ (5.2) is the correlations among the level 1 data, often referred to as intra class correlations (ICC); N_1 is the number of level 1 units per level 2 units; $N_2^{(0,0)}$ is the level 2 sample size in the $(X = 0, Z = 0)$ arm; Total $N = (1 + \lambda_{01} + \lambda_{10} + \lambda_{11}) N_1 N_2^{(0,0)}$ is the required total number of observations; $\varphi_{XZ(2)}$ is the statistical power to test $H_0 : \delta_{XZ(2)} = 0$.

5.5.2   Randomization at the First Level

When the random allocations of the intervention arms are assigned at the first level units for both X and Z, i.e., when the 2-by-2 factorial design is implemented within each second level unit, a mixed-effects linear regression model can be constructed as follows:

$$Y_{ij} = \beta_0 + \delta_{X(1)} X_{ij} + \delta_{Z(1)} Z_{ij} + \delta_{XZ(1)} X_{ij} Z_{ij} + u_i + e_{ij}. \tag{5.28}$$

All of the results from the second level randomization model can be extended to the first level randomization model.

5.5.2.1   Balanced Allocations

Here, we assume that $X_{ij} = X_j$ and $Z_{ij} = Z_j$ for all i and that $\Sigma_j X_j Z_j = N_1^{(1,1)}$ so that $i = 1, 2, \ldots, N_2$ and $j = 1, \ldots, 4N_1^{(1,1)}$. Specifically, N_2 is the level 2 sample size *in total*, and $N_1^{(1,1)}$ represents the level 1 sample size assigned to $(X = 1, Z = 1)$ arm, or $(1,1)$ cell in a 2-by-2 factorial table of X and Z. In a balanced design, where the sample sizes of all cells are identical, i.e., $N_1^{(1,1)} = N_1^{(1,0)} = N_1^{(0,1)} = N_1^{(0,0)}$, the required sample size for N_2 for testing $H_0 : \delta_{XZ(1)} = 0$ can be obtained using the following test statistic:

$$D_{XZ(1)} = \frac{\sqrt{N_2 N_1^{(0,0)}} \left\{ \left(\bar{Y}^{(1,1)} - \bar{Y}^{(1,0)}\right) - \left(\bar{Y}^{(0,1)} - \bar{Y}^{(0,0)}\right) \right\}}{\sigma \sqrt{4(1-\rho)}}.$$

Note that $\hat{\delta}_{XZ(1)} = \left(\bar{Y}^{(1,1)} - \bar{Y}^{(1,0)}\right) - \left(\bar{Y}^{(0,1)} - \bar{Y}^{(0,0)}\right)$ is an unbiased estimate of $\delta_{XZ(1)}$ and

$$Var\left(\hat{\delta}_{XZ(1)}\right) = \frac{4(1-\rho)\sigma^2}{N_2 N_1^{(0,0)}}.$$

The corresponding statistical power can be expressed as

$$\varphi_{XZ(1)} = \Phi \left\{ \Delta_{XZ(1)} \sqrt{N_2 N_1^{(0,0)} \big/ 4(1-\rho)} - \Phi^{-1}(1 - \alpha/2) \right\},$$

for $\Delta_{XZ(1)} = |\delta_{XZ(1)}/\sigma|$. It follows that

$$N_2 = \frac{4(1-\rho)z_{\alpha,\phi}^2}{N_1^{(0,0)} \Delta_{XZ(1)}^2}, \tag{5.29}$$

and

$$N_1^{(0,0)} = \frac{4(1-\rho)z_{\alpha,\phi}^2}{N_2 \Delta_{XZ(1)}^2}.$$

Again, this sample size is twice as large as $N_2^{(0)}$ for testing $H_0 : \delta_{(1)} = 0$ in model (5.8), which implies that testing $H_0 : \delta_{XZ(1)} = 0$ in model (5.28) requires four times the total number of measurements required for testing $H_0 : \delta_{(1)} = 0$.

5.5.2.2 Unbalanced Allocations

Under an unbalanced design where $N_1^{(0,1)} = \lambda_{01} N_1^{(0,0)}$, $N_1^{(1,0)} = \lambda_{10} N_1^{(0,0)}$, and $N_1^{(1,1)} = \lambda_{11} N_1^{(0,0)}$, the sample size N_2 and $N_1^{(0,0)}$ can be determined as:

$$N_2 = \frac{(1-\rho)(1 + 1/\lambda_{11} + 1/\lambda_{10} + 1/\lambda_{01}) z_{\alpha,\phi}^2}{N_1^{(0,0)} \Delta_{XZ(1)}^2} \tag{5.30}$$

and

$$N_1^{(0,0)} = \frac{(1-\rho)(1 + 1/\lambda_{11} + 1/\lambda_{10} + 1/\lambda_{01}) z_{\alpha,\phi}^2}{N_2 \Delta_{XZ(1)}^2}.$$

Depending on the values of λ_1 in Section 5.3.2.2 and the values of $\lambda = (\lambda_{01}\lambda_{10}, \lambda_{11})$ in this section, the total number of measurements required for testing $H_0 : \delta_{(1)}^{xz} = 0$ in model (5.28) is approximately or exactly four times larger than the total number of measurements required for testing $H_0 : \delta_{(1)} = 0$ in model (5.8). Furthermore, the statistical power $\varphi_{XZ(1)}$ is invariant over the product of $N_1^{(0,0)} N_2$ regardless of whether a design is balanced or unbalanced.

5.5.2.3 Summary

The sample size N_2 in (5.29) or in (5.30) required for a desired magnitude of statistical power decreases with increasing $N_1^{(0,0)}$, $\Delta_{XZ(1)}$, and ρ unlike the case with a 2-by-2 factorial design with second-level randomization. Table 5.6 displays examples of statistical power for testing the null hypothesis $H_0 : \delta_{XZ(1)} = 0$, which also shows that the invariance over the product of $N_1^{(0,0)} N_2$ property holds. The four-fold increase in sample size required for testing interaction effect versus the main effect can be confirmed by comparing the results in Tables 5.2 and 5.7.

5.5.3 Comparisons of Sample Sizes

The ratio between the first and the second level randomizations of the total numbers of observations required for testing the interaction effects of the two main factors is identical to the corresponding ratio for testing main effects: $R_{2.1} = f_1/(1 - \rho) = 1 + N_1\rho/(1 - \rho)$ in equation (5.3.9) This holds for balanced designs from equations (5.26) and (5.29) and unbalanced designs with equal values of λ's from equations (5.27) and (5.30) between the first and the second level randomizations. Thus, as we have shown above, trials with second level randomization require a large sample size compared to first level randomization for the same values of Δ, φ, α, and correlation ρ when $\rho \neq 0$. This is also evident when comparing the results in Tables 5.6 and 5.7.

5.6 Longitudinal Factorial Designs: Treatment Effects on Slopes

For trials using a 2×2 factorial design in which subjects are randomly assigned to one of the four treatment arms and then repeatedly evaluated for outcomes over time, it may be of interest to test whether the trends in the outcome over the study period (i.e., slopes) in the combined treatment arm is beyond what would be expected if the effects of X and Z are additive. This hypothesis can be evaluated by including a *three-way interaction term* between the two

TABLE 5.7

Sample size and power for detecting a two-way interaction XZ effect $\delta_{XZ(1)}$ in model (5.28) for a 2-by-2 factorial design when randomizations occur at the first level (two-sided significance level $\alpha = 0.05$)

$\lambda_{01}\lambda_{10}, \lambda_{11}$	$\Delta_{XZ(1)}$	ρ	$N_1^{(0,0)}$	N_2	Total N	$\varphi_{XZ(1)}$
1,1,1	0.4	0.1	10	18	720	0.807
			20	9	720	0.807
		0.2	10	16	640	0.807
			20	8	640	0.807
	0.5	0.1	10	12	480	0.823
			20	6	480	0.823
		0.2	10	12	480	0.865
			20	6	480	0.865
1.5,1.5,1.5	0.4	0.1	10	14	770	0.821
			20	7	770	0.821
		0.2	10	12	660	0.807
			20	6	660	0.807
	0.5	0.1	10	10	550	0.861
			20	5	550	0.861
		0.2	10	8	440	0.823
			20	4	440	0.823

Note: $N_1^{(0,1)} = \lambda_{01}N_1^{(0,0)}$; $N_1^{(1,0)} = \lambda_{10}N_1^{(0,0)}$; $N_1^{(1,1)} = \lambda_{11}N_1^{(0,0)}$; $\Delta_{XZ(1)} = |\delta_{XZ(1)}/\sigma|$; ρ (5.2) is the correlations among the level 1 data, often referred to as intra class correlations (ICC); $N_1^{(0,0)}$ is the level 1 sample size in the $(X = 0, Z = 0)$ arm within second-level units; N_2 is the number of level 2 units; Total $N = (1 + \lambda_{01} + \lambda_{10} + \lambda_{11})N_1^{(0,0)}N_2$ is the required total number of observations; $\varphi_{XZ(2)}$ is the statistical power to test $H_0 : \delta_{XZ(2)} = 0$.

treatments and time in a linear mixed-effects model as follows:

$$Y_{ij} = \beta_0 + \delta_X X_{ij} + \delta_Z Z_{ij} + \delta_T T_{ij} + \delta_{XZ} X_{ij} Z_{ij}$$
$$+ \delta_{XT} X_{ij} T_{ij} + \delta_{ZT} Z_{ij} T_{ij} + \delta_{XZT} X_{ij} Z_{ij} T_{ij} + \nu_i T_{ij} + u_i + e_{ijk} \quad (5.31)$$

The slopes can be considered fixed ($\sigma_\tau^2 = 0$) or random ($\sigma_\tau^2 > 0$), as in models (5.12) and (5.4.5), respectively, for the case of two-way interactions between X and T The slope parameters in model (5.31) are geometrically depicted in Figure 5.2.

5.6.1 Balanced Allocations

As in section (5.5.1.1), for a balanced design it is assumed that $\Sigma_i X_i Z_i = N_2^{(1,1)}$ so that $i = 1, 2, \ldots, 4N_2^{(1,1)}$ and $j = 1, \ldots, N_1$, where $N_2^{(1,1)}$ represents the level 2 sample size assigned to $(X = 1, Z = 1)$ group, or the (1,1) cell in a 2-by-2 factorial table of X and Z, and N_1 is the level 1 sample size per level

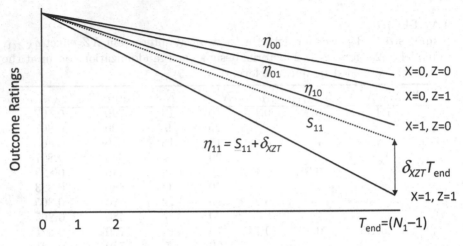

FIGURE 5.2

Geometrical representations of fixed parameters in model (5.31) for a 2-by-2 factorial longitudinal cluster randomized trial.

Note: $\eta_{00} = \delta_T;\; \eta_{01} = \eta_{00} + \delta_{ZT} = \delta_T + \delta_{ZT}\eta_{10} = \eta_{00} + \delta_{XT} = \delta_T + \delta_{XT};\; S_{11} = \eta_{00} + \delta_{XT} + \delta_{ZT} = \delta_T + \delta_{XT} + \delta_{ZT}$ represents a hypothetical slope for the $(X = 1, Z = 1)$ arm under the null hypothesis $H_0 : \delta_{XZT} = 0$ in model (5.31); and $\eta_{11} = \eta_{00} + \delta_{XT} + \delta_{ZT} + \delta_{XZT} = S_{11} + \delta_{XZT}$.

2 units. In a balanced design, the sample sizes of the other cells are identical to $N_2^{(1,1)}$, i.e., $N_2^{(1,1)} = N_2^{(1,0)} = N_2^{(0,1)} = N_2^{(0,0)}$ It is further assumed that $T_{ij} = T_j$ for all i, and the length of the intervals between time points is a unit increment from 0 (the baseline) to $T_{end} = N_1 - 1$ (the last time point). The relevant null hypothesis in model (5.31) is $H_0 : \delta_{XZT} = 0$, which can be evaluated using the following statistic:

$$D_{XZT} = \frac{\left\{\left(\hat{\eta}^{(1,1)} - \hat{\eta}^{(1,0)}\right) - \left(\hat{\eta}^{(0,1)} - \hat{\eta}^{(0,0)}\right)\right\}\sqrt{N_2^{(0,0)} N_1 Var_p(T)}}{\sqrt{4\{(1-\rho)\sigma^2 + N_1 Var_p(T)\sigma_\tau^2\}}},$$

where $\hat{\eta}^{(x,z)}$ is the OLS estimate of slope in the $(X = x, Z = z)$ cell; $\hat{\delta}_{XZT} = \left(\hat{\eta}^{(1,1)} - \hat{\eta}^{(1,0)}\right) - \left(\hat{\eta}^{(0,1)} - \hat{\eta}^{(0,0)}\right)$ is an unbiased estimate of δ_{XZT} and

$$Var\left(\hat{\delta}_{XZT}\right) = \frac{4\{(1-\rho)\sigma^2 + N_1 Var_p(T)\sigma_\tau^2\}}{N_2^{(0,0)} N_1 Var_p(T)}.$$

A power function can be derived from D_{XZT} as follows:

$$\varphi_{XZT} = \Phi\left\{\Delta_{XZT}\sqrt{\frac{N_2^{(0,0)} N_1 Var_p(T)}{4\{(1-\rho) + r_\tau N_1 Var_p(T)\}}} - \Phi^{-1}(1 - \alpha/2)\right\},$$

where $\Delta_{XZT} = |\delta_{XZT}/\sigma|$ and r_τ in Equation (5.17) is the ratio of the random slope variance to the sum of the other variances ($r_\tau = 0$ for the fixed slope models). It follows that the required sample size per arm for the second level unit N_2 can be obtained as [17]

$$N_2^{(0,0)} = \frac{4\{(1-\rho) + r_\tau N_1 Var_p(T)\} z_{\alpha,\phi}^2}{N_1 Var_p(T)\Delta_{XZT}^2}. \tag{5.32}$$

By comparing the required sample size $N_2^{(0,0)}$ in (5.32) with the corresponding sample sizes $N_2^{(0)}$ in (5.14) and (5.18), it is evident that testing $H_0 : \delta_{XZT} = 0$ under a fixed or random slope model requires four times the total number of observations required for testing the effect of the two-way interaction (XT), i.e., $H_0 : \delta_{(f)} = 0$ in model (5.12) or $H_0 : \delta_{(r)} = 0$ in model (5.4.5), respectively.

5.6.2   Unbalanced Allocations

Under unbalanced designs where $N_2^{(0,1)} = \lambda_{01} N_2^{(0,0)}$ $N_2^{(1,0)} = \lambda_{10} N_2^{(0,0)}$ and $N_2^{(1,1)} = \lambda_{11} N_2^{(0,0)}$, the sample size $N_2^{(0,0)}$ can be determined as

$$N_2^{(0,0)} = \frac{(1 + 1/\lambda_{11} + 1/\lambda_{10} + 1/\lambda_{01})\{(1-\rho) + r_\tau N_1 Var_p(T)\} z_{\alpha,\phi}^2}{N_1 Var_p(T)\Delta_{XZT}^2} \tag{5.33}$$

5.6.3   Summary

The sample size $N_2^{(0,0)}$ in (5.32) or in (5.33) required for a desired magnitude of statistical power decreases with increasing N_1, and Δ_{XZT}, and also decreases with increasing ρ. However, $N_2^{(0,0)}$ increases with increasing r_τ, which again has an enormous effect on the sample size. Table 5.7 displays examples of statistical power for testing the null hypothesis $H_0 : \delta_{XZT} = 0$. The four-fold increase in samples size can be confirmed by comparing the results in Table 5.3 ($r_\tau = 0$) and Table 5.4 ($r_\tau = 0.1$) with those in Table 5.8.

5.6.4   Comparison of Sample Sizes

The ratio $R_{r.f}(N_2^{(0,0)})$ of sample sizes $N_2^{(0,0)}$ in Equation (5.32) between fixed ($r_\tau = 0$) and random slope ($r_\tau > 0$) models for testing the significance of δ_{XZT} in model (5.31) is identical to the ratio $R_{r.f}(N_2)$ in Equation (5.20) for testing slope differences between arms. That is,

$$R_{r.f}(N_2^{(0,0)}; r_\tau, N_1, \rho) \equiv \frac{N_2^{(0,0)}|r_\tau > 0}{N_2^{(0,0)}|r_\tau = 0} = 1 + r_\tau N_1 Var_p(T)/(1-\rho)$$

$$= R_{r.f}(N_2; r_\tau, N_1, \rho)$$

This ratio $R_{r.f}(N_2^{(0,0)})$ holds for balanced designs and also for unbalanced designs with equal values of λ's. In summary, the same properties of $R_{r.f}(N_2)$

TABLE 5.8

Sample size and statistical power for detecting a three-way interaction XZT effect δ_{XZT} in model (5.31) for a 2-by-2 factorial design when randomizations occur at the second level (two-sided significance level $\alpha = 0.05$)

$\lambda_{01}\lambda_{10}, \lambda_{11}$	r_τ	Δ_{XZT}	ρ	N_1	$N_2^{(0,0)}$	Total N	φ_{XZT}
1,1,1	0	0.4/4	0.1	5	283	5660	0.801
			0.2	5	252	5040	0.801
		0.5/8	0.1	9	121	4356	0.801
			0.2	9	108	3888	0.803
	0.1	0.4/4	0.1	5	597	11940	0.800
			0.2	5	566	11320	0.801
		0.5/8	0.1	9	925	33300	0.800
			0.2	9	911	32796	0.800
1.5,1.5,1.5	0	0.4/4	0.1	5	212	5830	0.800
			0.2	5	189	5198	0.801
		0.5/8	0.1	9	91	4505	0.803
			0.2	9	81	4010	0.803
	0.1	0.4/4	0.1	5	448	12320	0.801
			0.2	5	424	11660	0.800
		0.5/8	0.1	9	694	34353	0.800
			0.2	9	684	33858	0.800

Note: $N_2^{(0,1)} = \lambda_{01} N_2^{(0,0)}$; $N_2^{(1,0)} = \lambda_{10} N_2^{(0,0)}$; $N_2^{(1,1)} = \lambda_{11} N_2^{(0,0)}$; $\Delta_{XZT} = |\delta_{XZT}/\sigma|$; ρ (5.2) is the correlations among the level 1 data, often referred to as intra class correlations (ICC); N_1 is the number of level 1 units per level 2 units; $N_2^{(0,0)}$ is the level 2 sample size in the $(X = 0, Z = 0)$ arm; Total $N = (1 + \lambda_{01} + \lambda_{10} + \lambda_{11}) N_1 N_2^{(0,0)}$ is the required total number of observations; φ_{XZT} is the statistical power to test $H_0 : \delta_{XZT} = 0$.

discussed in Section 5.4.3 also apply to $R_{r.f}(N_2^{(0,0)})$, which is evident by comparing the results in Table 5.8 for $r_\tau = 0$ and $r_\tau = 0.1$.

5.7　Sample Sizes for Binary Outcomes

In many trials, the primary outcome variable of interest is binary. Examples include remission status of certain symptoms like depression, termination (yes/no) of risky behaviors like smoking, and presence/absence of specific conditions or diseases like diabetes. If those binary outcomes have a two level data structure, correlations among the first level data units within the second level data units should be taken into account when designing the study. Approaches for sample size determination for trials with binary outcomes are essentially an extension of the methods for continuous outcomes presented earlier and are based on large sample normal theory approximations [18].

When randomization is at the mth ($m = 1, 2$) level, we consider the following mixed-effects logistic regression model for the binary outcome Y_{ij}:

$$\log\left(\frac{p_{ij}}{1 - p_{ij}}\right) = \beta_0 + \xi_{(m)}X_{ij} + u_i, \tag{5.34}$$

where $p_{ij} = E(Y_{ij}|X_{ij})$ and $u_i \sim N(0, \sigma_u^2)$. We further assume that $p_{ij} = p$ if $X_{ij} = 0$ and $p_{ij} = p_1$ if $X_{ij} = 1$. Therefore, $Var(Y_{ij}|X_{ij} = 0) = p(1 - p)$ and $Var(Y_{ij}|X_{ij} = 1) = p_1(1 - p_1)$. Under model (5.34), the correlation can be computed as [19]

$$\rho \equiv Corr(Y_{ij}, Y_{ij'}) = \frac{\sigma_u^2}{\sigma_u^2 + \pi^2/3} \tag{5.35}$$

The null hypothesis of interest can be expressed as $H_0 : \xi_{(m)} = 0$ or equivalently, $H_0 : p - p_1 = 0$. Sections 5.7.1 and 5.7.2 present sample size calculations when randomization occurs at the second ($m = 2$) and first ($m = 1$) levels, respectively. The comparison of sample sizes under the two approaches is discussed in Section 5.7.3.

5.7.1 Randomization at the Second Level

Let us denote the level 2 sample sizes for the control ($X = 0$) and intervention ($X = 1$) arms by $N_2^{(0)}$ and $N_2^{(1)}$ while N_1 represents the level 1 sample size per level 2 units.

5.7.1.1 Balanced Allocations

When randomization occurs at the second level in a balanced design, i.e., when $m = 2$ and $X_{ij} = X_i$ for all j, the indices i and j range as follows: $i = 1, 2, \ldots, 2N_2^{(0)}$, and $j = 1, \ldots, N_1$, yielding $\Sigma_i X_i = N_2^{(1)}$. Therefore, in this case $N_2 = N_2^{(0)} = N_2^{(1)}$ is the number of second level units per treatment arm, and N_1 is the sample size per second level unit. The null hypothesis to be tested is $H_0 : \xi_{(2)} = 0$ in model (5.34). Extending formula (5.5) for determining N_2 for testing $H_0 : \delta_{(2)} = 0$ in model (5.3) for continuous outcomes to the binary outcome case with unequal variances under the null and alternative hypotheses, an approximate sample size for N_2 for a two-sided type I error rate α and a statistical power $\varphi = \phi$ can be computed as [10, 18]

$$N_2 = \frac{f z_{\alpha,\phi,p}^2}{N_1 (p_1 - p_0)^2} \tag{5.36}$$

from a power function

$$\varphi = \Phi\left\{\frac{|p_1 - p_0|\sqrt{N_1 N_2/f} - \Phi^{-1}(1 - \alpha/2)\sqrt{2\bar{p}(1 - \bar{p})}}{\sqrt{p_0(1 - p_0) + p_1(1 - p_1)}}\right\}$$

where $\bar{p} = (p_0 + p_1)/2$, $f = 1 + (N_1 - 1)\rho$ in equation (5.3.2) and

$$z_{\alpha,\phi,p} = \Phi^{-1}\left(1 - \alpha/2\right)\sqrt{2\bar{p}(1 - \bar{p})} + \Phi^{-1}\left(\phi\right)\sqrt{p_0\left(1 - p_0\right) + p_1\left(1 - p_1\right)}.$$

The sample size for N_1 can be obtained by solving equation (5.36) for N_1. Another formula is presented in Equation (2.13) in Section (2.5.1).

5.7.1.2   Unbalanced Allocations

For an unbalanced design with $\Sigma_i X_i = \lambda N_2^{(0)}$ for $\lambda > 0$ and $i = 1, 2, \ldots, (1 + \lambda)N_2^{(0)}$, the sample size $N_2^{(0)}$ for the control arm can be obtained as [18]:

$$N_2^{(0)} = \frac{f z_{\alpha,\phi,\lambda,p}^2}{N_1\left(p_1 - p_0\right)^2} \tag{5.37}$$

from a power function

$$\varphi = \Phi\left\{\frac{|p_1 - p_0|\sqrt{N_1 N_2^{(0)}/f} - \Phi^{-1}(1 - \alpha/2)\sqrt{(1 + 1/\lambda)\,\bar{p}\,(1 - \bar{p})}}{\sqrt{p_0\left(1 - p_0\right) + p_1\left(1 - p_1\right)/\lambda}}\right\}$$

where

$$z_{\alpha,\phi,p,\lambda} = \Phi^{-1}\left(1 - \alpha/2\right)\sqrt{(1 + 1/\lambda)\,\bar{p}_\lambda(1 - \bar{p}_\lambda)}$$
$$+ \Phi^{-1}\left(\phi\right)\sqrt{p_0\left(1 - p_0\right) + p_1\left(1 - p_1\right)/\lambda},$$

and $\bar{p}_\lambda = (p_0 + \lambda p_1)/(1 + \lambda)$ It is easy to see that $z_{\alpha,\phi,p}$ is a special case of $z_{\alpha,\phi,p,\lambda}$ when $\lambda = 1$, i.e., when a design is balanced.

Shih [20] proposed the following formula for N_2 for the control arm based on a generalized estimating equation approach:

$$N_2 = \frac{f\left[\{p_0\left(1 - p_0\right)\}^{-1} + \{\lambda p_1\left(1 - p_1\right)\}^{-1}\right] z_{\alpha,\phi}^2}{N_1 b^2}$$

where $b = \log\left(p_0/(1 - p_0)\right) - \log\left(p_1/(1 - p_1)\right)$ and $z_{\alpha\varphi}$ is as defined in (5.6). Another sample size formula with unequal level 2 unit sizes is presented in Section (2.5.2).

5.7.1.3   Summary

The sample size N_2 in (5.36) or $N_2^{(0)}$ in (5.37) required for a desired magnitude of statistical power decreases with increasing N_1, and increases with increasing ρ. The sample size decreases in general with increasing $|p - p_1|$. However, for the same magnitude of $|p - p_1|$, the sample size depends on the specific combinations of p and p_1 since the variances of the difference in proportions are a function of both p and p_1. For example, smaller sample size is required when $p = 0.1$ and $p_1 = 0.2$ than when $p = 0.2$ and $p_1 = 0.3$. Table 5.9 displays examples of statistical power for testing the null hypothesis $H_0 : p - p_1 = 0$.

TABLE 5.9

Sample size and statistical power for detecting a main effect $|p_1 - p_0|$ on binary outcome in model with $m = 2$ (5.34) when randomizations occur at the second level (two-sided significance level $\alpha = 0.05$)

λ	p_0	p_1	ρ	N_1	$N_2^{(0)}$	Total N	φ
1	0.4	0.6	0.1	10	19	380	0.812
				20	15	600	0.826
			0.2	10	28	560	0.812
				20	24	960	0.812
		0.7	0.1	10	8	160	0.801
				20	7	280	0.854
			0.2	10	12	240	0.808
				20	11	440	0.834
1.5	0.4	0.5	0.1	10	16	400	0.817
				20	12	600	0.810
			0.2	10	23	575	0.807
				20	20	100	0.813
		0.6	0.1	10	7	175	0.823
				20	6	300	0.865
			0.2	10	10	250	0.811
				20	9	450	0.830

Note: λ represents the ratio of the number of level 2 units in the intervention arm to the corresponding number in the control arm; ρ (5.35) is the correlations among the level 1 data, often referred to as intra class correlations (ICC); N_1 is the number of level 1 units per level 2 units; $N_2^{(0)}$ is the level 2 sample size in the *control* arm; Total $N = (1 + \lambda) N_1 N_2^{(0)}$ is the required total number of observations; φ is the statistical power to test $H_0 : p_1 - p_0 = 0$.

5.7.2   Randomization at the First Level

Let us denote the level 1 sample sizes for the control ($X = 0$) and intervention ($X = 1$) arms within level 2 units by $N_1^{(0)}$ and $N_1^{(1)}$ while N_2 represents the *total* level 2 sample size.

5.7.2.1   Balanced Allocations

When random allocations were assigned at the first level for a balanced design, i.e., when $m = 1$ and $X_{ij} = X_j$ for all i the indices i and j range as follows: $i = 1, 2, \ldots, N_2$, and $j = 1, \ldots, 2N_1^{(0)}$, yielding $\Sigma_j X_j = N_1^{(1)}$. Therefore, N_2 in this case represents the total number of second level units and $N_1 = N_1^{(0)} = N_1^{(1)}$ is the number of first level units per arm in each second level unit. The null hypothesis to be tested is $H_0 : \xi_{(1)} = 0$. Again, the formula (5.9) for determining N_2 for testing $H_0 : \delta_{(1)} = 0$ in model (5.8) for continuous outcomes can be extended to the binary outcome case. An approximate sample size for N_2 for a two-sided type I error rate of α and statistical power $\varphi = \phi$ can be computed as

$$N_2 = \frac{(1 - \rho)\, z^2_{\alpha,\phi,p}}{N_1\, (p_1 - p_0)^2} \tag{5.38}$$

from the power function

$$\varphi = \Phi\left\{ \frac{|p_1 - p_0|\,\sqrt{N_1 N_2/(1 - \rho)} - \Phi^{-1}(1 - \alpha/2)\sqrt{2\bar{p}\,(1 - \bar{p})}}{\sqrt{p_0\,(1 - p_0) + p_1\,(1 - p_1)}} \right\}$$

From Equation (5.38), the sample size N_1 for a given N_2 can be immediately obtained.

5.7.2.2   Unbalanced Allocations

For an unbalanced design with $\Sigma_j X_j = \lambda N_1^{(0)}$ for $\lambda > 0$ and $i = 1, 2, \ldots, N_2$, and $j = 1, 2, \ldots (1 + \lambda)N_1^{(0)}$, the sample size N_2 for the total number of the second level data units can be obtained as

$$N_2 = \frac{(1 - \rho) z^2_{\alpha,\phi,p,\lambda}}{N_1^{(0)}\, (p_1 - p_0)^2} \tag{5.39}$$

from the power function

$$\varphi = \Phi\left\{ \frac{|p_1 - p_0|\,\sqrt{N_1^{(0)} N_2 \big/ (1 - \rho)} - \Phi^{-1}(1 - \alpha/2)\sqrt{(1 + 1/\lambda)\,\bar{p}\,(1 - \bar{p})}}{\sqrt{p_0\,(1 - p_0) + p_1\,(1 - p_1)/\lambda}} \right\}.$$

Here, again, the statistical power for testing $H_0 : p - -p_1 = 0$ is invariant over the product of $N_1^{(0)} N_2$ regardless of whether a design is balanced or unbalanced.

5.7.2.3   Summary

The sample size N_2 in (5.36) or $N_2^{(0)}$ in (5.37) required for a desired magnitude of statistical power decreases with increasing N_1, and increases with increasing ρ. The sample size decreases in general with increasing $|p - p_1|$. As noted in section 5.7.1.3, the sample size depends on the specific combinations of p and p_1 for the same magnitude of $|p - p_1|$. Table 5.10 displays examples of statistical power for testing the null hypothesis $H_0 : p - -p_1 = 0$, which also shows that the invariance over the product of $N_1^{(0)} N_2$ property holds.

5.7.3   Comparisons of Sample Sizes

The ratio between second level randomization to first level randomization of the total numbers of observations required for testing the treatment effects

TABLE 5.10

Sample size and statistical power for detecting a main effect $|p_1 - p_0|$ on binary outcome in model with $m = 1$ (5.34) when randomizations occur at the first level (two-sided significance level $\alpha = 0.05$)

λ	p_0	p_1	ρ	$N_1^{(0)}$	N_2	Total N	φ
1	0.4	0.6	0.1	10	10	200	0.851
				20	5	200	0.851
			0.2	10	8	160	0.812
				20	4	160	0.812
		0.7	0.1	10	4	80	0.823
				20	2	80	0.823
			0.2	10	4	80	0.866
				20	2	80	0.866
1.5	0.4	0.6	0.1	10	8	200	0.837
				20	4	200	0.837
			0.2	10	8	200	0.878
				20	4	200	0.878
		0.7	0.1	10	4	100	0.888
				20	2	100	0.888
			0.2	10	4	100	0.921
				20	2	100	0.921

Note: λ represents the ratio of the number of level 1 units in the intervention arm to the corresponding number in the control arm within level 2 units; ρ (5.35) is the correlations among the level 1 data, often referred to as intra class correlations (ICC); $N_1^{(0)}$ is the level 1 sample size in the *control* arm within level 2 units; N_2 is the total level 2 sample size; Total $N = (1 + \lambda)N_1^{(0)}N_2$ is the required total number of observations; φ is the statistical power to test $H_0 : p_1 - p_0 = 0$.

on binary outcomes is identical to $R_{2.1} = f_1/(1 - \rho) = 1 + N_1\rho/(1 - \rho)$ in equation (5.3.9) for continuous outcomes, which is also apparent by comparing (5.36) and (5.38). Again, unless $\rho = 0$, the second level randomization design requires a larger sample size than first level randomization for the same values of Δ, φ, α, and correlation ρ. This result holds for balanced or unbalanced designs (assuming a fixed λ) as shown by comparing (5.37) and (5.39), and also by comparing Tables 5.9 and 5.10

5.8   Further Readings

All of the aforementioned sample size approaches can be applied to non-inferiority trials by replacing the effect size Δ for superiority trials with the margin δ of non-inferiority or the maximum clinically acceptable difference in the primary outcome between the two treatments. A comprehensive review of the statistical aspects of non-inferiority trials can be found in Wellek [21]. Moerbeek proposed a method for choosing the level of randomization based on the potential for contamination bias and the cost of recruiting the units for each level [22]. A variation of the two-level study design is a stepped wedge design in which treatment randomization is assigned to clusters of subjects in a progressive manner (See Section 2.8 for more details). The implementation of the stepped wedge design is usually considered when logistical restrictions make it difficult to assign treatments in a parallel fashion [23]. For sample size approaches for stepped wedge designs, see Hussey and Hughes [24] and Woertman et al. [25]. The extension of sample size approached to models with different ICCs in outcome between treatment arms is possible following the approaches applied in Liang and Pulver [18]. Moerbeek [26] and Hedeker et al [27] proposed approaches for determining sample sizes under situations that factor in attrition rates and assume general covariance structures. Approaches based on generalized estimating equations are addressed in Pan[28] for binary outcomes with first-order auto regression, or AR(1), correlation structures. Gangnon and Kosorok [29] derived sample size formulas for clustered survival outcomes by extending Schoenfeld's approach [30] for independent survival outcomes. Optimal allocation of level 1 and level 2 units to treatment arms for different study designs and hypotheses are discussed in several papers [31–35]. As noted earlier, the effect of the magnitude and direction of the ICC on sample size requirements can be substantial; therefore the ICC should be carefully evaluated when designing a two-level randomized trial [36–39]. Murray [40] framed a general compressive modeling approach that accommodates more than two arms in a trial. In addition to general issues concerning the implementation of cluster randomized trials in real world settings, the sample size approaches for matched and stratified cluster randomized trials are discussed in Hayes and Moulton [41].

Bibliography

[1] A. Donner and N. Klar. *Design and Analysis of Cluster Randomization Trials in Health Research.* Arnold: London, 2000.

[2] D.M. Murray. *Design and Analysis of Group Randomized Trials.* Oxford University Press: New York, 1998.

[3] D.M. Murray, S.P. Varnell, and J.L. Blitstein. Design and analysis of group-randomized trials: A review of recent methodological developments. *American Journal of Public Health,* 94:423–432, 2004.

[4] S.P. Varnell, D.M. Murray, J.B. Janega, and J.L. Blitstein. Design and analysis of group-randomized trials: A review of recent practices. *American Journal of Public Health*, 94:393–399, 2004.

[5] S.W. Raudenbush and A.S. Bryk. *Hierarchical Linear Models: Application and Data Analysis Methods (2nd ed)*. SAGE: Thousand Oaks, 2006.

[6] A. Donner and J.J. Koval. The estimation of intraclass correlation in the analysis of family data. *Biometrics* 36:19–25, 1980.

[7] A. Donner, N. Birkett, and C. Buck. Randomization by cluster. Sample size requirements and analysis. *American Journal of Epidemiology* 114:906–914, 1981.

[8] J.M. Bland. Cluster randomised trials in the medical literature: two bibliometric surveys. *BMC Medical Research Methodology* 4:21, 2004.

[9] J. Cohen. *Statistical Power Analysis for the Behavioral Science*. Lawrence Erlbaum Associates: Hillsdale, NJ, 1988.

[10] P.J. Diggle, P. Heagerty, K.Y. Linag, and S.L. Zeger. *Analysis of Longitudinal data* (2nd ed).Oxford University Press: New York, 2002.

[11] E. Vierron and B. Giraudeau, Sample size calculation for multicenter randomized trial: taking the center effect into account. *Contemporary Clinical Trials* 28:451-458, 2007.

[12] S. Bellary, J.P. O'Hare, N.T. Raymond, A. Gumber, S. Mughal, A. Szczepura, S. Kumar, A.H. Barnett, and U.S. Grp. Enhanced diabetes care to patients of south Asian ethnic origin (the United Kingdom Asian Diabetes Study): a cluster randomised controlled trial. *Lancet* 371:1769-1776, 2008.

[13] N.T. Longford. *Random Coefficient Models*. Oxford University Press: New York, 1993.

[14] J.J. Schlesselman. Planning a longitudinal study: II. Frequency of measurement and study duratio. *Journal of Chronic Diseases* 26:561–570, 1973.

[15] J.L. Fleiss. *The Design and Analysis of Clinical Experiments*. Wiley & Sons: New York, 1986.

[16] A.C. Leon and M. Heo. Sample sizes required to detect interactions between two binary fixed-effects in a mixed-effects linear regression model. *Computational Statistics & Data Analysis* 53:603–608, 2009.

[17] M. Heo and A.C. Leon. Sample sizes required to detect two-way and three-way interactions involving slope differences in mixed-effects linear models. *Journal of Biopharmaceutical Statistics* 20:787–802, 2010.

[18] K.Y. Liang and A.E. Pulver. Analysis of case-control/family sampling design. *Genetic Epidemiology* 13:253–270, 1996.

[19] D. Hedeker and R.D. Gibbons. *Longitudinal Data Analysis*. Wiley: Hoboken, NJ, 2006.

[20] W.J. Shih. Sample size and power calculations for periodontal and other studies with clustered samples using the method of generalized estimating equations. *Biometrical Journal* 39:899–908, 1997.

[21] S. Wellek. *Testing Statistical Hypotheses of Equivalence and Noninferiority*. (2nd edn).Chapman & Hall/CRC: New York, 2010.

[22] M. Moerbeek. Randomization of clusters versus randomization of persons within clusters: Which is preferable? *American Statistician* 59:173–179, 2005.

[23] N.D. Mdege, M.S. Man, C.A. Taylor, and D.J. Torgerson. Systematic review of stepped wedge cluster randomized trials shows that design is particularly used to evaluate interventions during routine implementation. *Journal of Clinical Epidemiology* 64:936–948, 2011.

[24] M.A. Hussey and J.P. Hughes. Design and analysis of stepped wedge cluster randomized trials. *Contemporary Clinical Trials* 28:182–191, 2007.

[25] W. Woertman, E. de Hoop, M. Moerbeek, S.U. Zuidema, D.L. Gerritsen, and S. Teerenstra. Stepped wedge designs could reduce the required sample size in cluster randomized trials. *Journal of Clinical Epidemiology* 66:752–758, 2013.

[26] M. Moerbeek. Powerful and cost-efficient designs for longitudinal intervention studies with two treatment groups. *Journal of Educational and Behavioral Statistics* 33:41–6, 2008.

[27] D. Hedeker, R.D. Gibbons, and C. Waternaux. Sample size estimation for longitudinal designs with attrition: Comparing time-related contrasts between two groups. *Journal of Educational and Behavioral Statistics* 24:70–93, 1999.

[28] W. Pan. Sample size and power calculations with correlated binary data. *Controlled Clinical Trials* 22:211–227, 2001.

[29] R.E. Gangnon and M.R. Kosorok. Sample-size formula for clustered survival data using weighted log-rank statistics. *Biometrika* 91:263–275, 2004.

[30] D.A. Schoenfeld. Sample-size formula for the proportional-hazards regression model. *Biometrics* 39:499–503, 1983.

[31] S.W. Raudenbush and X.F. Liu. Statistical power and optimal design for multisite randomized trials. *Psychological Methods* 5:199–213, 2000.

[32] B. Winkens, H.J.A. Schouten, G.J.P. van Breukelen, and M.P.F Berger. Optimal designs for clinical trials with second-order polynomial treatment effects. *Statistical Methods in Medical Research* 16:523–537, 2007.

[33] F.E.S. Tan and M.P.F. Berger. Optimal allocation of time points for the random effects model. *Communications in Statistics-Simulation and Computation* 28:517–540, 1999.

[34] B. Winkens, H.J. Schouten, G.J. van Breukelen, and M.P. Berger. Optimal time-points in clinical trials with linearly divergent treatment effects. *Statistics in Medicine* 24:3743–3756, 2006.

[35] B. Winkens, H.J. Schouten, G.J. van Breukelen, and M.P. Berger. Optimal number of repeated measures and group sizes in clinical trials with linearly divergent treatment effects. *Contemporary Clinical Trials* 27:57–69, 2006.

[36] D.M. Murray and J.L. Blitstein. Methods to reduce the impact of intraclass correlation in group-randomized trials. *Evaluation Review* 27:79–103, 2003.

[37] M.K. Campbell, P.M. Fayers, and J.M. Grimshaw. Determinants of the intracluster correlation coefficient in cluster randomized trials: The case of implementation research. *Clinical Trials* 2:99–107, 2005.

[38] K. Resnicow, N.H. Zhang, R.D. Vaughan, S.P. Reddy, S. James, and D.M. Murray. When intraclass correlation coefficients go awry: A case study from a school-based smoking prevention study in outh Africa. *American Journal of Public Health* 100:1714–1718, 2010.

[39] S.M. Eldridge, O.C. Ukoumunne, and J.B. Carlin. The intra-cluster correlation coefficient in cluster randomized trials: A review of definitions. *International Statistical Review* 77:378–394, 2009.

[40] D.M. Murray. Statistical models appropriate for designs often used in group-randomized trials. *Statistics in Medicine* 20:1373–1385, 2001.

[41] R.J. Hayes and L.H. Moulton. *Cluster Randomized Trials.* CRC Press: Boca Raton, 2009.

6

Sample Size Determination for Correlated Outcomes from Three-Level Randomized Clinical Trials

6.1 Introduction

We now consider sample size approaches for randomized clinical trials with a three-level data structure. One example of a study with this structure is a large scale randomized trial to test a behavioral intervention to increase physical activity in school age children. In this design, students (level 1) are nested within classrooms (level 2) which are nested within schools (level 3). A second example is a longitudinal study of patients from multiple clinics in which the repeated measures comprise the first level, patients are the second level, and clinics are the third level in the hierarchy. The choice of units for each level depends on the study context and research setting. Furthermore, randomization can occur at any level. In this chapter, we present sample size approaches for trials with a three level data structure which are based on direct extensions of the statistical models described in chapter 5 for the two-level case.

6.2 Statistical Model for Continuous Outcomes

We will consider in this chapter the following three-level mixed-effects linear model for continuous outcome Y[1, 2]:

$$Y_{ijk} = h\left(X_{ijk}, T_{ijk}; \theta\right) + \nu_{j(i)}T_{ijk} + u_i + u_{j(i)} + e_{ijk}. \tag{6.1}$$

This is an extension of model (5.2.1) for the analysis of two-level data where Y_{ijk} denotes the observed outcome on the kth level 1 unit in the jth level 2 unit in the ith level 3 unit where: $i = 1, 2, \ldots, N_3, j = 1, 2, \ldots, N_2$, and $k = 1, 2, \ldots, N_1$. Depending on the study design, N_3, N_2, and N_1 may be defined below as the required number of units at each level per intervention arm or the required number per higher level data unit.

The variable X_{ijk} represents the indicator for treatment assignment. For longitudinal studies in which the first level data are repeated measures, the variable T_{ijk} denotes the fixed times at which the measurements were obtained. The fixed effect components are represented by the term h and a parameter vector θ, the specification of which depends on the study design and hypothesis of interest. However, for cross-sectional study designs, the fixed effect $h(X, T; \theta)$ will be reduced to $h(X; \theta)$, and the term νT will be removed from model (6.1).

The other terms in model (6.1) represent random effects. It is assumed that the error term e_{ijk} is normally distributed as $N\left(0, \sigma_e^2\right)$, the level 2 random intercept (i.e., subject-specific intercept) as $u_{j(i)} \sim N\left(0, \sigma_2^2\right)$, the level 3 random intercept (i.e., cluster-specific intercept) as $u_i \sim N\left(0, \sigma_3^2\right)$. If the slope of the outcome over time is considered random, the random slope coefficient is represented as $\nu_{j(i)} \sim N\left(0, \sigma_\tau^2\right)$. Among these random components, it is further assumed that $u_i \perp u_{j(i)} \perp e_{ijk} \perp \nu_{j(i)}$, i.e., these four random components are mutually independent. In addition, *conditional independence* is assumed for all $u_{j(i)}, \nu_{j(i)}$ and e_{ijk}, whereas the u_i are *unconditionally* independent. That is, both $u_{j(i)}$ and $\nu_{j(i)}$ are independent over j conditional on u_i, and e_{ijk} are independent over k conditional on $u_i, \nu_{j(i)}$ and $u_{j(i)}$.

Under model (6.1), it can be shown that the elements of the mean vector for the outcome are equal to

$$E(Y_{ijk}) = f\left(X_{ijk}, T_{ijk}; \theta\right)$$

and the elements of the covariance matrix are

$$
\begin{aligned}
Cov\left(Y_{ijk}, Y_{i'j'k'}\right) = {} & 1(i = i' \cap j = j' \cap k = k')\sigma_e^2 \\
& + 1(i = i' \cap j = j')\left(T_k T_{k'}\sigma_\tau^2 + \sigma_2^2\right) \\
& + 1(i = i')\sigma_3^2,
\end{aligned}
$$

where $1(.)$ is an indicator function and T_{ijk} is assumed to be equal to T_k for all i and j. It follows that

$$Var\left(Y_{ijk}\right) = Cov\left(Y_{ijk}, Y_{ijk}\right) = \sigma^2 + T_k^2 \sigma_\tau^2,$$

where

$$\sigma^2 \equiv \sigma_e^2 + \sigma_2^2 + \sigma_3^2,$$

which is also equal to the variance of Y under the fixed slope model with $\sigma_\tau^2 = 0$ or if time is not involved in the study design. Therefore, the correlations among the level 2 data, i.e., among outcomes from different second level clusters but the same third level cluster, can be expressed for $j \neq j'$ as follows:

$$Corr\left(Y_{ijk}, Y_{ij'k'}\right) = \frac{\sigma_3^2}{\sqrt{\sigma^2 + T_k^2 \sigma_\tau^2}\sqrt{\sigma^2 + T_k'^2 \sigma_\tau^2}}.$$

The correlations among the level one data, i.e., among outcomes measured at different time points on the same subject nested within clinics, can be expressed for $k \neq k'$ as

$$Corr\left(Y_{ijk}, Y_{ijk'}\right) = \frac{\sigma_2^2 + \sigma_3^2 + T_k T_{k'}\sigma_\tau^2}{\sqrt{\sigma^2 + T_k^2 \sigma_\tau^2}\sqrt{\sigma^2 + T_k'^2 \sigma_\tau^2}}.$$

Under the fixed slope model, i.e., when $\sigma_\tau^2 = 0$, the correlations reduce to the following, respectively:

$$\rho_2 = \sigma_3^2/\sigma^2 \qquad (6.2)$$

and

$$\rho_1 = (\sigma_2^2 + \sigma_3^2)/\sigma^2. \qquad (6.3)$$

As a result, ρ_1 is greater than or equal to ρ_2, that is, $\rho_1 \geq \rho_2$.

6.3 Testing Main Effects

When testing the main effects of an intervention is the primary goal of the study, the randomization can occur at any level of the hierarchical structure and may depend on several factors including the research settings, logistical considerations, the potential for contamination bias within units, and the number of potentially available units at each level. Sample sizes required to test a main effect will vary across the levels of randomization and are presented in Sections 6.3.1, 6.3.2, and 6.3.3 for randomizations at the first, second, and third levels, respectively. Comparisons of the required sample sizes between designs with randomizations at those different levels are discussed in Section 6.3.4, followed by a real trial example in Section 6.3.5.

6.3.1 Randomization at the Third Level

When the randomization scheme is applied to the third level units, the fixed effect term $h(X, T; \theta)$ in model (6.1) is reduced to

$$h(X, T; \theta) = \beta_0 + \delta_{(3)}X,$$

resulting in a model

$$Y_{ijk} = \beta_0 + \delta_{(3)}X_{ijk} + u_i + u_{j(i)} + e_{ijk} \qquad (6.4)$$

where β_0 represents the overall fixed intercept term and the parameter $\delta_{(3)}$ represents the intervention effect. The relevant null hypothesis of interest is $H_0 : \delta_{(3)} = 0$ in model (6.4). Let us denote the level 3 sample sizes for the control ($X = 0$) and intervention ($X = 1$) arms by $N_3^{(0)}$ and $N_3^{(1)}$ while N_2 and N_1 represent the level 2 sample size per level 3 units and the level 1 sample size per level 2 units, respectively.

6.3.1.1 Balanced Allocations

The intervention assignment indicator variable $X_{ijk} = 0$ if the ith level three unit is assigned to a control intervention and $X_{ijk} = 1$ if assigned to an experimental intervention; therefore $X_{ijk} = X_i$ for all j and k. For a balanced design it is assumed that $\Sigma_i X_i = N_3^{(1)}$ so that $i = 1, 2, \ldots, 2N_3^{(0)}, j = 1, \ldots, N_2$ and $k = 1, 2, \ldots, N_1$. In particular, $N_3 = N_3^{(0)} = N_3^{(1)}$ represents the level 3 sample size *per treatment arm*.

To test the significance of $\delta_{(3)}$ in model (6.4), the following test statistic can be used [3]:

$$D_3 = \frac{\sqrt{N_3 N_2 N_1} \left(\bar{Y}_1 - \bar{Y}_2 \right)}{\sigma \sqrt{2f_3}}.$$

Here, $\bar{Y}_g = \frac{1}{N_3 N_2 N_1} \sum_{i=1}^{N_3} \sum_{j=1}^{N_2} \sum_{k=1}^{N_1} Y_{ijk}$ $(g = 0, 1)$ is the mean of the outcome Y for the gth arm, for which $X_i = g$, and

$$f_3 = 1 + N_1(N_2 - 1)\rho_2 + (N_1 - 1)\rho_1 \tag{6.5}$$

is referred to as a *variance inflation factor* or *design effect* and does not depend on N_3. Note that $\hat{\delta}_{(3)} = \bar{Y}_1 - \bar{Y}_0$ is an unbiased estimate of $\delta_{(3)}$ and $Var\left(\bar{Y}_g\right) = f_3 \sigma^2 / (N_3 N_2 N_1)$, which yields

$$Var\left(\hat{\delta}_{(3)}\right) = \frac{2f_3 \sigma^2}{N_3 N_2 N_1}.$$

It can be shown that the test statistic D_3 is normally distributed with mean $\delta_{(3)}$ and variance 1. The power of the test statistic D_3, denoted by $\varphi_{(3)}$, can therefore be written as follows:

$$\varphi_{(3)} = \Phi\left\{ \Delta_{(3)} \sqrt{N_3 N_2 N_1 / 2f_3} - \Phi^{-1}(1 - \alpha/2) \right\}.$$

The parameter $\Delta_{(3)} = \left| \delta_{(3)} / \sigma \right|$ is referred to as the standardized effect size or Cohen's d, the mean difference in the outcome Y expressed in units of standard deviation σ. The cumulative distribution function (CDF) and the inverse CDF of a standard normal distribution are denoted by Φ and Φ^{-1}, respectively. A two-sided significance level is denoted by α. It follows that the required third level sample size N_3 per arm for a desired statistical power $\varphi_{(3)} = \phi$ can be expressed as [3-5]

$$N_3 = \frac{2f_3 z_{\alpha,\phi}^2}{N_2 N_1 \Delta_{(3)}^2}, \tag{6.6}$$

where

$$z_{\alpha,\phi} = \Phi^{-1}(1 - \alpha/2) + \Phi^{-1}(\phi). \tag{6.7}$$

Solving Equation (6.6) for sample sizes N_2 and N_1, we have

$$N_2 = \frac{2\left\{ 1 + (\rho_1 - \rho_2)N_1 - \rho_1 \right\} z_{\alpha,\phi}^2}{N_1 N_3 \Delta_{(3)}^2 - 2\rho_2 N_1 z_{\alpha,\phi}^2},$$

and

$$N_1 = \frac{2(1 - \rho_1)z_{\alpha,\phi}^2}{N_2 N_3 \Delta_{(3)}^2 - 2\left\{ (N_2 - 1)\rho_2 + \rho_1 \right\} z_{\alpha,\phi}^2}.$$

These sample size formulas were validated by simulation studies [3]. Recall, as noted in Section 5.3.1.1, that N_2 or N_1 for a desired power cannot be determined for a given N_3 if the corresponding denominators are not positive.

6.3.1.2   Unbalanced Allocations

Suppose that the random allocation needs to be unbalanced so that $\Sigma_i X_i = \lambda_3 N_3^{(0)}$ for $\lambda_3 > 0$ and $i = 1, 2, \ldots, (1 + \lambda_3)N_3^{(0)}$. The ratio λ_3 represents the ratio of the number of third level units in the intervention arm to that in the control arm, i.e. $\lambda_3 = N_3^{(1)}/N_3^{(0)}$. If $\lambda_3 > 1$, the sample size in the intervention arm is larger, which is usually the case in practice. Then, the sample size $N_3^{(0)}$ in the control arm can be expressed as

$$N_3^{(0)} = \frac{f_3\left(1 + 1/\lambda_3\right)z_{\alpha,\phi}^2}{N_2 N_1 \Delta_{(3)}^2}. \tag{6.8}$$

The sample size of the level 2 units given $N_3^{(0)}$ and N_1 is equal to

$$N_2 = \frac{(1 + 1/\lambda_3)\left\{1 + (\rho_1 - \rho_2)N_1 - \rho_1\right\}z_{\alpha,\phi}^2}{N_1 N_3^{(0)}\Delta_{(3)}^2 - (1 + 1/\lambda_3)\rho_2 N_1 z_{\alpha,\phi}^2},$$

and the sample size of the level 1 units given $N_3^{(0)}$ and N_2 is equal to

$$N_1 = \frac{(1 + 1/\lambda_3)(1 - \rho_1)z_{\alpha,\phi}^2}{N_2 N_3^{(0)}\Delta_{(3)}^2 - (1 + 1/\lambda_3)\left\{(N_2 - 1)\rho_2 + \rho_1\right\}z_{\alpha,\phi}^2}.$$

If the sample sizes N_2 and N_1 need to vary across clusters, that is, $j = 1, 2, .., n_i$ and $k = 1, 2, \ldots, n_{ij}$, then they can be replaced by $\tilde{N}_2 = \sum_{i=1}^{N_3^{(0)}+N_3^{(1)}} n_i/\left(N_3^{(0)}+N_3^{(1)}\right)$ and $\tilde{N}_1 = \sum_{i=1}^{N_3^{(0)}+N_3^{(1)}} \sum_{j=1}^{n_i} n_{ij}/\left\{\tilde{N}_2(N_3^{(0)}+N_3^{(1)})\right\}$, respectively.

6.3.1.3   Summary

The sample size N_3 in (6.6) or $N_3^{(0)}$ in (6.8) required for a desired magnitude of statistical power decreases with increasing N_1, N_2, and $\Delta_{(3)}$, and increases with increasing ρ_1 and ρ_2. Table 6.1 displays examples of statistical power for testing the null hypothesis $H_0 : \delta_{(3)} = 0$.

6.3.2   Randomization at the Second Level

When the randomization occurs at the second level, the fixed-effect term $h(X, T; \theta)$ in model (6.1) is reduced to

$$h(X, T; \theta) = \beta_0 + \delta_{(2)}X,$$

resulting in the model

$$Y_{ijk} = \beta_0 + \delta_{(2)}X_{ijk} + u_i + u_{j(i)} + e_{ijk} \tag{6.9}$$

where the parameter $\delta_{(2)}$ represents the intervention effect. The null hypothesis to be tested is $H_0 : \delta_{(2)} = 0$ in model (6.9). Let us denote the level 2 sample sizes for the control $(X = 0)$ and intervention $(X = 1)$ arms within

TABLE 6.1
Sample size and power for detecting a main effect $\delta_{(3)}$ in model (6.4) when randomizations occur at the third level with $\rho_2 = 0.05$ (two-sided significance level $\alpha = 0.05$)

λ_3	$\Delta_{(3)}$	ρ_1	N_1	N_2	$N_3^{(0)}$	Total N	$\varphi_{(3)}$
1	0.3	0.1	5	4	19	760	0.805
			5	8	14	1120	0.807
		0.2	5	4	23	920	0.813
			5	8	16	1280	0.813
	0.4	0.1	5	4	11	440	0.816
			5	8	8	640	0.813
		0.2	5	4	13	520	0.815
			5	8	9	720	0.813
1.5	0.3	0.1	5	4	16	800	0.809
			5	8	12	1200	0.818
		0.2	5	4	19	950	0.810
			5	8	13	1300	0.803
	0.4	0.1	5	4	9	450	0.809
			5	8	7	700	0.832
		0.2	5	4	11	550	0.821
			5	8	8	800	0.837

Note: λ_3 represents the ratio of the number of level 3 units in the intervention arm to the corresponding number in the control arm; $\Delta_{(3)} = |\delta_{(3)}/\sigma|$; ρ_2 (6.2) is the correlations among the level 2 data; ρ_1 (6.3) is the correlations among the level 1 data; N_1 is the number of level 1 units per level 2 unit2; N_2 is the number of level 2 units per level 3 units; $N_3^{(0)}$ is the level 3 sample size in the *control* arm; Total $N = (1 + \lambda_3)N_1 N_2 N_3^{(0)}$ is the required total number of observations; $\varphi_{(3)}$ is the statistical power to test $H_0 : \delta_{(3)} = 0$.

level 3 units by $N_2^{(0)}$ and $N_2^{(1)}$ while N_3 and N_1 represent the *total* level 3 sample size and the level 1 sample size per level 2 units, respectively.

6.3.2.1 Balanced Allocations

The intervention assignment indicator variable $X_{ijk} = 0$ if the jth level 2 unit is assigned to a control intervention and $X_{ijk} = 1$ if assigned to an experimental intervention; therefore $X_{ijk} = X_j$ for all i and k. For a balanced design it is assumed that $\Sigma_j X_j = N_2^{(1)}$ for all i so that $i = 1, 2, \ldots, N_3, j = 1, \ldots, 2N_2^{(0)}$ and $k = 1, 2, \ldots, N_1$. That is, $N_2 = N_2^{(0)} = N_2^{(1)}$ is the level 2 sample size per arm within the third level units.

To test the significance of $\delta_{(2)}$ in model (6.9), the following test statistic can be used [6]:

$$D_2 = \frac{\sqrt{N_3 N_2 N_1}\,(\bar{Y}_1 - \bar{Y}_2)}{\sigma\sqrt{2f_2}}.$$

Here, $\bar{Y}_g = \frac{1}{N_3 N_2 N_1} \sum_{i=1}^{N_3} \sum_{j=1}^{N_2} \sum_{k=1}^{N_1} Y_{ijk}$ $(g = 0, 1)$ is the mean of the outcome Y for the gth arm, for which $X_j = g$, and the design effect is

$$f_2 = 1 + (N_1 - 1)\rho_1 - N_1\rho_2, \tag{6.10}$$

which does not depend on N_3 or N_2. Note that $\hat{\delta}_{(2)} = \bar{Y}_1 - \bar{Y}_0$ is an unbiased estimate of $\delta_{(2)}$ and $Var\left(\bar{Y}_g\right) = f_2\sigma^2/(N_3 N_2 N_1)$ which yields

$$Var(\hat{\delta}_{(2)}) = \frac{2f_2\sigma^2}{N_3 N_2 N_1}.$$

It can be shown that the test statistic D_2 is again normally distributed with mean $\delta_{(2)}$ and variance 1. The power of the test statistic D_2 can therefore be written in terms of $\Delta_{(2)} = |\delta_{(2)}/\sigma|$ as follows [5, 6]:

$$\varphi_{(2)} = \Phi\left\{\Delta_{(2)}\sqrt{N_3 N_2 N_1/2f_2} - \Phi^{-1}(1 - \alpha/2)\right\}.$$

It follows that the required total number of level 3 units N_3 for a desired statistical power $\varphi_{(2)} = \phi$ can be expressed:

$$N_3 = \frac{2f_2 z_{\alpha,\phi}^2}{N_2 N_1 \Delta_{(2)}^2}. \tag{6.11}$$

Solving equation (6.11) for sample sizes N_2 and N_1, we have

$$N_2 = \frac{2f_2 z_{\alpha,\phi}^2}{N_1 N_3 \Delta_{(2)}^2},$$

and

$$N_1 = \frac{2(1 - \rho_1)z_{\alpha,\phi}^2}{N_2 N_3 \Delta_{(2)}^2 - 2(\rho_1 - \rho_2)z_{\alpha,\phi}^2}.$$

These sample size formulas were validated by simulation studies [6]. Here we note that the power $\varphi_{(2)}$ is invariant over the product of $N_3 N_2$.

6.3.2.2   Unbalanced Allocations

Suppose that the random allocation at the second level needs to be unbalanced so that $\Sigma_j X_j = \lambda_2 N_2^{(0)}$ for $\lambda_2 > 0$ and $j = 1, 2, \ldots, (1 + \lambda_2)N_2$. The ratio λ_2 represents the ratio of the number of second level units in the intervention arm per third level unit to that in the control arm, i.e, $\lambda_2 = N_2^{(1)}/N_2^{(0)}$. The *total* sample size N_3 can be expressed as

$$N_3 = \frac{f_2\left(1 + 1/\lambda_2\right)z_{\alpha,\phi}^2}{N_2^{(0)} N_1 \Delta_{(2)}^2} \tag{6.12}$$

The required sample size of second level units in the control arm per third level unit can be obtained as

$$N_2^{(0)} = \frac{f_2 \left(1 + 1/\lambda_2\right) z_{\alpha,\phi}^2}{N_3 N_1 \Delta_{(2)}^2};$$

and for given N_3 and $N_2^{(0)}$, N_1 can be computed as

$$N_1 = \frac{\left(1 - \rho_1\right)\left(1 + 1/\lambda_2\right) z_{\alpha,\phi}^2}{N_2^{(0)} N_3 \Delta_{(3)}^2 - \left(\rho_1 - \rho_2\right)\left(1 + 1/\lambda_2\right) z_{\alpha,\phi}^2}$$

Again, if sample sizes N_2 and N_1 need to vary across their higher level units, that is, $j = 1, 2, \ldots, n_i$ and $k = 1, 2, \ldots, n_{ij}$, then they can be replaced by $\tilde{N}_2 = \sum_{i=1}^{N_3^{(0)} + N_3^{(1)}} n_i / (N_3^{(0)} + N_3^{(1)})$ and $\tilde{N}_1 = \sum_{i=1}^{N_3^{(0)} + N_3^{(1)}} \sum_{j=1}^{n_i} n_{ij} / \{\tilde{N}_2 (N_3^{(0)} + N_3^{(1)})\}$, respectively.

6.3.2.3   Summary

The sample size N_3 in (6.11) or in (6.12) required for a desired magnitude of statistical power decreases with increasing $N_1, N_2^{(0)}, \Delta_{(2)}$, and ρ_1 but increases with increasing ρ_2. Table 6.2 displays examples of statistical power for testing the null hypothesis $H_0 : \delta_{(2)} = 0$, and shows that the statistical power $\varphi_{(2)}$ is invariant over the product of $N_3 N_2^{(0)}$, regardless of whether or not λ_2 is equal to 1.

6.3.3   Randomization at the First Level

When the randomization occurs at the first level units, the fixed-effect term $h(X, T; \theta)$ in model (6.1) is reduced to

$$h(X, T; \theta) = \beta_0 + \delta_{(1)} X,$$

resulting in the model

$$Y_{ijk} = \beta_0 + \delta_{(1)} X_{ijk} + u_i + u_{j(i)} + e_{ijk} \qquad (6.13)$$

where the parameter $\delta_{(1)}$ represents the intervention effect. The null hypothesis to be tested is $H_0 : \delta_{(1)} = 0$. Let us denote the level 1 sample sizes for the control ($X = 0$) and intervention ($X = 1$) arms within the level 2 units by $N_1^{(0)}$ and $N_1^{(1)}$ while N_3 and N_2 represent the *total* level 3 sample size and the level 2 sample size per level 3 units, respectively.

6.3.3.1   Balanced Allocations

The intervention assignment indicator variable $X_{ijk} = 0$ if the jth level 2 unit is assigned to a control intervention and $X_{ijk} = 1$ if assigned to an experimental intervention; therefore $X_{ijk} = X_k$ for all i and j. For a balanced design

TABLE 6.2
Sample size and power for detecting a main effect $\delta_{(2)}$ in model (6.9) when randomizations occur at the second level with $\rho_2 = 0.05$ (two-sided significance level $\alpha = 0.05$)

λ_2	$\Delta_{(2)}$	ρ_1	N_1	$N_2^{(0)}$	N_3	Total N	$\varphi_{(2)}$
1	0.3	0.1	5	4	12	480	0.865
			5	8	6	480	0.865
			10	4	8	640	0.893
			10	8	4	640	0.893
		0.2	5	4	14	560	0.814
			5	8	7	560	0.814
	0.4	0.1	5	4	6	240	0.824
			5	8	3	240	0.824
		0.2	5	4	8	320	0.820
			5	8	4	320	0.820
			10	4	6	480	0.824
			10	8	3	480	0.824
1.5	0.3	0.1	5	4	10	500	0.865
			5	8	5	500	0.865
			10	4	6	600	0.861
			10	8	3	600	0.861
		0.2	5	4	12	600	0.824
			5	8	6	600	0.824
	0.4	0.1	5	4	6	300	0.886
			5	8	3	300	0.886
		0.2	5	4	8	400	0.883
			5	8	4	400	0.883
			10	4	6	600	0.886
			10	8	3	600	0.886

Note: λ_2 represents the ratio of the number of level 2 units in the intervention arm to the corresponding number in the control arm within level 3 units; $\Delta_{(2)} = |\delta_{(2)}/\sigma|$; ρ_2 (6.2) is the correlations among the level 2 data; ρ_1 (6.3) is the correlations among the level 1 data; N_1 is the number of level 1 units per level 2 units; $N_2^{(0)}$ is the sample size in the *control* arm within level 3 units; N_3 is the *total* level 3 sample size; Total $N = (1 + \lambda_2)N_1 N_2^{(0)} N_3$ is the required total number of observations; $\varphi_{(2)}$ is the statistical power to test $H_0 : \delta_{(2)} = 0$.

it is assumed that $\Sigma_k X_k = N_1^{(1)}$ for all i and j so that $i = 1, 2, \ldots, N_3, j = 1, \ldots, N_2$ and $k = 1, 2, \ldots, 2N_1^{(0)}$. That is, $N_1 = N_1^{(0)} = N_1^{(1)}$ is the level 1 sample size per arm per level 2 units.

To test the significance of $\delta_{(1)}$ in model (6.13), the following test statistic can be used [6]:

$$D_1 = \frac{\sqrt{N_3 N_2 N_1}\,(\bar{Y}_1 - \bar{Y}_2)}{\sigma\sqrt{2f_1}}$$

Here, $\bar{Y}_g = \frac{1}{N_3 N_2 N_1} \sum_{i=1}^{N_3} \sum_{j=1}^{N_2} \sum_{k=1}^{N_1} Y_{ijk}$ $(g = 0, 1)$ is the mean of the outcome Y for the gth arm, for which $X_j = g$, and the design effect is

$$f_1 = (1 - \rho_1), \tag{6.14}$$

which does not depend on N_3, N_2, or N_1. Note that $\hat{\delta}_{(1)} = \bar{Y}_1 - \bar{Y}_0$ is an unbiased estimate of $\delta_{(1)}$ and $Var\left(\bar{Y}_g\right) = f_1 \sigma^2 / (N_3 N_2 N_1)$ which yields

$$Var(\hat{\delta}_{(1)}) = \frac{2\left(1 - \rho_1\right)\sigma^2}{N_3 N_2 N_1}.$$

It can be shown that the test statistic D_1 is again normally distributed with mean $\delta_{(2)}$ and variance 1. The power of the test statistic D_1 can therefore be written in terms of $\Delta_{(1)} = \left|\delta_{(1)}/\sigma\right|$ as follows [5, 6]:

$$\varphi_{(1)} = \Phi\left\{\Delta_{(1)}\sqrt{N_3 N_2 N_1 / 2 f_1} - \Phi^{-1}(1 - \alpha/2)\right\}.$$

It follows that the required total number of level 3 units N_3 for a desired statistical power $\varphi_{(1)} = \phi$ can be expressed as

$$N_3 = \frac{2 f_1 z_{\alpha,\phi}^2}{N_2 N_1 \Delta_{(1)}^2} = \frac{2(1 - \rho_1) z_{\alpha,\phi}^2}{N_2 N_1 \Delta_{(1)}^2}, \tag{6.15}$$

Solving equation (6.15) for each of sample sizes N_2 and N_1, we have

$$N_2 = \frac{2 f_1 z_{\alpha,\phi}^2}{N_1 N_3 \Delta_{(1)}^2},$$

and

$$N_1 = \frac{2 f_1 z_{\alpha,\phi}^2}{N_2 N_3 \Delta^2}.$$

These sample size formulas were validated by simulation studies [6]. Here we note that the power $\varphi_{(1)}$ is invariant over the product of $N_3 N_2 N_1$.

6.3.3.2   Unbalanced Allocations

Suppose that the random allocation at the second level needs to be unbalanced such that $\Sigma_j X_j = \lambda_1 N_1^{(0)}$ for $\lambda_1 > 0$ and $j = 1, 2, \ldots, (1 + \lambda_1) N_1^{(0)}$. The ratio λ_1 represents the ratio of the number of first level units in the intervention arm per second level unit to that in the control arm, i.e., $\lambda_1 = N_1^{(1)}/N_1^{(0)}$. The *total* sample size N_3 can be expressed as

$$N_3 = \frac{f_1\left(1 + 1/\lambda_1\right) z_{\alpha,\phi}^2}{N_2 N_1^{(0)} \Delta_{(1)}^2} \tag{6.16}$$

Owing to the invariance over product of $N_3 N_2 N_1$ property, the other sample sizes can be determined as follows:

$$N_2 = \frac{f_1 \left(1 + 1/\lambda_1\right) z_{\alpha,\phi}^2}{N_3 N_1^{(0)} \Delta_{(1)}^2}$$

and

$$N_1^{(0)} = \frac{f_1 \left(1 + 1/\lambda_1\right) z_{\alpha,\phi}^2}{N_3 N_2 \Delta_{(1)}^2}$$

which is the sample size of the first level units in the control arm per second level data unit.

6.3.3.3   Summary

The sample size N_3 in (6.15) or in (6.16) required for a desired magnitude of statistical power is not a function of ρ_2 and decreases with increasing $N_1^{(0)}$, N_2, $\Delta_{(1)}$, and ρ_1. Table 6.2 displays examples of statistical power for testing the null hypothesis $H_0 : \delta_{(1)} = 0$, and shows that the statistical power $\varphi_{(1)}$ is invariant over the product of $N_3 N_2 N_1^{(0)}$, or total number of observations $(1+\lambda_1) N_3 N_2 N_1^{(0)}$, regardless of whether or not λ_1 is equal to 1.

6.3.4   Comparisons of Sample Sizes

Either for balanced designs or for unbalanced designs with identical $\lambda_1 = \lambda_2 = \lambda_3$ values, the ratios of the total number of *observations*, that is $2N_3 N_2 N_1$, required for trials across randomizations at the three different levels are identical to those of the corresponding design effects: $f_3 = 1 + N_1(N_2-1)\rho_2 + (N_1-1)\rho_1$, $f_2 = 1 + (N_1-1)\rho_1 - N_1\rho_2$, and $f_1 = 1 - \rho_1$. The ratio denoted by $R_{3.2}$ between the third and the second level randomizations can be expressed as follows from Equations (6.6) and (6.11), and from (6.8) and (6.12):

$$R_{3.2} = f_3/f_2 = 1 + N_2 N_1 \rho_2 / f_2 \geq 1.$$

This is also apparent by comparing Tables 6.1 and 6.2. Likewise, the ratio denoted by $R_{3.1}$ between the third and the second level randomizations can be expressed as follows from Equations (6.6) and (6.15), and from (6.8) and (6.16):

$$R_{3.1} = f_3/f_1 = 1 + \{N_1 \left(N_2 - 1\right) \rho_2 + N_1 \rho_1\}/f_1 \geq 1.$$

This is evident by comparing Tables 6.1 and 6.3. Finally, the ratio denoted by $R_{2.1}$ between the third and the second level randomizations can be expressed as follows from equation (6.11) and (6.15), and from (6.12) and (6.16):

$$R_{2.1} = f_2/f_1 = 1 + N_1 \left(\rho_1 - \rho_2\right)/f_1 \geq 1$$

TABLE 6.3
Sample size and power for detecting a main effect $\delta_{(1)}$ in model (6.13) when randomizations occur at the first level with $\rho_2 = 0.05$ (two-sided significance level $\alpha = 0.05$)

λ_1	$\Delta_{(1)}$	ρ_1	$N_1^{(0)}$	N_2	N_3	Total N	$\varphi_{(1)}$
1	0.3	0.1	5	4	8	320	0.807
			5	8	4	320	0.807
			10	4	4	320	0.807
			10	8	2	320	0.807
		0.2	5	4	8	320	0.851
			5	8	4	320	0.851
	0.4	0.1	5	4	6	240	0.904
			5	8	3	240	0.904
		0.2	5	4	4	160	0.807
			5	8	2	160	0.807
			10	2	4	160	0.807
			10	4	2	160	0.807
1.5	0.3	0.1	5	4	8	400	0.873
			5	8	4	400	0.873
			10	4	4	400	0.873
			10	8	2	400	0.873
		0.2	5	4	6	300	0.812
			5	8	3	300	0.812
	0.4	0.1	5	4	4	200	0.832
			5	8	2	200	0.832
		0.2	5	4	4	200	0.873
			5	8	2	200	0.873
			10	2	4	200	0.873
			10	4	2	200	0.873

Note: λ_1 represents the ratio of the number of level 1 units in the intervention arm to the corresponding number in the control arm within level 2 units; $\Delta_{(1)} = |\delta_{(1)}/\sigma|$; ρ_2 (6.2) is the correlations among the level 2 data; ρ_1 (6.3) is the correlations among the level 1 data; $N_1^{(0)}$ is the level 1 sample size in the *control* arm units per level 2 units; N_2 is the number of level 2 unit per level 3 units; N_3 is the *total* level 3 sample size; Total $N = (1 + \lambda_1)N_1^{(0)}N_2N_3$ is the required total number of observations; $\varphi_{(1)}$ is the statistical power to test $H_0 : \delta_{(1)} = 0$.

and is also shown by comparing Tables 6.2 and 6.3. In summary, for the same magnitudes of Δ, φ, α, ρ_1 and ρ_2, the sample size requirements are largest for trials with third level randomization, followed by second level randomization and then first level randomization.

6.3.5 Example

Many schools in low-income urban communities lack physical education facilities. To assess the impact on physical activity levels of Moving Smart (MS), a daily 10-minute classroom-based physical activity program led by elementary school teachers that integrates academic objectives with physical activity during the school day, a cluster-randomized trial was planned. In this design, two intervention schools are randomly chosen to administer the MS program to their students three times daily, for 10 minutes each time, in addition to regular physical education classes. Two control schools have the students receive their usual physical education classes only. A pedometer, an objective measure of physical activity in children, is used to determine the mean number of steps each student takes during school hours for 5 consecutive days at baseline, and every 3 months during the 2 year study period. The overall goal is to evaluate whether the MS program increases physical activity in elementary school children. The baseline activity of the subset of kindergarten and first grade students are described in Reznik et al. [7].

To illustrate the sample size methods in this chapter for three-level hierarchical data, we consider the primary outcome in the trial to be the mean number of steps per day at the end of the intervention period. The pedometer measurements obtained on 5 consecutive days at the end of the intervention period are assumed to be averaged prior to analysis, yielding one measurement per subject and a three-level hierarchical data structure: subjects (level 1) are nested within classrooms (level 2) which in turn are nested within schools (level 3). We further assume a balanced design with third level randomization in which $N_3 = N_3^{(1)} = N_3^{(0)} = 2$ schools per arm ($\lambda_3 = 1$), $N_2 = 20$ classrooms per school, and $N_1 = 25$ students per classroom, resulting in $2 \times 20 \times 25 = 1,000$ subjects per arm, or 2,000 subjects in total. In addition, prior studies suggest school-level intraclass correlations for physical activity in the range of (0.01 − 0.06). Based on these values, we set $\rho_1 = 0.06$ for the correlation in measures among students from the same classroom, and $\rho_2 = 0.01$ for the correlation in measures between students from different classrooms but within the same school.

Under these assumptions and using the approach described in Section 6.3.1 for testing the main intervention effect with three-level balanced designs and randomization at the third level, the variance inflation factor or design effect in (6.3.2) is equal to $f_3 = 1 + N_1(N_2 - 1)\rho_2 + (N_1 - 1)\rho_1 = 1 + 25(20 - 1)(0.01) + (25 - 1)(0.06) = 7.19$ and the minimum detectable standardized effect size is $\Delta_{(3)} = 0.34$ with 80% power ($\varphi_{(3)} = 0.809$) at a two-sided type I error rate of 5%. On the original scale, this effect size corresponds to a minimum difference between the intervention and control groups in the mean number of steps per day of approximately 340 steps assuming a standard deviation of 1,000 steps from prior studies.

Now consider the case where randomization to the MS program or control intervention could occur at the classroom (second) level rather than at the school (third) level. With the second level randomization design, the variance inflation factor is equal to $f_2 = 1 + (N_1 - 1)\rho_1 - N_1\rho_2 = 1 + (25 - 1)(0.06) - 25(0.01) = 2.19$. Comparing the two designs, the results in Section 6.3.4 indicate that the third level randomization design would require $R_{3.2} = f_3/f_2 = 7.19/2.19 = 3.3$ times the sample size required for a second level randomization design for the same magnitude of power and effect size. In other words, a total sample size of $1000/3.3 = \sim 300$ per arm would be needed with second level randomization as opposed to 1000 subjects per arm with third level randomization. Furthermore, since the power $\varphi_{(2)}$ with second level randomization is invariant over the product of $N_3 N_2^{(0)}$, the smaller total sample size could be achieved by either reducing $N_2 = N_2^{(0)} = N_2^{(1)}$, the number of classes per arm per school ($\lambda_2 = 1$), or reducing N_3, the total number of schools, as long as the product $N_3 N_2^{(0)}$ is equal to (300 subjects/N_1) $= 300/25 = 12$ and all other factors being equal. For example, the total number of schools could be maintained at 4 and $N_2^{(0)}$ reduced to 3 classrooms per arm per school, or the number of schools could be reduced to 2 and $N_2^{(0)}$ reduced to 6 classrooms per arm per school. In either case, $\varphi_{(2)} = 0.803$ with $\Delta_{(2)} = 0.34$ and 300 subjects per arm, or 600 subjects in total.

6.4   Testing Slope Differences

A longitudinal cluster randomized trial (longitudinal CRT) is a three-level design in which the repeated measurements obtained during follow-up are nested within subjects who are nested within a higher structure such as clinic. For example, primary care clinics are randomly assigned to either experimental or control intervention and subjects within clinics are repeatedly measured over time for outcomes such as blood pressure or glucose level. Under this type of design, the primary goal of the longitudinal CRT is to test whether the outcome trends over the follow-up period will be different between experimental and control intervention arms, that is, to test the significance of an interaction effect between time and intervention. The outcome trends can be modeled as subject-specific slopes, i.e., the regression coefficients of the time variable for the repeatedly measured study outcomes and these slopes can be considered fixed or random. In addition, the randomization can occur at the third cluster level across clusters (Section 6.4.1) or at the second subject level within clusters (Section 6.4.2). Sample size requirements for fixed and random slope models are compared in Section 6.4.3 and for testing main effects at the end of study are addressed in Section 6.4.5.

6.4.1   Randomization at the Third Level

Let us denote the level 3 sample sizes for the control ($X = 0$) and intervention ($X = 1$) arms by $N_3^{(0)}$ and $N_3^{(1)}$ while N_2 and N_1 represent the level 2 sample size per level 3 units and the level 1 sample size per level 2 unit, respectively.

6.4.1.1   Fixed Slope Model

When the subject-specific slopes of the outcome over time are assumed to be fixed as opposed to random, the fixed-effect term $h(X, T; \theta)$ in model (6.1) can be specified as

$$h(X, T; \theta) = \beta_0 + \xi X + \tau T + \delta_{(f)} XT,$$

and the variance of the random slopes is set to $\nu = 0$. This results in the following model:

$$Y_{ijk} = \beta_0 + \xi X_{ijk} + \tau T_{ijk} + \delta_{(f)} X_{ijk} T_{ijk} + u_i + u_{j(i)} + e_{ijk}. \qquad (6.17)$$

6.4.1.1.1   Balanced Allocations

Under a balanced design with $\Sigma_i X_i = N_3^{(1)}$, the third level unit is indexed by index $i = 1, 2, \ldots, 2N_3^{(0)}$. In addition, $j = 1, \ldots, N_2$, is the index for the level 2 unit nested within each i; and $k = 1, 2, \ldots, N_1$, is the index for the level 1 unit within each j. Therefore, $N_3 = N_3^{(0)} = N_3^{(1)}$ represents the number of third level units per arm. It is further assumed that $T_{ijk} = T_k$ for all i and j, and that time increases from 0 (the baseline) to $T_{end} = N_1 - 1$ (the last time point) by unit increments. Again, however, these assumptions are not required in general.

The parameter ξ represents the intervention effect at baseline, which can be non-zero, and the parameter τ represents the slope of the time effect, that is, the change in outcome over time. The overall fixed intercept is denoted by β_0. The intervention by time effect $\delta_{(f)}$ represents the difference in slopes between treatment arms. The null hypothesis to be tested is $H_0 : \delta_{(f)} = 0$. Figure 5.1 depicts a geometrical relationship among the fixed parameters in model (6.17) as well.

To test the significance of $\delta_{(f)}$ in model (6.17), the following test statistic can be used [8, 9]:

$$D_f = \frac{\sqrt{N_3 N_2 N_1 Var_p(T)}\,(\hat{\eta}_1 - \hat{\eta}_0)}{\sigma \sqrt{2(1 - \rho_1)}},$$

where $\hat{\eta}_g (g = 0, 1)$ is an ordinary least square (OLS) estimate of the slope for the outcome Y in the gth arm, in which $X_i = g$, so that $\hat{\delta}_{(f)} = \hat{\eta}_1 - \hat{\eta}_0$, which is an unbiased estimate of $\delta_{(f)}$. When $\nu = 0$, this OLS is identical to the maximum likelihood estimate. Specifically, for the gth arm,

$$\hat{\eta}_g = \sum_{i=1}^{N_3} \sum_{j=1}^{N_2} \sum_{k=1}^{N_1} \left(T_k - \bar{T}\right) \left(Y_{ijk} - \bar{Y}_g\right) \Big/ \sum_{i=1}^{N_3} \sum_{j=1}^{N_2} \sum_{k=1}^{N_1} \left(T_k - \bar{T}\right)^2$$

$$= \sum_{i=1}^{N_3} \sum_{j=1}^{N_2} \sum_{k=1}^{N_1} \left(T_k - \bar{T}\right) \left(Y_{ijk} - \bar{Y}_g\right) \Big/ N_3 N_2 N_1 Var_p(T)$$

(6.18)

where: (1) $\bar{Y}_g (g = 0, 1)$ is the overall mean of the outcome Y for the gth arm; (2) $\bar{T} = \sum_{k=1}^{N_1} T_k / N_1$ is the "mean" time point; and (3) $Var_p(T) = \sum_{k=1}^{N_1} \left(T_k - \bar{T}\right)^2 / N_1$ is the "population variance" of the time variable T. The variance of $\hat{\delta}_{(f)}$ can be obtained as

$$Var(\hat{\delta}_{(f)}) = Var\left(\hat{\eta}_1 - \hat{\eta}_0\right) = \frac{2(1 - \rho_1)\sigma^2}{N_3 N_2 N_1 Var_p(T)}.$$

(6.19)

It is worth noting that the variance of $\hat{\delta}$ depends only on the residual variance σ_e^2, and not on σ_3^2, σ_2^2, or ρ_2. Therefore, *for a given total variance* σ^2, the variance of $\hat{\delta}_{(f)}$ decreases with decreasing σ_e^2 or increasing ρ_1, the correlation among the first level data.

The power of the test statistic D_f can be written as follows:

$$\varphi_{(f)} = \Phi \left\{ \Delta_{(f)} \sqrt{N_3 N_2 N_1 Var_p(T)/2(1 - \rho_1)} - \Phi^{-1}(1 - \alpha/2) \right\},$$

where the slope difference is expressed in terms of a standardized effect size, that is, $\Delta_{(f)} = \left|\delta_{(f)}/\sigma\right|$. It follows that the third level sample size N_3 per arm for a desired statistical power $\varphi_{(f)} = \phi$ can be obtained as

$$N_3 = \frac{2(1 - \rho_1)z_{\alpha,\phi}^2}{N_2 N_1 Var_p(T)\Delta_{(f)}^2}.$$

(6.20)

The sample size N_2 for the second level can be immediately determined from Equation (6.20) as follows since N_2 has a reciprocal relationship with N_3:

$$N_2 = \frac{2(1 - \rho_1)z_{\alpha,\phi}^2}{N_3 N_1 Var_p(T)\Delta_{(f)}^2}.$$

Here we note that the power $\varphi_{(f)}$ is invariant over the product of $N_3 N_2$. The sample size N_1 for the first level should, however, be determined in an iterative manner because $Var_p(T)$ is a function of N_1. Specifically, an iterative solution of (6.20) for N_1 must satisfy the following equation:

$$N_1 = \frac{2(1 - \rho_1)z_{\alpha,\phi}^2}{N_3 N_2 Var_p(T)\Delta^2}.$$

6.4.1.1.2   Unbalanced Allocations

Again, suppose that the random allocation needs to be unbalanced such that $\Sigma_i X_i = \lambda_3 N_3^{(0)}$ for $\lambda_3 = N_3^{(1)}/N_3^{(0)} > 0$ and $i = 1, 2, \ldots, (1 + \lambda_3)N_3^{(0)}$. Then, the sample size $N_3^{(0)}$ in the control arm can be expressed as

$$N_3^{(0)} = \frac{(1 - \rho_1)\,(1 + 1/\lambda_3)\,z_{\alpha,\phi}^2}{N_2 N_1 Var_p(T)\Delta_{(f)}^2}. \tag{6.21}$$

And subsequently, due to the invariance of product of $N_3 N_2$ we have

$$N_2 = \frac{(1 - \rho_1)\,(1 + 1/\lambda_3)\,z_{\alpha,\phi}^2}{N_3^{(0)} N_1 Var_p(T)\Delta^2}.$$

But, N_1 must be solved iteratively to satisfy the following:

$$N_1 = \frac{(1 - \rho_1)\,(1 + 1/\lambda_3)\,z_{\alpha,\phi}^2}{N_3^{(0)} N_2 Var_p(T)\Delta^2}.$$

If the sample size N_2 needs to vary across the third level units, that is, $j = 1, 2, \ldots, n_i$, then it can be replaced by $\tilde{N}_2 = \sum_{i=1}^{N_3^{(0)}+N_3^{(1)}} n_i/(N_3^{(0)} + N_3^{(1)})$.

6.4.1.1.3   Summary

The sample size N_3 in (6.20) or $N_3^{(0)}$ in (6.21) required for a desired magnitude of statistical power is not a function of ρ_2 and decreases with increasing N_1, N_2, $\Delta_{(f)}$, and ρ_1. Table 6.4 displays examples of statistical power for testing the null hypothesis $H_0 : \delta_{(f)} = 0$, and shows that the statistical power $\varphi_{(f)}$ is invariant over the product of $N_3 N_2^{(0)}$, regardless of whether or not λ_3 is equal to 1.

6.4.1.2   Random Slope Model

When the subject-specific slopes of the outcome measurements over time are assumed to be random, the fixed-effect term $h(X, T; \theta)$ in model (6.1) is specified as

$$h(X, T; \theta) = \beta_0 + \xi X + \tau T + \delta_{(r)} XT,$$

and the variance of the random slopes is assumed to be distributed as $\nu_{j(i)} \sim N\left(0, \sigma_\tau^2\right)$. These specifications result in the following model

$$Y_{ijk} = \beta_0 + \xi X_{ijk} + \tau T_{ijk} + \delta_{(r)} X_{ijk} T_{ijk} + \nu_{j(i)} T_{ijk} + u_i + u_{j(i)} + e_{ijk}. \tag{6.22}$$

TABLE 6.4

Sample size and power for detecting an effect $\delta_{(f)}$ on slope differences in a three-level fixed-slope model (6.17) with $r_\tau = 0$ when randomizations occur at the third level (two-sided significance level $\alpha = 0.05$)

λ_3	ρ_1	$\Delta_{(f)}$	N_1	N_2	$N_3^{(0)}$	Total N	$\varphi_{(f)}$
1	0.1	0.3/4	5	8	32	2560	0.807
			5	32	8	2560	0.807
		0.3/8	9	8	21	3024	0.845
			9	21	8	3024	0.845
	0.2	0.3/4	5	8	28	2240	0.801
		0.3/8	9	8	19	2736	0.808
	0.1	0.4/4	5	8	18	1440	0.807
		0.4/8	9	8	12	1728	0.807
	0.2	0.4/4	5	8	16	1280	0.807
			5	16	8	1280	0.807
		0.4/8	9	8	11	1584	0.819
			9	11	8	1584	0.819
1.5	0.1	0.3/4	5	8	27	2700	0.812
			5	27	8	2700	0.812
		0.3/8	9	8	18	3240	0.812
			9	18	8	3240	0.812
	0.2	0.3/4	5	8	24	2400	0.812
		0.3/8	9	8	16	2880	0.812
	0.1	0.4/4	5	8	15	1500	0.807
		0.4/8	9	8	10	1800	0.807
	0.2	0.4/4	5	8	14	1400	0.826
			5	14	8	1400	0.826
		0.4/8	9	8	9	1620	0.812
			9	9	8	1620	0.812

Note: λ_3 represents the ratio of the number of level 3 units in the intervention arm to the corresponding number in the control arm; $\Delta_{(f)} = \left|\delta_{(f)}/\sigma\right|$; ρ_2 (6.2) is the correlations among the level 2 data; ρ_1 (6.3) is the correlations among the level 1 data; N_1 is the number of level 1 units per level 2 units; N_2 is the number of level 2 units per level 3 units; $N_3^{(0)}$ is the level 3 sample size in the *control* arm; Total $N = (1 + \lambda_3)N_1 N_2 N_3^{(0)}$ is the required total number of observations; $\varphi_{(f)}$ is the statistical power to test $H_0 : \delta_{(f)} = 0$.

6.4.1.2.1   Balanced Allocations

To test the significance of $\delta_{(r)}$ in model (6.22) in the balanced case, the following test statistic can be used [10]:

$$D_r = \frac{(\hat{\eta}_1 - \hat{\eta}_0)\sqrt{N_3 N_2 N_1 Var_p(T)}}{\sqrt{2\left\{(1 - \rho_1)\sigma^2 + N_1 Var_p(T)\sigma_\tau^2\right\}}},$$

where $\hat{\eta}_g (g = 0, 1)$ is an ordinary least square (OLS) estimate of the slope for the outcome Y as in Equation (6.18) so that $\hat{\delta}_{(r)} = \hat{\eta}_1 - \hat{\eta}_0$, which is an unbiased estimate of $\delta_{(r)}$. For the random slope with $\nu_{j(i)} \sim N\left(0, \sigma_\tau^2\right)$ this OLS is not necessarily identical to the maximum likelihood estimate (MLE). However, the test statistic can be constructed based on the OLS estimate. In addition, the variance of $\hat{\delta}_{(r)}$ can be obtained as

$$Var(\hat{\delta}_{(r)}) = Var\left(\hat{\eta}_1 - \hat{\eta}_0\right) = \frac{2\left\{(1 - \rho_1)\sigma^2 + N_1 Var_p(T)\sigma_\tau^2\right\}}{N_3 N_2 N_1 Var_p(T)}, \qquad (6.23)$$

which is again not a function of ρ_2. The power of the test statistic D_r can therefore be expressed as follows:

$$\varphi_{(r)} = \Phi\left\{\Delta_{(r)}\sqrt{\frac{N_3 N_2 N_1 Var_p(T)}{2\left\{(1 - \rho_1) + r_\tau N_1 Var_p(T)\right\}}} - \Phi^{-1}(1 - \alpha/2)\right\},$$

where test $\Delta_{(r)} = \left|\delta_{(r)}/\sigma\right|$ and

$$r_\tau = \sigma_\tau^2 / (\sigma_2^2 + \sigma_3^2 + \sigma_e^2) = \sigma_\tau^2/\sigma^2 \qquad (6.24)$$

is the ratio of the random slope variance to the sum of the other variances. Note that this is an extension of r_τ (5.4.6) in section 5.4.2.1 for the two-level continuous outcome analysis. It follows that the required sample size per arm for the third level N_3 can be obtained as [8, 10]:

$$N_3 = \frac{2\left\{(1 - \rho_1) + r_\tau N_1 Var_p(T)\right\} z_{\alpha,\phi}^2}{N_2 N_1 Var_p(T)\Delta_{(r)}^2}. \qquad (6.25)$$

Here we note that the power $\varphi_{(r)}$ is also invariant over the product of $N_3 N_2$. It follows that N_2 per third level units can be expressed as

$$N_2 = \frac{2\left\{(1 - \rho_1) + r_\tau N_1 Var_p(T)\right\} z_{\alpha,\phi}^2}{N_3 N_1 Var_p(T)\Delta_{(r)}^2}.$$

Nevertheless, the sample size for the level 1 data, N_1, needs to be determined in an iterative manner because $Var_p(T)$ is a function of N_1 and the solution for N_1 must satisfy the following equation:

$$N_1 = \frac{(1 - \rho_1)}{Var_p(T)\left\{N_3 N_2 \Delta_{(r)}^2 \Big/ \left(2 z_{\alpha,\phi}^2\right) - r_\tau\right\}}.$$

6.4.1.2.2 Unbalanced Allocations

Again, suppose that the random allocation needs to be unbalanced such that $\Sigma_i X_i = \lambda_3 N_3^{(0)}$ for $\lambda_3 = N_3^{(1)}/N_3^{(0)} > 0$ and $i = 1, 2, \ldots, (1 + \lambda_3) N_3^{(0)}$. Then, the sample size $N_3^{(0)}$ in the control arm can be expressed as

$$N_3^{(0)} = \frac{(1 + 1/\lambda_3)\left\{(1 - \rho_1) + r_\tau N_1 Var_p(T)\right\} z_{\alpha,\phi}^2}{N_2 N_1 Var_p(T) \Delta_{(r)}^2} \qquad (6.26)$$

Due to the invariance over product property, N_2 can be immediately determined as

$$N_2 = \frac{(1 + 1/\lambda_3)\left\{(1 - \rho_1) + r_\tau N_1 Var_p(T)\right\} z_{\alpha,\phi}^2}{N_3^{(0)} N_1 Var_p(T) \Delta_{(r)}^2},$$

but N_1 should be determined iteratively so that it can satisfy the following equation

$$N_1 = \frac{(1 - \rho_1)}{Var_p(T)\left[N_3^{(0)} N_2 \Delta_{(r)}^2 \Big/ \left\{(1 + 1/\lambda_3) z_{\alpha,\phi}^2\right\} - r_\tau\right]}.$$

If the sample size N_2 needs to vary across the third level units, that is, $j = 1, 2, \ldots, n_i$, then it can be replaced by $\tilde{N}_2 = \sum_{i=1}^{N_3^{(0)} + N_3^{(1)}} n_i / (N_3^{(0)} + N_3^{(1)})$.

6.4.1.2.3 Summary

The sample size N_3 in (6.25) or $N_3^{(0)}$ in (6.26) required for a desired magnitude of statistical power is not a function of ρ_2 and decreases with increasing N_2, $\Delta_{(f)}$, and ρ_1, but increases with increasing r_τ. Table 6.6 displays examples of statistical power for testing the null hypothesis $H_0 : \delta_{(r)} = 0$, and shows that the statistical power $\varphi_{(r)}$ is invariant over the product of $N_3 N_2^{(0)}$, regardless of whether or not λ_3 is equal to 1.

6.4.2 Randomization at the Second Level

One of the important assumptions of the size approaches discussed so far is that the randomization occurs at the third level, e.g., clinic, so that all the subjects within the cluster receive the same intervention, experimental or control. But in many settings, the randomization needs to be made at the subject level (i.e., second level) within clusters instead of at the third level. In this case, the slope estimates in the two intervention arms within clusters might not necessarily be independent because data from subjects who are randomized to different interventions within the same cluster are correlated, and this correlation may affect sample size calculations. The sample size requirements for testing main effects are quite different across randomizations at different levels as discussed in Sections 6.3.1 to 6.3.4.

It can be shown, however, that both $Var(\hat{\delta}_{(f)})$ (6.19) and $Var(\hat{\delta}_{(r)})$ (6.23) will be unaffected even when randomization occurs at the subject (second)

TABLE 6.5

Sample size and power for detecting an effect $\delta_{(r)}$ on slope differences in a three-level random-slope model (6.22) with $r_\tau = 0.1$ when randomizations occur at the third level (two-sided significance level $\alpha = 0.05$)

λ_3	ρ_1	$\Delta_{(r)}$	N_1	N_2	$N_3^{(0)}$	Total N	$\varphi_{(r)}$
1	0.1	0.3/4	5	8	67	5360	0.804
			5	67	8	5360	0.804
		0.3/8	9	8	161	23184	0.801
			9	161	8	23184	0.801
	0.2	0.3/4	5	8	63	5040	0.801
		0.3/8	9	8	159	22896	0.802
	0.1	0.4/4	5	8	38	3040	0.807
		0.4/8	9	8	91	13104	0.803
	0.2	0.4/4	5	8	36	2880	0.807
			5	36	8	2880	0.807
		0.4/8	9	8	89	12816	0.800
			9	89	8	12816	0.800
1.5	0.1	0.3/4	5	8	56	5600	0.805
			5	56	8	5600	0.805
		0.3/8	9	8	134	24120	0.801
			9	134	8	24120	0.801
	0.2	0.3/4	5	8	53	5300	0.805
		0.3/8	9	8	132	23760	0.801
	0.1	0.4/4	5	8	32	3200	0.811
		0.4/8	9	8	76	13680	0.804
	0.2	0.4/4	5	8	30	3000	0.807
			5	30	8	3000	0.807
		0.4/8	9	8	75	13500	0.805
			9	75	8	13500	0.805

Note: λ_3 represents the ratio of the number of level 3 units in the intervention arm to the corresponding number in the control arm; $\Delta_{(r)} = \left| \delta_{(r)}/\sigma \right|$; ρ_2 (6.2) is the correlations among the level 2 data; ρ_1 (6.3) is the correlations among the level 1 data; N_1 is the number of level 1 units per level 2 units; N_2 is the number of level 2 units per level 3 units; $N_3^{(0)}$ is the level 3 sample size in the *control* arm; Total $N = (1 + \lambda_3)N_1 N_2 N_3^{(0)}$ is the required total number of observations; $\varphi_{(r)}$ is the statistical power to test $H_0 : \delta_{(r)} = 0$.

level within the third/cluster level units [11]. In other words, even if the correlation between slope estimates within clusters is expected to be non-zero, it can be shown that $Cov\,(\hat{\eta}_0, \hat{\eta}_1) = 0$ regardless of whether the slopes are assumed to be fixed or random. It follows that the sample size formulas (6.20) and (6.25) under fixed and random slope models, respectively, are identical to those when randomization occurs at the second level. This property also applies when the randomization is unbalanced, that is, the formulas (6.21) and

(6.26) will remain identical as well, the only differences being the definitions of the sample sizes for the three-levels: N_3, N_2, and N_1. For trials with randomization at the third level, N_3 or $N_3^{(0)}$ = number of clusters for the control arm; N_2 = number of subjects per cluster; N_1 = number of observations per subject. For trials with randomization at the second level, N_3 = total number of clusters, N_2 or $N_2^{(0)}$ = number of subjects for the control arm within clusters and N_1 is defined as before. Nonetheless, the total number of observations, $(1 + \lambda_3)N_3^{(0)}N_2N_1 = (1 + \lambda_2)N_3N_2^{(0)}N_1$, remains the same for trials with third or second level randomization. We call this property invariance over level of randomization, which was validated by simulation studies [11]. This property along with the invariance over product enables clinical trialists and statisticians to flexibly plan longitudinal CRTs in regard to the level of randomizations and allocations of level 3 and level 2 units.

6.4.3   Comparison of Sample Sizes

When the randomization occurs at the third cluster level, the ratio $R_{r.f}(N_3)$ of sample size N_3 required for trials with random slopes to that required for fixed slopes for balanced trials can be expressed as follows from Equations (6.20) and (6.25) for fixed Δ, φ, α, and ρ:

$$R_{\mathrm{r.f}}(N_3; r_\tau, N_1, \rho_1) \equiv \frac{N_3 \,|r_\tau > 0}{N_3 \,|r_\tau = 0} = 1 + r_\tau N_1 Var_p(T)/(1 - \rho_1). \qquad (6.27)$$

This ratio holds even for unbalanced trials with identical $\lambda = (\lambda_{11}, \lambda_{10}, \lambda_{01})$ values between random and fixed slopes as shown by comparing Equations (6.21) and (6.26). The ratio $R_{r.f}(N_3)$ does not depend on N_2 or ρ_2, and therefore, if ρ_1 in Equation (6.3) is equal to ρ in Equation (5.2.2), $R_{r.f}(N_3)$ is Equal to $R_{r.f}(N_2)$ in Equation (5.4.9) in Section 5.4.3. Again, under the assumption that the value of T increases from 0 to $T_{end} = N_1 - 1$ by unit time increments, $R_{r.f}(N_3)$ can be expressed as follows:

$$R_{\mathrm{r.f}}(N_3) = 1 + r_\tau N_1(N_1^2 - 1)/\{12(1 - \rho_1)\}.$$

Thus, $R_{r.f}(N_3)$ is an increasing function of r_τ, ρ_1 and N_1. Like $R_{r.f}(N_2)$, the effect of N_1 on the sample size ratio $R_{r.f}(N_3)$ is greatest for larger r_τ because the variance of the outcome increases quadratically with N_1 and the magnitude of the increase in the variance is larger for larger r_τ. The effect of r_τ on $R_{r.f}(N_3)$ is again enormous in that $R_{r.f}(N_3)$ is as high as 7.7 even for $r_\tau = 0.1$ for $N_1 = 9$, and $\rho_1 = 0.1$, as shown by comparing the results in Tables 6.4 and 6.5. Furthermore, the ratios above should be identical to the sample size ratios between random and fixed models when randomization occurs at the second level since the sample size determinations remain unchanged between third and second level randomizations.

6.4.4 Testing Main Effects at the End of Follow-Up

When the hypothesis focuses on the intervention effect at the end of follow-up (i.e., when $T_k = T_{end}$) under model (6.17) and assuming fixed slopes with third level randomization, a contrast representing that intervention effect can be constructed on a shifted scale of the time variable $T'_k = T_k - T_{end}$ (increasing from $-T_{end}(= 1 - N_1)$ to 0 by 1) as follows:

$$Y_{ijk} = \beta'_0 + \delta_{(e)} X_i + \tau T'_k + \delta_{(f)} X_i T'_k + u_i + u_{j(i)} + e_{ijk}, \qquad (6.28)$$

where $\beta'_0 = \beta_0 + \tau T_{end}$ and $\delta_{(e)} = \xi + \delta_{(f)} T_{end}$. Since $T_k = T_{end}$ is equivalent to $T'_k = 0$, the reparameterized intervention effect $\delta_{(e)}$ in model (6.28) represents the intervention effect at the end of the study, which is geometrically depicted in Figure 5.1. Here again, we assume for the purpose of deriving the power function that the parameter ξ in model (6.17) representing the intervention effect at baseline is zero which is plausible due to randomization.

6.4.4.1 Balanced Allocations

The null hypothesis to be tested is $H_0 : \delta_{(e)} = 0$. Because subjects are randomized to the two intervention arms, we assume no mean difference in outcome Y between the two arms at baseline, that is, $\xi = 0$ in model (6.17). Under this assumption, the null hypothesis can be tested with

$$D_e = \frac{\sqrt{N_3 N_2 N_1} \hat{\delta}_{(e)}}{\sigma \sqrt{2 f_3 C_{(3)}}},$$

where $f_3 = 1 + N_1(N_2 - 1)\rho_2 + (N_1 - 1)\rho_1$, $\hat{\delta}_{(e)} = (\bar{Y}_1 - \bar{Y}_0) - \hat{\delta}_{(f)} \bar{T}'$, $\bar{T}' = \sum_{k=1}^{N_1} T'_k / N_1$, $\hat{\delta}_{(f)} = \hat{\eta}_1 - \hat{\eta}_0$ and $C_{(3)} = 1 + (1 - \rho_1) CV^{-2}(T') / f_3$; $CV(T') = SD_p(T') / \bar{T}'$ is the coefficient of variation (CV) of the time variable T', and $SD_p(T')$ is a "population" standard deviation or the square root of $Var_p(T')$. The variance of $\hat{\delta}_{(e)}$ can be computed as

$$Var(\hat{\delta}_{(e)}) = Var\left(\bar{Y}_1 - \bar{Y}_0\right) + \bar{T}'^2 Var(\hat{\delta}_{(f)}) = \frac{2 f \sigma^2}{N_3 N_2 N_1} C_{(3)}.$$

Accordingly, the statistical power of the test statistic D_e can be written as follows:

$$\varphi_{(e)} = \Phi \left\{ \Delta_{(e)} \sqrt{\frac{N_3 N_2 N_1}{2 f_3 C_{(3)}}} - \Phi^{-1}(1 - \alpha/2) \right\},$$

where $\Delta_{(e)} = |\delta_{(e)}/\sigma|$ is a standardized effect size, also known as Cohen's d [12].

The third level unit sample size N_3 per arm for a desired statistical power $\varphi_{(e)} = \phi$ can be obtained as [13]

$$N_3 = \frac{2 f_3 C_{(3)} z_{\alpha,\phi}^2}{N_2 N_1 \Delta_{(e)}^2}. \qquad (6.29)$$

The sample size N_2 of the level 2 data units can be obtained by solving for it from equation (6.29) as follows:

$$N_2 = \frac{2\left\{1 - \rho_1 + (\rho_1 - \rho_2)N_1 + (1 - \rho_1)CV^{-2}(T')\right\} z_{\alpha,\phi}^2}{N_1 N_3 \Delta_{(e)}^2 - 2\rho_2 N_1 z_{\alpha,\phi}^2}.$$

The sample size N_1 for the level 1 data should, however, be an iterative solution that must satisfy the following equation:

$$N_1 = \frac{2\left(1 - \rho_1\right)\left(1 + CV^{-2}(T')\right) z_{\alpha,\phi}^2}{N_2 N_3 \Delta_{(e)}^2 - 2\left\{(N_2 - 1)\rho_2 + \rho_1\right\} z_{\alpha,\phi}^2}.$$

Note that elongation of the time intervals will not affect the sample size (6.29) so that the required sample sizes with time intervals $t = T_k - T_{k-1}$ for all k will be the same as those with time intervals wt for any $w > 0$.

6.4.4.2   Unbalanced Allocations

Again, suppose that the random allocation needs to be unbalanced such that $\Sigma_i X_i = \lambda_3 N_3^{(0)}$ for $\lambda_3 = N_3^{(1)}/N_3^{(0)} > 0$ and $i = 1, 2, \ldots, (1 + \lambda_3)N_3^{(0)}$. Then, the sample size $N_3^{(0)}$ in the control arm can be expressed as

$$N_3^{(0)} = \frac{f_3 C_{(3)}\left(1 + 1/\lambda_3\right) z_{\alpha,\phi}^2}{N_2 N_1 \Delta_{(e)}^2}. \tag{6.30}$$

The sample size N_2 can be expressed as

$$N_2 = \frac{(1 + 1/\lambda_3)\left\{1 - \rho_1 + (\rho_1 - \rho_2)N_1 + (1 - \rho_1)CV^{-2}(T')\right\} z_{\alpha,\phi}^2}{N_1 N_3^{(0)} \Delta_{(e)}^2 - (1 + 1/\lambda_3)\rho_2 N_1 z_{\alpha,\phi}^2}.$$

An iterative solution of Equation (6.30) for N_1 must satisfy the following equation:

$$N_1 = \frac{(1 + 1/\lambda_3)\left(1 - \rho_1\right)\left(1 + CV^{-2}(T')\right) z_{\alpha,\phi}^2}{N_2 N_3^{(0)} \Delta_{(e)}^2 - (1 + 1/\lambda_3)\left\{(N_2 - 1)\rho_2 + \rho_1\right\} z_{\alpha,\phi}^2}.$$

If the sample size N_2 needs to vary across the level 3 units, that is, $j = 1, 2, \ldots, n_i$, then it can be replaced by $\tilde{N}_2 = \sum_{i=1}^{N_3^{(0)}+N_3^{(1)}} n_i/(N_3^{(0)} + N_3^{(1)})$.

6.4.4.3   Summary

The sample size N_3 in (6.29) or $N_3^{(0)}$ in (6.30) required for a desired magnitude of statistical power decreases with increasing N_1, N_2, and $\Delta_{(e)}$. Table 6.6 displays examples of statistical power for testing the null hypothesis $H_0 : \delta_{(e)} = 0$.

TABLE 6.6

Sample size and power for detecting a main effect $\delta_{(e)}$ at the end of study in a three-level fixed-slope model (6.28) when randomizations occur at the third level (two-sided significance level $\alpha = 0.05$)

λ_3	$\Delta_{(e)}$	ρ_1	N_1	N_2	$N_3^{(0)}$	$\varphi_{(e)}$
1	0.3	0.1	5	4	35	0.806
			5	8	22	0.807
		0.2	5	4	37	0.809
			5	8	23	0.809
	0.4	0.1	5	4	20	0.812
			5	8	13	0.826
		0.2	5	4	21	0.812
			5	8	13	0.811
1.5	0.3	0.1	5	4	29	0.804
			5	8	18	0.800
		0.2	5	4	31	0.811
			5	8	19	0.806
	0.4	0.1	5	4	17	0.820
			5	8	11	0.832
		0.2	5	4	17	0.801
			5	8	11	0.817

Note: λ_3 represents the ratio of the number of level 3 units in the intervention arm to the corresponding number in the control arm; $\Delta_{(e)} = |\delta_{(e)}/\sigma|$; ρ_2 (6.2) is the correlations among the level 2 data; ρ_1 (6.3) is the correlations among the level 1 data; N_1 is the number of level 1 units per level 2 units; N_2 is the number of level 2 units per level 3 units; $N_3^{(0)}$ is the level 3 sample size in the *control* arm; $\varphi_{(e)}$ is the statistical power to test $H_0 : \delta_{(e)} = 0$.

6.5 Cross-Sectional Factorial Designs: Interactions between Treatments

In a 2×2 factorial design with experimental treatments X and Z, it is often of interest to test the significance of the effect of the interaction between X and Z on the study outcome. The principles for determining the sample size for interaction effects presented in the last chapter for two level data still hold for three-level data, i.e., the sample sizes required for testing the interaction effect are four times as large as those required for testing a main effect. When randomizations are assigned at the mth level, one can test the interaction effect using the following mixed-effects linear model:

$$Y_{ijk} = \beta_0 + \delta_{X(m)} X_{ijk} + \delta_{Z(m)} Z_{ijk} + \delta_{XZ(m)} X_{ijk} Z_{ijk} + u_i + u_{j(i)} + e_{ijk} \quad (6.31)$$

Sections 6.5.1, 6.5.2, and 6.5.3 present sample size determinations when randomization occurs at the third ($m = 3$), second ($m = 2$), and first ($m =$

1) levels, respectively. Comparisons of sample sizes across different levels of randomizations are discussed in Section 6.3.4.

6.5.1 Randomization at the Third Level

For trials with $m = 3$, we define the intervention assignment indicator for treatment X as $X_{ijk} = 0$ if the ith level 3 unit is not assigned to receive X and $X_{ijk} = 1$ if assigned to receive X; therefore $X_{ijk} = X_i$ for all j and k. Likewise, we define the assignment indicator for treatment Z as $Z_{ijk} = 0$ if the ith level 3 unit is not assigned to receive Z and $Z_{ijk} = 1$ if assigned to receive Z; therefore $Z_{ijk} = Z_i$ for all j and k. Here, the null hypothesis of interest. Here, the null hypothesis of interest is $H_0 : \delta_{XZ(3)} = 0$ in model (6.31) for $m = 3$.

6.5.1.1 Balanced Allocations

For a balanced design it is assumed that $\Sigma_i X_i Z_i = N_3^{(1,1)}$ so that $i = 1, 2, \ldots, 4N_3^{(1,1)}, j = 1, \ldots, N_2$ and $k = 1, 2, \ldots, N_1$. Specifically, $N_3^{(1,1)}$ represents the level 3 sample size assigned to $(X = 1, Z = 1)$ arm, or the (1,1) cell in a 2-by-2 factorial table of X and Z, N_2 is the level 2 sample size per 3 level unit2, and N_1 is the level 1 sample size per level two unit2. In a balanced design, the sample sizes of the other cells are identical to $N_3^{(1,1)}$, i.e., $N_3^{(1,1)} = N_3^{(1,0)} = N_3^{(0,1)} = N_3^{(0,0)}$. The following test statistic can be used to test $H_0 : \delta_{XZ(3)} = 0$:

$$D_{XZ(3)} = \frac{\sqrt{N_3^{(0,0)} N_2 N_1} \left\{ \left(\bar{Y}^{(1,1)} - \bar{Y}^{(1,0)}\right) - \left(\bar{Y}^{(0,1)} - \bar{Y}^{(0,0)}\right) \right\}}{\sigma \sqrt{4f_3}}.$$

where $\bar{Y}^{(x,z)}$ is the mean of Y in the $(X = x, Z = z)$ cells and $f_3 = 1 + N_1(N_2 - 1)\rho_2 + (N_1 - 1)\rho_1$. It can be shown that $\hat{\delta}_{XZ(3)} = \left(\bar{Y}^{(1,1)} - \bar{Y}^{(1,0)}\right) - \left(\bar{Y}^{(0,1)} - \bar{Y}^{(0,0)}\right)$ is an unbiased estimate of $\delta_{XZ(3)}$ and

$$Var(\hat{\delta}_{XZ(3)}) = \frac{4f_3 \sigma^2}{N_3^{(0,0)} N_2 N_1}.$$

The corresponding statistical power can be expressed as

$$\varphi_{XZ(3)} = \Phi \left\{ \Delta_{XZ(3)} \sqrt{N_3^{(0,0)} N_2 N_1 / 4f_3} - \Phi^{-1}(1 - \alpha/2) \right\},$$

where $\Delta = |\delta_{XZ(3)}/\sigma|$. Accordingly, the corresponding required sample size is

$$N_3^{(0,0)} = \frac{4f_3 z_{\alpha,\phi}^2}{N_2 N_1 \Delta_{XZ(3)}^2}. \tag{6.32}$$

The sample sizes of the second and first level units can be determined by solving (6.32) for N_2 and N_1, respectively.

Note that when Equations (6.32) and (6.6) are compared, $N_3^{(0,0)} = 2N_3$, where N_3 is the required sample size for testing $H_0 : \delta_{(3)} = 0$ in model (6.4) in Section 6.3.1. If follows that the total number of measurements, that is $4N_3^{(0,0)}N_2N_1$, required for testing $H_0 : \delta_{XZ(3)} = 0$ is four times larger than the total number of measurements, that is $2N_3N_2N_1$, required for testing $H_0 : \delta_{(3)} = 0$. In other words, the required sample size for testing an interaction is four times larger than that for testing a main effect. This finding can also be applied when randomization is assigned at the second or the first level as discussed further next.

6.5.1.2   Unbalanced Allocations

Under an unbalanced design for which $N_3^{(1,1)} = \lambda_{11} N_3^{(0,0)}$, $N_3^{(1,0)} = \lambda_{10} N_3^{(0,0)}$, $N_3^{(0,1)} = \lambda_{01} N_3^{(0,0)}$, the sample size $N_3^{(0,0)}$ can be determined as

$$N_3^{(0,0)} = \frac{f_3 \left(1 + 1/\lambda_{11} + 1/\lambda_{10} + 1/\lambda_{01}\right) z_{\alpha,\phi}^2}{N_2 N_1 \Delta_{XZ(3)}^2}. \tag{6.33}$$

6.5.1.3   Summary

The sample size $N_3^{(0,0)}$ in (6.32) or in (6.33) required for a desired magnitude of statistical power decreases with increasing N_1, N_2, and $\Delta_{XZ(3)}$ but increases with increasing ρ_1 and ρ_2. Table 6.7 displays examples of statistical power for testing the null hypothesis $H_0 : \delta_{XZ(3)} = 0$. When compared to N_3 in Table 6.1, it can be seen under balanced designs that $N_3^{(0,0)} = 2N_3^{(0)}$ and $N_3^{(0,0)}N_2N_1 = 4N_3^{(0)}N_2N_1$ for the same magnitude of power $\varphi_{(3)} = \varphi_{XZ(3)}$. Under unbalanced designs, $N_3^{(0,0)}$ is approximately twice as large.

6.5.2   Randomization at the Second Level

For trials with $m = 2$ in model (6.31), the intervention assignment indicator variable $X_{ijk} = 0$ if the jth level 2 unit is not assigned to receive X and $X_{ijk} = 1$ if assigned to receive X; therefore $X_{ijk} = X_j$ for all i and k. Likewise, we define the assignment indicator for treatment Z as $Z_{ijk} = 0$ if the jth level 2 unit is not assigned to receive Z and $Z_{ijk} = 1$ if assigned to receive Z; therefore $Z_{ijk} = Z_j$ for all i and k. Here, the null hypothesis to be tested is $H_0 : \delta_{XZ(2)} = 0$ in model (6.31) for $m = 2$.

6.5.2.1   Balanced Allocations

For a balanced design it is assumed that $\Sigma_j X_j Z_j = N_2^{(1,1)}$ so that $i = 1, 2, \ldots, N_3, j = 1, \ldots, 4N_2^{(1,1)}$ and $k = 1, 2, \ldots, N_1$. Specifically, $N_2^{(1,1)}$

TABLE 6.7
Sample size and power for detecting a two-way interaction XZ effect $\delta_{XZ(3)}$ in model with $m = 3$ (6.31) for a 2-by-2 factorial design when randomizations occur at the third level with $\rho_2 = 0.05$ (two-sided significance level $\alpha = 0.05$)

λ	Δ	ρ_1	N_1	N_2	$N_3^{(0,0)}$	Total N	φ
1	0.3	0.1	5	4	38	3040	0.805
			5	8	28	4480	0.807
		0.2	5	4	45	3600	0.805
			5	8	31	4960	0.801
	0.4	0.1	5	4	22	1760	0.816
			5	8	16	2560	0.813
		0.2	5	4	26	2080	0.815
			5	8	18	2880	0.813
1.5,1.5,1.5	0.3	0.1	5	4	29	3190	0.812
			5	8	21	4620	0.807
		0.2	5	4	34	3740	0.807
			5	8	24	5280	0.813
	0.4	0.1	5	4	16	1760	0.804
			5	8	12	2640	0.813
		0.2	5	4	19	2090	0.805
			5	8	14	3080	0.827

Note: $N_3^{(0,1)} = \lambda_{01} N_3^{(0,0)}$; $N_3^{(1,0)} = \lambda_{10} N_3^{(0,0)}$; $N_3^{(1,1)} = \lambda_{11} N_3^{(0,0)}$; $\Delta_{XZ(3)} = |\delta_{XZ(3)}/\sigma|$; ρ_2 (6.2) is the correlations among the level 2 data; ρ_1 (6.3) is the correlations among the level 1 data; N_1 is the number of level 1 units per level 2 unit; N_2 is the number of level 2 units per level 3 units; $N_3^{(0,0)}$ is the level 3 sample size in the $(X = 0, Z = 0)$ arm; Total $N = (1 + \lambda_{01} + \lambda_{10} + \lambda_{11}) N_1 N_2 N_3^{(0,0)}$ is the required total number of observations; $\varphi_{XZ(3)}$ is the statistical power to test $H_0 : \delta_{XZ(3)} = 0$.

represents the level 2 sample size assigned to the $(X = 1, Z = 1)$ arm, or the $(1,1)$ cell, in a 2-by-2 factorial table of X and Z within the third level units, N_3 is the level 3 sample size in total, and N_1 is the level 1 sample size per level 2 units. In a balanced design, the sample sizes of the other cells are identical to $N_2^{(1,1)}$, i.e., $N_2^{(1,1)} = N_2^{(1,0)} = N_2^{(0,1)} = N_2^{(0,0)}$. The following test statistic can be used to test $H_0 : \delta_{XZ(2)} = 0$:

$$D_{XZ(2)} = \frac{\sqrt{N_3 N_2^{(0,0)} N_1} \left\{ \left(\bar{Y}^{(1,1)} - \bar{Y}^{(1,0)} \right) - \left(\bar{Y}^{(0,1)} - \bar{Y}^{(0,0)} \right) \right\}}{\sigma \sqrt{4 f_2}},$$

where $\bar{Y}^{(x,z)}$ is the mean of Y in $(X = x, Z = z)$ cells and $f_2 = 1 + (N_1 - 1)\rho_1 - N_1 \rho_2$. It can be shown that $\hat{\delta}_{XZ(2)} = \left(\bar{Y}^{(1,1)} - \bar{Y}^{(1,0)} \right) - \left(\bar{Y}^{(0,1)} - \bar{Y}^{(0,0)} \right)$ is an unbiased estimate of $\delta_{(2)}^{xz}$ and

$$Var(\hat{\delta}_{XZ(2)}) = \frac{4f_2\sigma^2}{N_3 N_2^{(0,0)} N_1}.$$

The corresponding statistical power can be expressed as

$$\varphi_{XZ(2)} = \Phi\left\{\Delta_{XZ(2)}\sqrt{N_3 N_2^{(0,0)} N_1 / 4f_2} - \Phi^{-1}(1 - \alpha/2)\right\},$$

where $\Delta_{XZ(2)} = |\delta_{XZ(2)}/\sigma|$. Accordingly, the corresponding required sample size is

$$N_3 = \frac{4f_2 z_{\alpha,\phi}^2}{N_2^{(0,0)} N_1 \Delta_{XZ(2)}^2}. \tag{6.34}$$

The sample sizes of the second and first level units can be determined by solving Equation (6.34) for $N_2^{(0,0)}$ and N_1, respectively. In fact, the power $\varphi_{XZ(2)}$ is invariant over the product of $N_3 N_2^{(0,0)}$, which makes the solution of $N_2^{(0,0)}$ immediate from Equation (6.34).

Note that when Equations (6.34) and (6.11) are compared, $N_2^{(0,0)} = 2N_2$, where N_2 is the required sample size of the second level for testing $H_0 : \delta_{(2)} = 0$ in model (6.9) in Section 6.3.2. If follows that the total number of measurements, that is, $4N_3 N_2^{(0,0)} N_1$ required for testing $H_0 : \delta_{XZ(2)} = 0$ is four times larger than the total number of measurements, $2N_3 N_2 N_1$, required for testing $H_0 : \delta_{(2)} = 0$.

6.5.2.2 Unbalanced Allocations

Under an unbalanced design for which $N_2^{(1,1)} = \lambda_{11} N_2^{(0,0)}$, $N_2^{(1,0)} = \lambda_{10} N_2^{(0,0)}$, $N_2^{(0,1)} = \lambda_{01} N_2^{(0,0)}$, the *total* sample size N_3 can be determined as

$$N_3 = \frac{f_2 \left(1 + 1/\lambda_{11} + 1/\lambda_{10} + 1/\lambda_{01}\right) z_{\alpha,\phi}^2}{N_2^{(0,0)} N_1 \Delta^2}. \tag{6.35}$$

6.5.2.3 Summary

The sample size $N_3^{(0,0)}$ in (6.34) or in (6.35) required for a desired magnitude of statistical power decreases with increasing N_1, N_2, $\Delta_{XZ(2)}$ and ρ_2 but increases with increasing ρ_1. Table 6.8 displays examples of statistical power for testing the null hypothesis $H_0 : \delta_{XZ(2)} = 0$ and shows that the statistical power $\varphi_{XZ(2)}$ is invariant over the product of $N_3 N_2^{(0,0)}$, regardless of whether the design is balanced. When total sample sizes are compared between Tables 6.2 and 6.8, it can be seen under balanced designs that total N in Table 6.8 is approximately four times larger than total N in Table 6.2 for the same magnitude of power $\varphi_{(2)} = \varphi_{XZ(2)}$.

TABLE 6.8
Sample size and power for detecting a two-way interaction XZ effect $\delta_{XZ(2)}$ in model with $m = 2$ (6.31) for a 2-by-2 factorial design when randomizations occur at the second level with $\rho_2 = 0.05$ (two-sided significance level $\alpha = 0.05$)

λ	Δ	ρ_1	N_1	$N_2^{(0,0)}$	N_3	Total N	φ
1	0.3	0.1	5	4	22	1760	0.835
			5	8	11	1760	0.835
			10	4	14	2240	0.851
			10	8	7	2240	0.851
		0.2	5	4	28	2240	0.814
			5	8	14	2240	0.814
	0.4	0.1	5	4	12	960	0.824
			5	8	6	960	0.824
		0.2	5	4	16	1280	0.820
			5	8	8	1280	0.820
			10	4	12	1920	0.824
			10	8	6	1920	0.824
1.5,1.5,1.5	0.3	0.1	5	4	16	1760	0.824
			5	8	8	1760	0.824
			10	4	10	2200	0.833
			10	8	5	2200	0.833
		0.2	5	4	21	2310	0.814
			5	8	11	2420	0.831
	0.4	0.1	5	4	10	1100	0.861
			5	8	5	1100	0.861
		0.2	5	4	12	1320	0.820
			5	8	6	1320	0.820
			10	4	10	2200	0.861
			10	8	5	2200	0.861

Note: $N_2^{(0,1)} = \lambda_{01} N_2^{(0,0)}$; $N_2^{(1,0)} = \lambda_{10} N_2^{(0,0)}$; $N_2^{(1,1)} = \lambda_{11} N_2^{(0,0)}$; $\Delta_{XZ(2)} = |\delta_{XZ(2)}/\sigma|$; ρ_2 (6.2) is the correlations among the level 2 data; ρ_1 (6.3) is the correlations among the level 1 data; N_1 is the number of level 1 units per level 2 units; $N_2^{(0,0)}$ is the level 2 sample size in the $(X = 0, Z = 0)$ arm within level 3 units; N_3 is the total number of level 3 units; Total $N = (1 + \lambda_{01} + \lambda_{10} + \lambda_{11}) N_1 N_2^{(0,0)} N_3$ is the required total number of observations; $\varphi_{XZ(2)}$ is the statistical power to test $H_0 : \delta_{XZ(2)} = 0$.

6.5.3   Randomization at the First Level

For trials with $m = 1$, the intervention assignment indicator variable $X_{ijk} = 0$ if the kth level 1 unit is not assigned to receive X and $X_{ijk} = 1$ if assigned to receive X; therefore $X_{ijk} = X_j$ for all i and j. Likewise, the indicator variable $Z_{ijk} = 0$ if the kth level 1 unit is not assigned to receive Z and $Z_{ijk} = 1$ if assigned to receive Z; therefore $Z_{ijk} = Z_j$ for all i and j. Here, the null hypothesis to be tested is $H_0 : \delta_{(1)}^{xz} = 0$ in model (6.31) for $m = 1$.

6.5.3.1   Balanced Allocations

For a balanced design, it is assumed that $\Sigma_k X_k Z_k = N_1^{(1,1)}$ so that $i = 1, 2, \ldots, N_3, j = 1, \ldots, N_2$ and $k = 1, 2, \ldots, 4N_1^{(1,1)}$. Specifically, $N_1^{(1,1)}$ represents the level 1 sample size assigned to the $(X = 1, Z = 1)$ arm, or the $(1,1)$ cell, in a 2-by-2 factorial table of X and Z within the second level units, N_3 is the level 3 sample size in total, and N_2 is the level 2 sample size per third-level units. In a balanced design, the sample sizes of the other cells are identical to $N_1^{(1,1)}$, i.e., $N_1^{(1,1)} = N_1^{(1,0)} = N_1^{(0,1)} = N_1^{(0,0)}$. The following test statistic can be used to test $H_0 : \delta_{XZ(1)} = 0$:

$$D_{XZ(1)} = \frac{\sqrt{N_3 N_2 N_1^{(0,0)}} \left\{ \left(\bar{Y}^{(1,1)} - \bar{Y}^{(1,0)} \right) - \left(\bar{Y}^{(0,1)} - \bar{Y}^{(0,0)} \right) \right\}}{\sigma \sqrt{4(1 - \rho_1)}},$$

where $\bar{Y}^{(x,z)}$ is the mean of Y in the $(X = x, Z = z)$ cells and ρ_1 is the correlation among the first level data units (6.3). It can be shown that $\hat{\delta}_{XZ(1)} = \left(\bar{Y}^{(1,1)} - \bar{Y}^{(1,0)} \right) - \left(\bar{Y}^{(0,1)} - \bar{Y}^{(0,0)} \right)$ is an unbiased estimate of $\delta_{XZ(1)}$ and

$$Var(\hat{\delta}_{XZ(1)}) = \frac{4(1 - \rho_1)\sigma^2}{N_3 N_2 N_1^{(0,0)}}.$$

The corresponding statistical power can be expressed as

$$\varphi_{XZ(1)} = \Phi \left\{ \Delta_{XZ(1)} \sqrt{N_3 N_2 N_1^{(0,0)} \Big/ 4(1 - \rho_1)} - \Phi^{-1}(1 - \alpha/2) \right\},$$

where $\Delta_{XZ(1)} = \left| \delta_{XZ(1)}/\sigma \right|$. Accordingly, the corresponding required sample size is

$$N_3 = \frac{4(1 - \rho_1) z_{\alpha,\phi}^2}{N_2 N_1^{(0,0)} \Delta_{XZ(1)}^2}. \tag{6.36}$$

The sample sizes of the second and first level units can be determined by solving (6.36) for N_3 and $N_1^{(0,0)}$, respectively. The determinations are immediate since the power $\varphi_{XZ(1)}$ is invariant over the product of $N_3 N_2 N_1^{(0,0)}$

Again, when Equations (6.36) and (6.15) are compared, $N_1^{(0,0)} = 2N_1$, where N_1 is the required sample size of the first level data units for testing $H_0 : \delta_{(1)} = 0$ in model (6.13) in Section 6.3.3. If follows that the total number measurements, that is, $4N_3 N_2 N_1^{(0,0)}$ required for testing $H_0 : \delta_{(1)}^{xz} = 0$ is four times larger than the total number of measurements, $2N_3 N_2 N_1$, required for testing $H_0 : \delta_{(1)} = 0$.

6.5.3.2   Unbalanced Allocations

Under an unbalanced design for which $N_1^{(0,1)} = \lambda_{01} N_1^{(0,0)}$, $N_1^{(1,0)} = \lambda_{10} N_1^{(0,0)}$, and $N_1^{(1,1)} = \lambda_{11} N_1^{(0,0)}$, the *total* sample size N_3 can be determined as

$$N_3 = \frac{(1 - \rho_1)(1 + 1/\lambda_{11} + 1/\lambda_{10} + 1/\lambda_{01}) z_{\alpha,\phi}^2}{N_2 N_1^{(0,0)} \Delta^2}. \tag{6.37}$$

6.5.3.3   Summary

The sample size $N_3^{(0,0)}$ in (6.34) or in (6.35) required for a desired magnitude of statistical power is not a function of ρ_2 and decreases with increasing N_1, N_2, $\Delta_{XZ(2)}$ and ρ_1. Table 6.9 displays examples of statistical power for testing the null hypothesis $H_0 : \delta_{XZ(1)} = 0$ and shows that the statistical power $\varphi_{XZ(1)}$ is invariant over the product of $N_3 N_2 N^{(0,0)}$, or total number of observations $(1+\lambda_{01}+\lambda_{10}+\lambda_{11})N_1^{(0,0)} N_3 N_2 N^{(0,0)}$, regardless of whether the design is balanced. When total sample sizes are compared between Tables 6.9 and 6.3, it can be seen under balanced designs that total N in Table 6.9 is approximately four times larger than total N in Table 6.3 for the same magnitude of power $\varphi_{(1)} = \varphi_{XZ(1)}$.

6.5.4   Comparison of Sample Sizes

The ratio of the total numbers of observations required for testing the interaction effects of the two main factors across randomizations at the three different levels are identical to those of $R_{3.2}$, $R_{3.1}$ and $R_{2.1}$ presented in Section (6.7) as evident by comparing Equations (6.32) and (6.34); Equations (6.32) and (6.36); and Equations (6.34) and (6.36). These ratios hold for unbalanced trial designs with identical $\lambda = (\lambda_{11}, \lambda_{10}, \lambda_{01})$ values as shown by comparing Equations (6.33) and (6.35); Equations (6.33) and (6.37); and Equations (6.35) and (6.37). Compare the results on Table 6.7, Table 6.8, and Table 6.9. Therefore, again, for the same values of Δ, φ, α, and correlation ρ, the sample size requirements are greatest for trials with the third level randomization, followed by second level randomization, followed by first level randomization

6.6   Longitudinal Factorial Designs: Treatment Effects on Slopes

For trials using a 2×2 factorial design in which subjects are randomly assigned to one of the four treatment combinations and then repeatedly evaluated for outcomes over time, it may be of interest to test whether the trends in the outcome over the study period (i.e., slopes) is beyond what would be expected if the effects of X and Z are additive. This hypothesis can be evaluated by including a three-way interaction between the two treatments and time in a linear mixed-effects linear model as follows:

$$Y_{ijk} = \beta_0 + \delta_X X_{ijk} + \delta_Z Z_{ijk} + \delta_T T_{ijk} + \delta_{XZ} X_{ijk} Z_{ijk} + \delta_{XT} X_{ijk} T_{ijk}$$
$$+ \delta_{ZT} Z_{ijk} T_{ijk} + \delta_{XZT} X_{ijk} Z_{ijk} T_{ijk} + \nu_{j(i)} T_{ijk} + u_i + u_{j(i)} + e_{ijk}$$
$$(6.38)$$

TABLE 6.9

Sample size and power for detecting a two-way interaction XZ effect $\delta_{XZ(1)}$ in model with $m = 1$ (6.31) for a 2-by-2 factorial design when randomizations occur at the first level with $\rho_2 = 0.05$ (two-sided significance level $\alpha = 0.05$)

λ	Δ	ρ_1	$N_1^{(0,0)}$	N_2	N_3	Total N	φ
1	0.3	0.1	5	4	16	1280	0.807
			5	8	8	1280	0.807
			10	4	8	1280	0.807
			10	8	4	1280	0.807
		0.2	5	4	14	1120	0.801
			5	8	7	1120	0.801
	0.4	0.1	5	4	10	800	0.846
			5	8	5	800	0.846
		0.2	5	4	8	640	0.807
			5	8	4	640	0.807
			10	4	4	640	0.807
			10	8	2	640	0.807
1.5,1.5,1.5	0.3	0.1	5	4	12	1320	0.807
			5	8	6	1320	0.807
			10	4	6	1320	0.807
			10	8	3	1320	0.807
		0.2	5	4	12	1320	0.851
			5	8	6	1320	0.851
	0.4	0.1	5	4	8	880	0.868
			5	8	4	880	0.868
		0.2	5	4	6	660	0.807
			5	8	3	660	0.807
			10	2	6	660	0.807
			10	4	3	660	0.807

Note: $N_1^{(0,1)} = \lambda_{01} N_1^{(0,0)}$; $N_1^{(1,0)} = \lambda_{10} N_1^{(0,0)}$; $N_1^{(1,1)} = \lambda_{11} N_1^{(0,0)}$; $\Delta_{XZ(1)} = |\delta_{XZ(1)}/\sigma|$; ρ_2 (6.2) is the correlations among the level 2 data; ρ_1 (6.3) is the correlations among the level 1 data; $N_1^{(0,0)}$ is the level 1 sample size in the $(X = 0, Z = 0)$ arm within level 2 units; N_2 is the number of level 2 units per level 3 units; N_3 is the total number of level 3 units; Total $N = (1 + \lambda_{01} + \lambda_{10} + \lambda_{11}) N_1^{(0,0)} N_2 N_3$ is the required total number of observations; $\varphi_{XZ(1)}$ is the statistical power to test $H_0 : \delta_{XZ(1)} = 0$.

The slopes can be considered fixed ($\sigma_\tau^2 = 0$) or random ($\sigma_\tau^2 > 0$) in this model, which is an extension of testing the two-way interaction between X and T in models (6.17) and (6.22), respectively. Section 6.6.1 presents sample size approaches for designs with third level randomization. Brief remarks on sample size requirements for studies with second level randomization are provided in Section 6.6.2. Comparisons of sample sizes between fixed and random slope models are discussed in Section 6.6.3. Figure 5.2 geometrically depicts the slope parameters in model (6.38).

6.6.1   Randomizations at the Third Level

When randomization occurs at the third level, all subjects within the same third cluster unit will uniformly receive one of the four (X, Z) treatment combinations and then are repeatedly evaluated for outcomes over time. Therefore, the outcomes of subjects taking different treatment combinations are uncorrelated since the subjects belong to different clusters which are independent.

6.6.1.1   Balanced Allocations

It is assumed that $\Sigma_i X_i Z_i = N_3^{(1,1)}$ so that $i = 1, 2, \ldots, 4N_3^{(1,1)}$, $j = 1, \ldots, N_2$ and $k = 1, 2, \ldots, N_1$. Specifically, $N_3^{(1,1)}$ represents the level 3 sample size assigned to $(X = 1, Z = 1)$ arm, or the $(1,1)$ cell in a 2-by-2 factorial table of X and Z, N_2 is the level 2 sample size per third level units, and N_1 is the level 1 sample size per level two units. In a balanced design, the sample sizes of the other cells are identical to $N_3^{(1,1)}$, i.e., $N_3^{(1,1)} = N_3^{(1,0)} = N_3^{(0,1)} = N_3^{(0,0)}$. It is further assumed that $T_{ijk} = T_k$ for all i and j, and that time increases from 0 (the baseline) to $T_{end} = N_1 - 1$ (the last time point) by unit increments. The null hypothesis of interest is $H_0 : \delta_{XZT} = 0$ in model (6.38) which can be evaluated using the following test statistic:

$$D_{XZT} = \frac{\left\{\left(\hat{\eta}^{(1,1)} - \hat{\eta}^{(1,0)}\right) - \left(\hat{\eta}^{(0,1)} - \hat{\eta}^{(0,0)}\right)\right\} \sqrt{N_3^{(0,0)} N_2 N_1 Var_p(T)}}{\sqrt{4\left\{(1 - \rho_1)\sigma^2 + N_1 Var_p(T)\sigma_\tau^2\right\}}}$$

where $\hat{\eta}^{(x,z)}$ is the OLS estimate of the slope in the $(X = x, Z = z)$ cell. It can be shown that $\hat{\delta}_{XZT} = \left(\hat{\eta}^{(1,1)} - \hat{\eta}^{(1,0)}\right) - \left(\hat{\eta}^{(0,1)} - \hat{\eta}^{(0,0)}\right)$ is an unbiased estimate of δ_{XZT} and

$$Var\left(\hat{\delta}_{XZT}\right) = \frac{4\left\{(1 - \rho_1)\sigma^2 + N_1 Var_p(T)\sigma_\tau^2\right\}}{N_3^{(0,0)} N_2 N_1 Var_p(T)}.$$

The statistical power of this test can be expressed as

$$\varphi_{XZT} = \Phi \left\{ \Delta_{XZT} \sqrt{\frac{N_3^{(0,0)} N_2 N_1 Var_p(T)}{4\left\{(1 - \rho_1) + r_\tau N_1 Var_p(T)\right\}}} - \Phi^{-1}(1 - \alpha/2) \right\},$$

where $\Delta_{XZT} = |\delta_{XZT}/\sigma|$ and again r_τ (6.23) is the ratio of the random slope variance to the sum of the other variances (equal to 0 for the fixed slope models). It follows that the required sample size per arm for the third level unit N_3 can be expressed as

$$N_3^{(0,0)} = \frac{4\left\{(1 - \rho_1) + r_\tau N_1 Var_p(T)\right\} z_{\alpha,\phi}^2}{N_2 N_1 Var_p(T)\Delta_{XZT}^2}. \tag{6.39}$$

This equation also shows that testing $H_0 : \delta_{XZT} = 0$ in model (6.38) requires four times the total number of observations required for testing H_0 : $\delta_{(f)} = 0$ with fixed-slope models (6.17) and also for testing H_0 : $\delta_{(r)} = 0$ with random-slope models (6.22). The former can be seen by comparing the sample size in Equation (6.39) with $r_\tau = 0$ with that in Equation (6.19) and the latter by comparing the sample size in Equation (6.39) with $r_\tau > 0$ with that in Equation (6.24).

6.6.1.2   Unbalanced Allocations

With unbalanced designs for which $N_3^{(1,1)} = \lambda_{11} N_3^{(0,0)}$, $N_3^{(1,0)} = \lambda_{10} N_3^{(0,0)}$, $N_3^{(0,1)} = \lambda_{01} N_3^{(0,0)}$, the sample size $N_3^{(0,0)}$ can be determined as

$$N_3^{(0,0)} = \frac{(1 + 1/\lambda_{11} + 1/\lambda_{10} + 1/\lambda_{01})\{(1 - \rho_1) + r_\tau N_1 Var_p(T)\} z_{\alpha,\phi}^2}{N_2 N_1 Var_p(T) \Delta_{XZT}^2}$$

(6.40)

6.6.1.3   Summary

The sample size $N_3^{(0,0)}$ in (6.39) or in (6.40) required for a desired magnitude of statistical power is not a function of ρ_2 and decreases with increasing N_2, N_1, ρ_1, and Δ_{XZT}. However, $N_3^{(0,0)}$ increases with increasing r_τ, which again has a large effect on the sample size which is again enormous. Table 6.10 displays examples of statistical power for testing the null hypothesis $H_0 : \delta_{XZT} = 0$ and shows that the statistical power φ_{XZT} is invariant over the product of $N_3^{(0,0)} N_2$. The four-fold increase in sample size for interaction effects relative to main effects for the same statistical power can be confirmed by comparing results in Table 6.4 ($r_\tau = 0$) and Table 6.5 ($r_\tau = 0.1$) with those in Table 6.10.

6.6.2   Randomizations at the Second Level

When randomization occurs at the second level, subjects in each cluster are randomly assigned to one of the four (X, Z) treatment combinations and then repeatedly evaluated for outcomes over time. Therefore, the subjects could receive different treatment combinations even if they are nested within the same clusters. However the sample size formulas (6.39) and (6.40) under balanced and unbalanced designs, respectively, can still be applied to trials with second level randomization since the correlations among slope estimates will be zero for the reasons discussed in Section 6.4.4. Again, the only differences are the definitions of the sample sizes for trials with the third level randomization: $N_3^{(0,0)}$, N_2, and N_1; the corresponding sample sizes for trials with second level randomization should be denoted by N_3, $N_2^{(0,0)}$, and N_1. Therefore, the sample sizes required for detecting a three-way interaction are the same regardless of whether randomization occurs at the third or second level or whether

TABLE 6.10

Sample size and power for detecting a three-way interaction XZT effect δ_{XZT} in model (6.38) for a 2-by-2 factorial design when randomizations occur at the third level (two-sided significance level $\alpha = 0.05$)

λ_3	r_τ	Δ_{XZT}	ρ_1	N_1	N_2	$N_3^{(0,0)}$	Total N	φ_{XZT}
1	0	0.3/4	0.1	5	8	63	10080	0.801
				5	63	8	10080	0.801
			0.2	5	8	56	8960	0.801
		0.3/8	0.1	9	8	42	12096	0.801
			0.2	9	8	38	10944	0.808
				9	38	8	10944	0.808
	0.1	0.3/4	0.1	5	8	133	21280	0.801
				5	133	8	21280	0.801
			0.2	5	8	126	20160	0.801
		0.3/8	0.1	9	8	321	92448	0.801
			0.2	9	8	317	91296	0.801
				9	317	8	91296	0.801
1.5	0	0.3/4	0.1	5	8	48	10560	0.807
				5	48	8	10560	0.807
			0.2	5	8	42	9240	0.801
		0.3/8	0.1	9	8	32	12672	0.807
			0.2	9	8	28	11088	0.801
				9	28	8	11088	0.801
	0.1	0.3/4	0.1	5	8	100	22000	0.802
				5	100	8	22000	0.802
			0.2	5	8	95	20900	0.803
		0.3/8	0.1	9	8	241	95436	0.800
			0.2	9	8	238	94248	0.801
				9	238	8	94248	0.801

Note: $N_3^{(0,1)} = \lambda_{01} N_3^{(0,0)}$; $N_3^{(1,0)} = \lambda_{10} N_3^{(0,0)}$; $N_3^{(1,1)} = \lambda_{11} N_3^{(0,0)}$; $\Delta_{XZT} = |\delta_{XZT}/\sigma|$; ρ_2 (6.2) is the correlations among the level 2 data; ρ_1 (6.3) is the correlations among the level 1 data; N_1 is the number of level 1 units per level 2 unit; N_2 is the number of level 2 units per level 3 unit; $N_3^{(0,0)}$ is the level 3 sample size in the $(X = 0, Z = 0)$ arm; Total $N = (1 + \lambda_{01} + \lambda_{10} + \lambda_{11}) N_1 N_2 N_3^{(0,0)}$ is the required total number of observations; φ_{XZT} is the statistical power to test $H_0 : \delta_{XZT} = 0$.

subject-level slopes are considered fixed or random. This result has been validated by simulation studies [11].

6.6.3   Comparison of Sample Sizes

When randomization occurs at the third level, the ratio $R_{r.f}(N_3^{(0,0)})$ of sample sizes $N_3^{(0,0)}$ in Equation (6.39) between fixed ($r_\tau = 0$) and random slope

$(r_\tau > 0)$ models for testing the significance of δ_{XZT} in model (6.38) is identical to the ratio $R_{r.f}(N_3)$ in equation (6.27) for testing slope differences between arms. That is,

$$R_{r.f}(N_3^{(0,0)}; r_\tau, N_1, \rho_1) \equiv \frac{N_3^{(0,0)} \mid r_\tau > 0}{N_3^{(0,0)} \mid r_\tau = 0} = 1 + r_\tau N_1 Var_p(T)/(1 - \rho_1)$$

$$= R_{r.f}(N_3; r_\tau, N_1, \rho_1)$$

This ratio $R_{r.f}(N_3^{(0,0)})$ holds also for unbalanced designs with equal values of λ's. In summary, $R_{r.f}(N_3^{(0,0)}) = R_{r.f}(N_3) = R_{r.f}(N_2) = R_{r.f}(N_2^{(0,0)})$ in Equation (5.4.9), if ρ_1 in Equation (6.3) is equal to ρ in Equation (5.2.2). Thus the properties of $R_{r.f}(N_3)$ discussed in Section 6.4.2 hold also for $R_{r.f}(N_3^{(0,0)})$ as evident by comparing the results in Table 6.10 for $r_\tau = 0$ and $r_\tau = 0.1$. Furthermore, it can be inferred that the ratios above should be identical to the sample size ratios between random and fixed models when randomization occurs at the second level since the sample size requirements remain unchanged between third and second level randomization (section 6.6.2), i.e., due to the invariance over level of randomization property.

6.7 Sample Sizes for Binary Outcomes

Sample size determinations for main effects in clinical trials involving binary outcomes with three-level data should take into account correlations among the second level data units within the third level data units, and those among the first level data units within the second level data units. The approach for sample size determination is again based on large sample normal theory approximations [14]. When randomization occurs at the mth ($m = 1, 2, 3$) level, a statistical model for binary outcome $Y_{ijk} = 0$ or 1 can be formulated as

$$\log\left(\frac{p_{ijk}}{1 - p_{ijk}}\right) = \beta_0 + \xi_{(m)} X_{ijk} + u_i + u_{j(i)}, \tag{6.41}$$

where $p_{ijk} = E(Y_{ijk}|X_{ijk})$ and $u_i \sim N\left(0, \sigma_3^2\right)$ further $u_{j(i)} \sim N\left(0, \sigma_3^2\right)$. We further assume that $p_{ijk} = p$ if $X_{ijk} = 0$ and $p_{ijk} = p_1$ if $X_{ijk} = 1$ so that β_0 and $\xi_{(m)}$ can accordingly be determined. Therefore, $Var(Y_{ijk}|X_{ijk} = 0) = p(1 - p)$ and $Var(Y_{ijk}|X_{ijk} = 1) = p_1(1 - p_1)$. Under model (6.41), the correlations can be computed as

$$\rho_2 \equiv Corr\left(Y_{ij'k'}, Y_{ijk'}\right) = \frac{\sigma_3^2}{\sigma_2^2 + \sigma_3^2 + \pi^2/3} \quad \text{for } j \neq j', \tag{6.42}$$

and

$$\rho_1 \equiv Corr\left(Y_{ijk}, Y_{ijk'}\right) = \frac{\sigma_2^2 + \sigma_3^2}{\sigma_2^2 + \sigma_3^2 + \pi^2/3} \quad for \ k \neq k'. \tag{6.43}$$

The null hypothesis to be tested is $H_0 : \xi_{(m)} = 0$ or equivalently $H_0 :$ $p - p_1 = 0$. Sections 6.7.1, 6.7.2, and 6.7.3 present sample size approaches when randomization occurs at the third $(m = 3)$, second $(m = 2)$, and first $(m = 1)$ levels, respectively, and these are compared in Section 6.7.4.

6.7.1   Randomization at Third Level

Let us denote the level 3 sample sizes for the control $(X = 0)$ and intervention $(X = 1)$ arms by $N_3^{(0)}$ and $N_3^{(1)}$ while N_2 and N_1 represent the level 2 sample size per level 3 unit and the level 1 sample size per level 2 units, respectively.

6.7.1.1   Balanced Allocations

When randomization occurs at the third level, i.e., when $m = 3$ and $X_{ijk} = X_i$ for all j and k a balanced design is assumed so that $\Sigma_i X_i = N_3^{(1)} i = 1, 2, \ldots, 2N_3^{(0)}, j = 1, \ldots, N_2$ and $k = 1, 2, \ldots, N_1$. The null hypothesis to be tested is $H_0 : \xi_{(3)} = 0$. Extending the formula (6.6) for determining $N_3 = N_3^{(0)} = N_3^{(1)}$ per treatment arm for testing $H_0 : \delta_{(3)} = 0$ in model (6.4) for a continuous outcome to the case of a binary outcome with unequal variances between the null and alternative hypotheses, an approximate sample size for N_3 per arm for two-sided α and statistical power $\varphi = \phi$ can be computed as

$$N_3 = \frac{f_3 z_{\alpha,\phi,p}^2}{N_2 N_1 \left(p_1 - p_0\right)^2}, \tag{6.44}$$

from a power function

$$\varphi = \Phi\left\{ \frac{|p_1 - p_0|\sqrt{N_1 N_2 N_3 / f_3} - \Phi^{-1}(1 - \alpha/2)\sqrt{2\bar{p}\left(1 - \bar{p}\right)}}{\sqrt{p_0\left(1 - p_0\right) + p_1\left(1 - p_1\right)}} \right\}$$

where $\bar{p} = (p_0 + p_1)/2$, $f_3 = 1 + N_1(N_2 - 1)\rho_2 + (N_1 - 1)\rho_1$ (6.3.2) and

$$z_{\alpha,\phi,p} = \Phi^{-1}\left(1 - \alpha/2\right)\sqrt{2\bar{p}(1 - \bar{p})} + \Phi^{-1}\left(\phi\right)\sqrt{p_0\left(1 - p_0\right) + p_1\left(1 - p_1\right)}.$$

Sample sizes for N_2 and N_1 can be obtained by solving for them using Equation (6.44).

6.7.1.2   Unbalanced Allocations

For an unbalanced design with $\Sigma_i X_i = \lambda N_3^{(0)}$ for $\lambda > 0$ and $i = 1, 2, \ldots, (1 + \lambda)N_3^{(0)}$, the sample size $N_3^{(0)}$ for the control arm with $X_i = 0$ can be obtained as

$$N_3 = \frac{f_3 z_{\alpha,\phi,p,\lambda}^2}{N_2 N_1 (p_1 - p_0)^2}, \qquad (6.45)$$

from a power function

$$\varphi = \Phi \left\{ \frac{|p_1 - p_0| \sqrt{N_1 N_2 N_3^{(0)}/f_3} - \Phi^{-1}(1 - \alpha/2)\sqrt{(1 + 1/\lambda)\,\bar{p}\,(1 - \bar{p})}}{\sqrt{p_0(1 - p_0) + p_1(1 - p_1)/\lambda}} \right\}$$

where

$$\begin{aligned} z_{\alpha,\phi,p,\lambda} &= \Phi^{-1}(1 - \alpha/2)\sqrt{(1 + 1/\lambda)\,\bar{p}_\lambda(1 - \bar{p}_\lambda)} \\ &\quad + \Phi^{-1}(\phi)\sqrt{p_0(1 - p_0) + p_1(1 - p_1)/\lambda} \end{aligned} \qquad (6.46)$$

and $\bar{p}_\lambda = (p_0 + \lambda p_1)/(1 + \lambda)$.

We note that Teerenstra et al. [15] proposed the following formula for N_3 for the control arm based on a generalized estimating equation approach:

$$N_3 = \frac{f_3 \left[\{p_0(1 - p_0)\}^{-1} + \{\lambda p_1(1 - p_1)\}^{-1} \right] \left(t_{1-\alpha/2, N_3-2} + t_{\phi, N_3-2} \right)^2}{N_2 N_1 b^2}, \qquad (6.47)$$

where $b = \log(p_0/(1 - p_0)) - \log(p_1/(1 - p_1))$ and $t_{\alpha df}$ is the $100\alpha\%$ percentile of a t-distribution with df degrees of freedom. This result was based on an extension of Shih [16] introduced in Section 5.7.1.2. Since both the $100(1 - \alpha/2)\%$ and $100\varphi\%$ percentiles of the t distribution in Equation (6.47) are a function of N_3, an iterative solution to the equation is necessary. For this reason, Teerenstra et al. [15] suggested replacing the t-percentiles by the corresponding z-percentiles and then multiplying the result by $(N_1 + 1)/(N_1 - 1)$. The sample size determined with this approach is always greater than that determined by Equation (6.45).

6.7.1.3 Summary

Sample size $N_3(6.44)$ or $N_3^{(0)}(6.45)$ required for a desired magnitude of statistical power decreases with increasing N_1 and N_2, but increases with increasing ρ_1 and ρ_2. The sample size decreases in general with increasing $|p - p_1|$. However, for the same magnitude of $|p - p_1|$, the sample size depends on the specific combinations of p and p_1 since the variances of the difference in proportions are a function of both p and p_1. For example, a smaller sample size is required when $p = 0.1$ and $p_1 = 0.2$ than when $p = 0.2$ and $p_1 = 0.3$. Table 6.11 displays examples of statistical power for testing the null hypothesis $H_0 : p - p_1 = 0$.

6.7.2 Randomization at the Second Level

Let $N_2^{(0)}$ and $N_2^{(1)}$ denote the level 2 sample sizes within level 3 units for the control ($X = 0$) and intervention ($X = 1$) arms, respectively, while N_3 and

TABLE 6.11
Sample size and statistical power for detecting a main effect $|p_1 - p_0|$ on binary outcome in model with $m = 3$ (6.41) when randomizations occur at third level (two-sided significance level $\alpha = 0.05$)

λ	p_0	p_1	ρ_1	N_1	N_2	$N_3^{(0)}$	Total N	φ
1	0.4	0.5	0.1	5	4	42	1680	0.803
				5	8	31	2480	0.806
			0.2	5	4	50	2000	0.805
				5	8	35	2800	0.807
		0.6	0.1	5	4	11	440	0.821
				5	8	8	640	0.818
			0.2	5	4	13	520	0.820
				5	8	9	720	0.818
1.5	0.4	0.5	0.1	5	4	35	1750	0.803
				5	8	26	2600	0.808
			0.2	5	4	42	2100	0.807
				5	8	29	2900	0.804
		0.6	0.1	5	4	9	450	0.815
				5	8	7	700	0.837
			0.2	5	4	11	550	0.826
				5	8	8	800	0.842

Note: λ represents the ratio of the number of level 3 units in the intervention arm to the corresponding number in the control arm; ρ_2 (6.7.2) is the correlations among the level 2 data; ρ_1 (6.7.3) is the correlations among the level 1 data; N_1 is the number of level 1 units per level 2 units; N_2 is the number of level 2 units per level 3 units; $N_3^{(0)}$ is the level 3 sample size in the *control* arm; Total $N = (1+\lambda)N_1 N_2 N_3^{(0)}$ is the required total number of observations; φ is the statistical power to test $H_0 : p_1 - p_0 = 0$.

N_1 represent the *total* level 3 sample size and the level 1 sample size per level 2 units, respectively.

6.7.2.1 Balanced Allocations

When random allocations are assigned at the second level, i.e., when $m = 2$ and $X_{ijk} = X_j$ for all i and k a balanced design is assumed so that $\Sigma_i X_j = N_2^{(1)}$ where $i = 1, 2, \ldots, N_3, j = 1, 2, \ldots, 2N_2^{(0)}$, and $k = 1, 2, \ldots, N_1$. Here, $N_2 = N_2^{(0)} = N_2^{(1)}$ represents sample size per intervention arm within third level clusters. The null hypothesis to be tested is $H_0 : \xi_{(2)} = 0$. Again, extending the formula (6.10) for determining the total sample size N_3 for testing $H_0 : \delta_{(2)} = 0$ in model (6.9) for continuous outcome to the binary outcome, an approximate sample size for N_3 for a two-sided α and a statistical power

φ can be computed as

$$N_3 = \frac{f_2 z_{\alpha,\phi,p}^2}{N_2 N_1 (p_1 - p_0)^2} \tag{6.48}$$

from the power function

$$\varphi = \Phi \left\{ \frac{|p_1 - p_0| \sqrt{N_1 N_2 N_3 / f_2} - \Phi^{-1}(1 - \alpha/2) \sqrt{2\bar{p}(1 - \bar{p})}}{\sqrt{p_0(1 - p_0) + p_1(1 - p_1)}} \right\}$$

From Equation (6.48), sample size N_2 for given N_1, and N_3 can be immediately obtained as

$$N_2 = \frac{f_2 z_{\alpha,\phi,p}^2}{N_3 N_1 (p_1 - p_0)^2}$$

since $f_2 = 1 + (N_1 - 1)\rho_1 - N_1\rho_2$ in (6.10) is not a function of N_2. In other words, the power is invariant over the product of $N_3 N_2$ Equation (6.48) should be solved to obtain N_1.

6.7.2.2   Unbalanced Allocations

For an unbalanced design with $\Sigma_j X_j = \lambda N_2^{(0)}$ for $\lambda > 0$ and $i = 1, 2, \ldots, N_3, j = 1, 2, \ldots (1 + \lambda) N_2^{(0)}$, and $k = 1, 2, \ldots, N_1$, the total sample size N_3 can be obtained as

$$N_3 = \frac{f_2 z_{\alpha,\phi,p,\lambda}^2}{N_2^{(0)} N_1 (p_1 - p_0)^2} \tag{6.49}$$

from the power function

$$\varphi = \Phi \left\{ \frac{|p_1 - p_0| \sqrt{N_1 N_2^{(0)} N_3 / f_2} - \Phi^{-1}(1 - \alpha/2) \sqrt{(1 + 1/\lambda)\bar{p}(1 - \bar{p})}}{\sqrt{p_0(1 - p_0) + p_1(1 - p_1)/\lambda}} \right\}$$

6.7.2.3   Summary

The sample size N_3 in (6.48) or in (6.49) required for a desired magnitude of statistical power decreases with $N_1, N_2^{(0)}$, and ρ_2, but increases with increasing ρ_1. The sample size decreases in general with increasing $|p - p_1|$ and for the same magnitude of $|p - p_1|$, the sample size depends on specific combinations of p and p_1. Table 6.12 displays examples of statistical power for testing the null hypothesis $H_0 : p - p_1 = 0$, and shows that the statistical power φ is invariant over the product of $N_3 N_2^{(0)}$, regardless of whether or not λ is equal to 1.

TABLE 6.12
Sample size and statistical power for detecting a main effect $|p_1 - p_0|$ on binary outcome in model with $m = 2$ (6.41) when randomizations occur at second level (two-sided significance level $\alpha = 0.05$)

λ	p_0	p_1	ρ_1	N_1	$N_2^{(0)}$	N_3	Total N	φ
1	0.4	0.5	0.1	5	4	24	960	0.829
				5	8	12	960	0.829
				10	4	14	1120	0.813
				10	8	7	1120	0.813
			0.2	5	4	32	1280	0.825
				5	8	16	1280	0.825
		0.6	0.1	5	4	6	240	0.829
				5	8	3	240	0.829
			0.2	5	4	8	320	0.824
				5	8	4	320	0.824
				10	4	6	480	0.829
				10	8	3	480	0.829
1.5	0.4	0.5	0.1	5	4	20	1000	0.828
				5	8	10	1000	0.828
				10	4	12	1200	0.823
				10	8	6	1200	0.823
			0.2	5	4	26	1300	0.815
				5	8	13	1300	0.815
		0.6	0.1	5	4	6	300	0.891
				5	8	3	300	0.891
			0.2	5	4	8	400	0.888
				5	8	4	400	0.888
				10	4	6	600	0.891
				10	8	3	600	0.891

Note: λ represents the ratio of the number of level 3 units in the intervention arm to the corresponding number in the control arm; ρ_2 (6.7.2) is the correlations among the level 2 data; ρ_1 (6.7.3) is the correlations among the level 1 data; N_1 is the number of level 1 units per level 2 units; $N_2^{(0)}$ is sample size in the *control* arm within level 3 units; N_3 is the *total* level 3 sample size; Total $N = (1 + \lambda)N_1 N_2^{(0)} N_3$ is the required total number of observations; φ is the statistical power to test $H_0 : p_1 - p_0 = 0$.

6.7.3   Randomization at the First Level

Let $N_1^{(0)}$ and $N_1^{(1)}$ denote the level 1 sample sizes within level 2 units for the control ($X = 0$) and intervention ($X = 1$) arms, respectively, and N_3 and N_2 represent the *total* level 3 sample size and the level 2 sample size per level 3 units, respectively.

6.7.3.1 Balanced Allocations

When randomization occurs at the first level, i.e., when $m = 1$ and $X_{ijk} = X_k$ for all i and j a balanced design is assumed so that $\Sigma_k X_k = N_1^{(1)}$ and $i = 1, 2, \ldots, N_3, j = 1, 2, \ldots, N_2$, and $k = 1, 2, \ldots, 2N_1^{(0)}$. Here, $N_1 = N_1^{(0)} = N_1^{(1)}$ represents the sample size per intervention arm within the second level data units. The null hypothesis to be tested is $H_0 : \xi_{(1)} = 0$. Again, extending the formula (6.14) for determining N_3 for testing $H_0 : \delta_{(1)} = 0$ in model (6.13) for a continuous outcome to the case of a binary outcome, an approximate sample size for N_3 for a two-sided α and statistical power $\varphi = \phi$ can be computed as

$$N_3 = \frac{f_1 z_{\alpha,\phi,p}^2}{N_2 N_1 (p_1 - p_0)^2} = \frac{(1 - \rho_1) z_{\alpha,\phi,p}^2}{N_2 N_1 (p_1 - p_0)^2} \qquad (6.50)$$

from the power function

$$\varphi = \Phi \left\{ \frac{|p_1 - p_0| \sqrt{N_1 N_2 N_3 / (1 - \rho_1)} - \Phi^{-1}(1 - \alpha/2) \sqrt{2\bar{p}(1 - \bar{p})}}{\sqrt{p_0 (1 - p_0) + p_1 (1 - p_1)}} \right\}$$

From Equation (6.50), N_2 and N_1 can be immediately obtained as

$$N_2 = \frac{(1 - \rho_1) z_{\alpha,\phi,p}^2}{N_3 N_1 (p_1 - p_0)^2}$$

and

$$N_1 = \frac{(1 - \rho_1) z_{\alpha,\phi,p}^2}{N_3 N_2 (p_1 - p_0)^2}$$

since the statistical power is invariant over the product of $N_3 N_2 N_1$

6.7.3.2 Unbalanced Allocations

For an unbalanced design with $\Sigma_k X_k = \lambda N_1^{(0)}$ for $\lambda > 0$ and $i - 1, 2, \ldots, N_3$, $j = 1, 2, \ldots N_2$, and $k = 1, 2, \ldots, (1 + \lambda) N_1^{(0)}$, the total sample size N_3 can be obtained as

$$N_3 = \frac{f_1 z_{\alpha,\phi,p,\lambda}^2}{N_2 N_1 (p_1 - p_0)^2} \qquad (6.51)$$

from the power function

$$\varphi = \Phi \left\{ \frac{|p_1 - p_0| \sqrt{N_1^{(0)} N_2 N_3 / (1 - \rho_1)} - \Phi^{-1}(1 - \alpha/2) \sqrt{(1 + 1/\lambda) \bar{p}(1 - \bar{p})}}{\sqrt{p_0 (1 - p_0) + p_1 (1 - p_1)/\lambda}} \right\}$$

6.7.3.3 Summary

The sample size N_3 in (6.50) or in (6.51) required for a desired magnitude of statistical power is not a function of ρ_2, and decreases with increasing $N_1^{(0)}, N_2$, and ρ_1. The sample size decreases in general with increasing $|p - p_1|$. Again, for the same magnitude of $|p - p_1|$, the sample size depends on specific combinations of p and p_1. Table 6.13 displays examples of statistical power for testing the null hypothesis $H_0 : p - p_1 = 0$, and shows that the statistical power φ is invariant over the product of $N_3 N_2 N_1^{(0)}$, or total number of observations $(in 1 + \lambda) N_3 N_2 N_1^{(0)}$, regardless of whether or not λ is equal to 1.

6.7.4 Comparisons of Sample Sizes

The ratios of the total numbers of observations required for testing the treatment effect on a binary outcome with randomization at the third level compared to the second level, randomization at the third level compared to the first level, and randomization at the second level compared to the first level are equal to $R_{3.2}$, $R_{3.1}$, and $R_{2.1}$, respectively, as defined in Section 6.3.4; compare Equations (6.44) and (6.48); Equations (6.44) and (6.50); and Equations (6.48) and (6.50). These ratios also hold for unbalanced trial designs with identical $\lambda = (\lambda_{11}, \lambda_{10}, \lambda_{01})$ values; compare Equations (6.45) and (6.49); Equations (6.45) and (6.51); and Equations (6.49) and (6.51)). These results are evident by comparing Tables 6.11, 6.12, and 6.13. Therefore, again, for the same values of Δ, φ, α, and correlation ρ, the required sample size is greatest with third level randomization, followed by second level randomization and then first level randomization. In summary, the ratios of these sample sizes are identical regardless of whether the outcome is continuous or binary; one is testing main effects or interaction effects; or the designs are balanced or unbalanced.

6.8 Further Readings

The reading list suggested in Section 5.8 should also be applied to this chapter. As before, the sample size requirements for three-level non-inferiority trials can be computed using the same formulas presented in this chapter by substituting the non-inferiority margin for the effect size in superiority trials [17]. In addition, sample size approaches with comprehensive general models can be found in Murray et al [8] including randomization to more than two arms. Preisser et al.[18] proposed approaches for determining sample sizes by applying generalized estimating equation methods. Roy et al. [19] proposed sample size determination approaches that take into account anticipated attrition rates, and Heo [20] examined the impact of subject attrition with different

TABLE 6.13

Sample size and statistical power for detecting a main effect $|p_1 - p_0|$ on binary outcome in model with $m = 1$ (6.41) when randomizations occur at first level (two-sided significance level $\alpha = 0.05$)

λ	p_0	p_1	ρ_1	$N_1^{(0)}$	N_2	N_3	Total N	φ
1	0.4	0.5	0.1	5	4	20	800	0.851
				5	8	10	800	0.851
				10	2	10	800	0.851
				10	4	5	800	0.851
			0.2	5	4	16	640	0.813
				5	8	8	640	0.813
		0.6	0.1	5	4	6	240	0.909
				5	8	3	240	0.909
			0.2	5	4	4	160	0.812
				5	8	2	160	0.812
				10	2	4	160	0.812
				10	4	2	160	0.812
1.5	0.4	0.5	0.1	5	4	16	800	0.836
				5	8	8	800	0.836
				10	4	8	800	0.836
				10	8	4	800	0.836
			0.2	5	4	14	700	0.830
				5	8	7	700	0.830
		0.6	0.1	5	4	4	200	0.837
				5	8	2	200	0.837
			0.2	5	4	4	200	0.878
				5	8	2	200	0.878
				10	2	4	200	0.878
				10	4	2	200	0.878

Note: λ represents the ratio of the number of level 3 units in the intervention arm to the corresponding number in the control arm; ρ_2 (6.7.2) is the correlations among the level 2 data; ρ_1 (6.7.3) is the correlations among the level 1 data; $N_1^{(0)}$ is the level 1 sample size in the *control* arm units per level 2 units; N_2 is the number of level 2 unit per level 3 units; N_3 is the *total* level 3 sample size; Total $N = (1 + \lambda)N_1^{(0)}N_2N_3$ is the required total number of observations; φ is the statistical power to test $H_0 : p_1 - p_0 = 0$.

mechanisms—namely, attrition completely at random, attrition at random, and attrition not at random—on the sample size determinations under model (6.22) for longitudinal cluster-RCT. Sample size requirements with different ICCs between clusters can be computed by extending Liang and Pulver's approaches [14] for two level data structures to three-level structures.

In summary, the sample size formulas presented in Chapters 5 and 6 can be expressed in the general form of $\chi z_{\alpha,\phi}^2 / \Delta^2$ for continuous outcome or

$\chi z^2_{\alpha,\phi,p,\lambda}/(p_1 - p_0)^2$ for binary outcomes, where the quantity of the multiplication factor χ depends on the number of levels, levels of randomizations, hypotheses of interest, and types of study design (crosssectional or longitudinal). For continuous outcomes, Δ represents the standardized effect size or Cohen's d [12] and $z^2_{\alpha,\phi}$ is defined in Equation (6.7). For binary outcomes, $p_1 - p$ represents a difference in proportions or rates between the null and alternative hypotheses and $z^2_{\alpha,\phi,p,\lambda}$ is defined in Equation (6.46).

Bibliography

[1] D. Hedeker and R.D. Gibbons. *Longitudinal Data Analysis*. Wiley: Hoboken, NJ, 2006.

[2] S.W. Raudenbush and A.S. Bryk. *Hierarchical Linear Models: Application and Data Analysis Methods* (2nd ed) SAGE: Thousand Oaks, 2006.

[3] M. Heo and A.C. Leon. Statistical power and sample size requirements for three-level hierarchical cluster randomized trials. *Biometrics* 64:1256–1262, 2008.

[4] S. Teerenstra, M. Moerbeek, T. van Achterberg, B.J. Pelzer, and G.F. Borm. Sample size calculations for 3-level cluster randomized trials. *Clinical Trials* 5:486–495, 2008.

[5] M. Moerbeek, G.J.P. van Breukelen, and M.P.F. Berger. Design issues for experiments in multilevel populations. *Journal of Educational and Behavioral Statistics* 25:271–284, 2000.

[6] M.J. Fazzari, M.Y. Kim, and M. Heo. Sample size determination for three-level randomized clinical trials with randomization at the first or second level. *Journal of Biopharmaceutical Statistics* 24:579–599, 2014.

[7] M. Reznik, J. Wylie-Rosett, M. Kim, and P.O. Ozuah. Physical activity during school in urban minority kindergarten and first-grade students. *Pediatrics* 131:E81–E87, 2013.

[8] D.M. Murray, J.L. Blitstein, P.J. Hannan, W.L. Baker, and L.A. Lytle. Sizing a trial to alter the trajectory of health behaviours: Methods, parameter estimates, and their application. *Statistics in Medicine* 26:2297–2316, 2007.

[9] M. Heo and A.C. Leon. Sample size requirements to detect an intervention by time interaction in longitudinal cluster randomized clinical trials. *Statistics in Medicine* 28:1017–1027, 2009.

[10] M. Heo, X. Xue, and M.Y. Kim. Sample size requirements to detect an intervention by time interaction in longitudinal cluster randomized clinical trials with random slopes. *Computational Statistics and Data Analysis* 60:169–178, 2013.

[11] M. Heo, X. Xue, and M.Y. Kim. Sample size requirements to detect a two- or three-way interaction in longitudinal cluster randomized clinical trials with second-level randomization. *Clinical Trials* 11:503–507, 2014.

[12] J. Cohen. *Statistical Power Analysis for the Behavioral Science.* Lawrence Erlbaum Associates: Hillsdale, NJ, 1988.

[13] M. Heo, Y. Kim, X.N. Xue, and M.Y. Kim. Sample size requirement to detect an intervention effect at the end of follow-up in a longitudinal cluster randomized trial. *Statistics in Medicine* 29:382–390, 2010.

[14] K.Y. Liang and A.E. Pulver. Analysis of case-control/family sampling design. *Genetic Epidemiology* 13:253–270, 1996.

[15] S. Teerenstra, B. Lu, J.S. Preisser, T. van Achterberg, and G. F. Borm. Sample size considerations for GEE analyses of three-level cluster randomized trials. *Biometrics* 66:1230–1237, 2010.

[16] W.J. Shih. Sample size and power calculations for periodontal and other studies with clustered samples using the method of generalized estimating equations. *Biometrical Journal* 39:899–908, 1997.

[17] S. Wellek. *Testing Statistical Hypotheses of Equivalence and Noninferiority* (2nd ed) Chapman & Hall/CRC: New York, 2010.

[18] J.S. Preisser, M.L. Young, D.J. Zaccaro , and M. Wolfson. An integrated population-averaged approach to the design, analysis and sample size determination of cluster-unit trials. *Statistics in Medicine* 22:1235–1254, 2003.

[19] A. Roy, D.K. Bhaumik, S. Aryal, and R.D. Gibbons. Sample size determination for hierarchical longitudinal designs with differential attrition rates. *Biometrics* 63:699-707, 2007.

[20] M. Heo. Impact of subject attrition on sample size determinations for longitudinal cluster randomized clinical trials. *Journal of Biopharmaceutical Statistics* **24**:507–522, 2014.

Index

Note: Page numbers ending in "f" refer to figures. Page numbers ending in "t" refer to tables.

A

Analysis of covariance (ANCOVA), 64–65, 71, 79
Analysis of variance (ANOVA) estimate, 25–35, 39–40, 46–48, 64–65, 69, 78
Autoregressive (AR) correlation structures, 62–63, 115–116

B

Binary outcomes
clustered outcomes, 28–34, 38–51
design parameters for, 122
equal cluster size, 46–47
equal weights to clusters, 32, 49
equal weights to observations, 31–32, 48–49
examples of, 4–5, 28–29, 33, 44–45, 50, 123
minimum variance weights, 32–33, 50
nonparametric methods, 45–51
one-sample clustered binary outcomes, 28–34, 45–51
precision analysis for, 4–5
primary endpoint for, 6
relative efficiency of estimators, 51
sample sizes for, 119–123, 176–181, 223–224
slope for, 119–123, 176–181

stratified cluster randomization for, 42–45
time-averaged difference for, 123–126
unequal cluster size, 47–48
Brief Psychiatric Rating Scale (BPRS), 68, 72–73

C

Cancer clinical trials, 13–14, 18, 38
Cancer drugs, 4–6
Cancer risk intake system (CRIS) trial, 38
Cancer studies, 4–6, 13–14, 18, 24, 28, 38, 101
Cardiovascular disease, 34
CKD, 44–45
Clinical attachment loss (CAL), 28
Clinical trials, 5 6, 19–20, 149–185, 187–223. *See also* Randomized clinical trials
Cluster randomization trials, 34–36, 39–41, 52–54, 74. *See also* Randomized clinical trials
Clustered outcomes
binary outcomes, 28–34, 38–51
cluster size, 19–20, 24–28, 38–42, 46–47
cluster size estimation, 41–42
continuous outcomes, 24–28, 34–38
equal cluster size, 24–26, 38–39, 46–47

equal weights to clusters, 32, 49
equal weights to observations,
     31–32, 48–49
examples of, 24, 28–29, 33, 37,
     40–41, 44–45, 50
explanation of, 23–24
fixed number of clusters, 41–42
minimum variance weights,
     32–33, 50
nonparametric methods, 45–51
one-sample clustered binary
     outcomes, 28–34, 45–51
one-sample clustered outcomes,
     24–34, 45–51
relative efficiency of estimators,
     51
sample size calculation for,
     19–20, 33
sample size determination for,
     23–60
split-mouth trials, 23, 51–52
stratified cluster randomization,
     42–45
two-sample clustered outcomes,
     34–41
unequal cluster size, 26–28,
     35–37, 39–40, 47–48
Compound symmetry (CS)
     correlation structures,
     53–54, 62–63, 115
Congestive heart failure (CHF),
     16–17
CONSORT statement, 78
Continuous outcomes
     clustered outcomes, 24–28,
     34–38
     example of, 3–4
     models for, 150–151, 187–189
     precision analysis for, 3–4
     primary endpoint for, 6
     sample size calculation for, 90
     slope for, 90–110
     statistical models for, 150–151,
     187–189
     time-averaged difference for,
     110–119
Controlled clinical trials, 19–20, 61,
     65–66. *See also* Randomized
     clinical trials
Correlated outcome measurements
     autoregressive correlation,
     115–116
     balanced allocations, 152–153,
     155–156, 159–163, 165–170,
     173–175, 177–180, 189–190,
     192–196, 201–202, 204–205,
     209–210, 212–215, 217,
     220–221, 224, 226–227, 229
     for binary outcomes, 176–181
     compound symmetry
     correlation, 115
     for continuous outcomes,
     150–151
     cross-sectional designs for,
     167–172, 211–218
     design parameters for, 122
     examples of, 107–110, 123,
     199–200
     explanation of, 83–85, 149–150,
     187
     factorial designs for, 167–176,
     211–223
     financial constraints, 105–107
     on fixed-slope model, 159–167
     GEE for, 83–147
     geometrical relationships,
     159–161, 160f
     impact of, 93–97, 95f, 97f,
     103–105
     independent missing patterns,
     95–99, 97f, 101f, 110t
     K treatment groups, 100–105,
     116–119
     longitudinal designs for,
     158–167, 172–176, 218–223
     marginal observant probabilities,
     108–109, 109f
     missing data impact, 95–97, 97f,
     101f, 113–118, 139–141

models for, 150–151, 187–189
monotone missing patterns,
    96–99, 97f, 101f, 110t
quasi-score test approach, 88–90
randomization at first level,
    154–157, 170–172, 179–180,
    194–197, 216–218, 228–230
randomization at second level,
    151–154, 168–170, 177–179,
    191–194, 206–208, 213–216,
    225–228
randomization at third level,
    189–191, 201–206, 212–213
on random-slope model,
    161–164, 203–206
sample size calculation for, 85,
    91–93, 116, 119–122,
    136–137
sample size determination for,
    83–147, 95f, 149–185,
    187–233
sample size formula for, 111–114
slope among, 100–105, 119–123
statistical models for, 150–151,
    187 189, 223
testing main effects for,
    151–158, 189–200, 209
testing slope differences for,
    158–167, 200–211
testing treatment effects for,
    165–166
from three-level randomized
    clinical trials, 187–233
time-averaged difference for,
    110–119
trade-offs within, 98–100,
    105–106, 114–116
treatment effects on slopes,
    172–176, 218–223
for treatment interactions,
    167–172, 211–218
from two-level randomized
    clinical trials, 149–185, 160f
unbalanced allocations, 153–154,
    156, 161, 163–164, 166, 169,

    171–172, 175, 178, 180, 191,
    193–194, 196–197, 203,
    205–206, 210, 213, 215, 217,
    221, 224–225, 227, 229
Wald test approach, 87–90, 127,
    137
Correlation structures
    exponential family of, 53–54,
        62–63, 93–97, 95f, 97f
    marginal observant probabilities,
        108–109, 109f
    missing data impact, 61–63,
        93–101, 97f, 101f
    sample size determination for,
        93–97, 95f, 97f, 101f
Count outcomes
    design factors for, 129, 132
    K treatment groups, 132–134
    sample size calculation for, 127,
        132–134
    slope for, 126–130
    time-averaged difference for,
        126–133

D
Diabetes, 44–45, 157–158, 176
Drug testing study
    anti-epileptic drug, 127
    cancer drugs, 4–6
    for congestive heart failure,
        16–17
    experimental drugs, 105, 118,
        123
    sample size formula for, 40–41,
        75, 79
    superiority test, 7–10
    for systolic blood pressure, 13

E
Effective sample size (ESS), 23–28
Enzymatic diagnostic test, 33
Ethical Guidelines of the American
        Statistical Association, 5
European Medicines Agency, 9

F
Food and Drug Administration
        (FDA), 9

G
Generalized estimating equation
        (GEE)
    for binary outcomes, 119–123,
        178
    for continuous outcomes, 90–110
    for correlated outcomes, 83–147,
        225, 230
    explanation of, 83–87
    for repeated measurement
        outcomes, 64, 76–78
    review of, 85–90
    sample size determination for,
        83–147, 225, 230
Generalized least squares (GLS),
        65–66, 70, 72–73
Generalized linear mixed model
        (GLMM), 64
GENISOS observations, 123
Geometrical relationships, 159–161,
        160f, 201
Geometrical representations, 160,
        173–174, 174f
GoodNEWS trial, 34

H
Health promotion program, 34, 37
Heart failure, 16–17
Hypertension, 44–45, 83
Hypothesis testing
    for binary outcomes, 120
    bioequivalence test for, 9
    for clustered outcomes, 35–36,
        41–42
    for continuous outcomes, 90–91,
        111
    for correlated outcomes, 120
    equivalence test for, 7–9
    error types for, 9–10
    hypothesis of interest, 6–8
    method of, 2–3, 6–15

non-inferiority test for, 8–9
null hypothesis, 2–3, 5–14,
        28–31, 35–40
    one-sample test for means,
        11–14, 12f
    one-sample test for proportions,
        13–14
    quasi-score test approach, 88–90
    sample size estimation for, 6–9
    superiority test for, 7–9
    two-sample test for means,
        15–16
    two-sample test for proportions,
        17–18
    Wald test approach, 87–90

I
Independent missing (IM) pattern,
        95–99, 97f, 101f, 110t
Independent outcomes
    explanation of, 1–2
    one-sample test for, 11–14
    power analysis for, 5–18
    precision analysis for, 2–5
    sample size calculation for,
        18–20
    sample size determination for,
        1–22
    sample size estimation, 11–14,
        12f
    two-sample test for, 15–16
Infection, proportion of, 33–34,
        33t
Intervention studies, 19, 136
Intracluster correlation coefficient
        (p), 26–28

L
Linear exponent autoregressive
        (LEAR) correlation, 63
Linear mixed model (LMM), 65, 138

M
Marginal observant probabilities,
        108–109, 109f

Measurements per subject (*m*)
  for continuous outcomes, 111
  for correlated outcomes, 85,
    98–100, 105–106, 111,
    114–116
  K treatment groups, 105
  sample size estimation for, 63,
    66, 76–79
  trade-offs, 98–100, 105–106,
    114–116
Mental retardation, 13
Minimally invasive periodontal
    surgery (MIPS), 28
Missing completely at random
    (MCAR), 79, 92, 117, 125,
    135–136
Missing data impact, 93–101, 97f,
    101f, 113–118, 139–141
Mixed-effect model (MM), 84
Monotone missing (MM) pattern,
    96–99, 97f, 101f, 110t
Morbidity, 158
Mortality, 158

N
Nonparametric methods, 45–51
Null hypothesis (*H0*), 2–3, 5–14,
    28–31, 35–40
Number of subjects (*n*)
  for clustered outcomes, 37,
    41–42
  for continuous outcomes, 111
  for correlated outcomes, 85,
    98–100, 105–106, 111,
    114–116
  K treatment groups, 105
  for randomization trials, 156,
    166
  sample size determination for,
    1–2, 5–11, 13–20
  sample size estimation for,
    52–53, 62–63, 76, 208
  trade-offs, 98–100, 105–106,
    114–116

O
Ordinary least squares (OLS)
  on fixed-slope model, 160–163,
    201, 205, 219–220
  on random-slope model, 219–220
  sample size estimation for,
    65–66, 70–73

P
Periodontal disease, 28
Physical activity, 34, 37, 187, 199
Physical activity recall (PAR), 37
Poisson model, 86, 127–129, 132
Positron emission tomography
    (PET), 19
Power analysis
  examples of, 13–14, 16–17
  information needed for, 5–11
  one-sample test for, 11–14, 12f
  sample size estimation for, 5–18
  test drug study, 13
  two-sample test for means,
    15–16
  two-sample test for proportions,
    17–18
Power estimation, 37, 52, 78. *See
    also* Sample size estimation
Precision analysis
  for binary outcomes, 4–5
  for continuous outcomes, 3–4
  example of, 3–5
  sample size estimation for, 2–5
Psychotic disorders, 68
Pulmonary fibrosis, 123

Q
Quasi-score test approach, 88–90

R
Randomized clinical trials
  balanced allocations, 152–153,
    155–156, 159–161, 165–170,
    173–175, 177–180, 189–190,
    192–196, 201–202, 204–205,

209–210, 212–215, 217,
220–221, 224, 226–227, 229
for binary outcomes, 176–181
for continuous outcomes,
150–151
controlled clinical trials, 19–20,
61, 65–66
for correlated outcome
measurements, 149–185
cross-sectional designs for,
167–172, 211–218
for detecting main effect,
154–157, 154t, 157t, 167t,
179t, 181t, 192t, 195t, 198t,
226t, 228t, 231t
examples of, 157–158, 199–200
explanation of, 149–150, 187
factorial designs for, 167–176,
211–223
at first level, 154–155
fixed parameters in, 159–161,
160f, 173–174, 174f
on fixed-slope model, 159–167,
162t, 167t, 201–202, 204t,
211t
longitudinal designs for,
158–167, 172–176, 218–223
models for, 150–151, 187–189
randomization at first level,
154–157, 170–172, 179–180,
194–197, 216–218, 228–230
randomization at second level,
151–154, 168–170, 177–179,
191–194, 206–208, 213–216,
225–228
randomization at third level,
189–191, 201–206, 212–213
on random-slope model,
161–164, 164t, 203–206,
207t
sample size calculation for, 74
sample size determination for,
149–185
sample size estimation for, 5–6,
53–54

statistical models for, 150–151,
187–189
summary statistics for, 65
testing main effects for,
151–158, 189–200, 209
testing slope differences for,
158–167, 200–211
testing treatment effects for,
165–166
three-level randomized clinical
trials, 187–233, 192t, 195t,
198t, 204t, 207t, 211t, 214t,
216t, 219t, 222t, 226t, 228t,
231t
on three-way interaction, 176t,
222t
treatment effects on slopes,
172–176, 218–223
treatment interactions in,
167–172, 211–218
two-level randomized trials,
149–185, 154t, 157t,
159–161, 160f, 170t,
173–174, 173t, 174f
on two-way interaction, 170t,
173t, 214t, 216t, 219t
unbalanced allocations, 153–154,
156, 161, 163–164, 166, 169,
171–172, 175, 178, 180, 191,
193–194, 196–197, 203,
205–206, 210, 213, 215, 217,
221, 224–225, 227, 229
Relative efficiency (RE), 51
Repeated measurement outcomes
binary outcomes, 77–78
change rates, 69–75
difference score analysis, 66–68
examples of, 68, 72–73, 75
explanation of, 61–62
financial constraints, 76
multiple treatment groups,
76–77
sample size determination for,
61–82

sample size estimation for, 9,
  18–19, 76, 78
score analysis, 66–68
time-averaged differences, 73–78
using summary statistics, 61–82

S
Sample size calculation
  for binary outcomes, 119–123,
    176–181, 223–224
  for clustered outcomes, 19–20,
    33
  for continuous outcomes, 90
  for correlated outcomes, 85,
    91–93, 116, 136–137
  for count outcomes, 127
  examples of, 10–15
  explanation of, 1–2, 6–8
  hypothesis of interest and, 6–9
  for independent outcomes, 18–20
  one-sample test for means,
    11–14, 12f
  one-sample test for proportions,
    13–14
  for randomized clinical trials, 74
  two-sample test for means,
    15–16
  two-sample test for proportions,
    17–18
Sample size comparisons
  for three-level randomizations,
    197–198, 208, 218, 222–223,
    230
  for two-level randomizations,
    156–157, 164–165, 172,
    175–176, 180–181
Sample size determination
  balanced allocations, 152–153,
    155–156, 159–163, 165–170,
    173–175, 177–180, 189–190,
    192–196, 201–202, 204–205,
    209–210, 212–215, 217,
    220–221, 224, 226–227, 229
  for binary outcomes, 119–123,
    176–181, 223–224

for clustered outcomes, 23–60
comparisons of sample sizes,
    156–157, 164–165, 172,
    175–176, 180–181, 197–198,
    208, 218, 222–223, 230
for correlated outcome
    measurements, 83–147, 95f,
    97f, 101f, 149–185, 187–233
for correlation structures, 93–97,
    95f, 97f, 101f
for detecting main effect,
    154–157, 154t, 157t, 167t,
    179t, 181t, 192t, 195t, 198t,
    226t, 228t, 231t
example of, 199–200
explanation of, 1–2
on fixed-slope model, 159–167,
    162t, 167t, 201–202, 204t,
    211t
GEE for, 83–147, 225, 230
geometrical relationships,
    159–161, 160f
for independent outcomes, 1–22
marginal observant probabilities,
    108–109, 109f
missing data impact, 93–101,
    97f, 101f
one-sample test for, 11–14
power analysis for, 5–18
precision analysis for, 2–5
on random-slope model,
    161–164, 164t, 203–206,
    207t
for repeated measurement
    outcomes, 61–82
sample size estimation, 1–5,
    11–14, 12f, 62–64
from three-level randomized
    clinical trials, 187–233
on three-way interaction, 176t,
    222t
from two-level randomized
    clinical trials, 149–185, 160f
two-sample test for, 15–16

on two-way interaction, 170t,
 173t, 214t, 216t, 219t
unbalanced allocations, 153–154,
 156, 161, 163–164, 166, 169,
 171–172, 175, 178, 180, 191,
 193–194, 196–197, 203,
 205–206, 210, 213, 215, 217,
 221, 224–225, 227, 229
using summary statistics, 61–82
under various scenarios, 110t
Sample size estimation
effective sample sizes, 23–28
examples of, 3–5, 13–17
financial constraints on, 76
for hypothesis testing, 6–9
for independent outcomes,
 11–14, 12f
information needed for, 5–11,
 18–19, 62–64
one-sample test for means,
 11–14, 12f
one-sample test for proportions,
 13–14
for power analysis, 5–18
power estimation and, 37, 52, 78
for precision analysis, 2–5
for randomization trials, 5–6,
 53–54, 65–66
for repeated measurement
 outcomes, 9, 18–19, 76, 78
sample size determination and,
 1–5
statistical method for, 10–11
two-sample test for means,
 15–16
two-sample test for proportions,
 17–18
Schizophrenia, 68
Scleroderma, 61, 123
Split-mouth trials, 23, 51–52
Statistical methods
for continuous outcomes,
 150–151
for data analysis, 7–10, 23, 62

nonparametric approach to,
 45–46
for sample size estimation,
 10–11, 18
Statistical models, 150–151, 187–189,
 223
Stepped wedge design, 52–53, 53t
Summary statistics
for controlled clinical trials, 65
explanation of, 64
for randomized clinical trials, 65
for repeated measurement
 outcomes, 61–82
Survival outcomes, 6, 182

T
Test drugs
anti-epileptic drug, 127
cancer drugs, 4–6
for congestive heart failure,
 16–17
experimental drugs, 105, 118,
 123
sample size formula for, 40–41,
 75, 79
superiority test, 7–10
for systolic blood pressure, 13
Three-level randomized clinical trials
balanced allocations, 189–190,
 192–196, 201–202, 204–205,
 209–210, 212–215, 217,
 220–221, 224, 226–227, 229
for correlated outcome
 measurements, 187–233
cross-sectional designs for,
 211–218
for detecting main effect, 192t,
 195t, 198t, 226t, 228t, 231t
example of, 199–200
explanation of, 187
factorial designs for, 211–223
on fixed-slope model, 201–202,
 204t, 211t
longitudinal designs for, 218–223

models for, 187–189
randomization at first level,
194–197, 216–218, 228–230
randomization at second level,
191–194, 206–208, 213–216,
225–228
randomization at third level,
189–191, 201–206, 212–213
on random-slope model,
203–206, 207t
sample size determination for,
187–233
statistical models for, 187–189
testing main effects for,
189–200, 209
testing slope differences for,
200–211
three-level hierarchical studies,
54, 199
on three-way interaction, 222t
treatment effects on slopes,
218–223
treatment interactions in,
211–218
on two-way interaction, 214t,
216t, 219t
unbalanced allocations, 191,
193–194, 196–197, 203,
205–206, 210, 213, 215, 217,
221, 224–225, 227, 229
Time averaged difference (TAD)
for binary outcomes, 123–126
for continuous outcomes,
110–119
for count outcomes, 126–133
K treatment groups and,
132–134
testing, 73–78, 110–119,
123–126, 130–134
Tumors, 5, 24. *See also* Cancer
studies
Two-level randomized clinical trials
balanced allocations, 152–153,
155–156, 159–163, 165–170,
173–175, 177–180

for binary outcomes, 176–181
for continuous outcomes,
150–151
for correlated outcome
measurements, 149–185,
160f
cross-sectional designs for,
167–172
for detecting main effect,
154–157, 154t, 157t, 167t,
179t, 181t
example of, 157–158
explanation of, 149–150
factorial designs for, 167–176
at first level, 154–155
fixed parameters in, 173–174,
174f
on fixed-slope model, 159–167,
162t, 167t
geometrical relationships,
159–163, 160f
longitudinal designs for,
158–167, 172–176
models for, 150–151
randomization at first level,
154–157, 170–172, 179–180
randomization at second
level, 151–154, 168–170,
177–179
on random-slope model,
161–164, 164t
sample size determination for,
149–185, 154t, 157t, 160f
statistical models for, 150–151
testing main effects for,
151–158
testing slope differences for,
158–167
testing treatment effects for,
165–166
on three-way interaction, 176t
treatment effects on slopes,
172–176
treatment interactions in,
167–172

on two-way interaction, 170t,
173t
unbalanced allocations, 153–154,
156, 161, 163–164, 166, 169,
171–172, 175, 178, 180

W
Wald test approach, 87–90, 127, 137
Wedge design, 52–53, 53

Printed in the United States
by Baker & Taylor Publisher Services